Advances in Experimental Medicine and Biology

Volume 1246

Advances in Experimental Medicine and Biology provides a platform for scientific contributions in the main disciplines of the biomedicine and the life sciences. This series publishes thematic volumes on contemporary research in the areas of microbiology, immunology, neurosciences, biochemistry, biomedical engineering, genetics, physiology, and cancer research. Covering emerging topics and techniques in basic and clinical science, it brings together clinicians and researchers from various fields.

Advances in Experimental Medicine and Biology has been publishing exceptional works in the field for over 40 years, and is indexed in SCOPUS, Medline (PubMed), Journal Citation Reports/Science Edition, Science Citation Index Expanded (SciSearch, Web of Science), EMBASE, BIOSIS, Reaxys, EMBiology, the Chemical Abstracts Service (CAS), and Pathway Studio.

2018 Impact Factor: 2.126.

More information about this series at http://www.springer.com/series/5584

Maurice B. Hallett

Editor

Molecular and Cellular Biology of Phagocytosis

 Springer

Editor
Maurice B. Hallett
School of Medicine
Cardiff University
Cardiff, UK

ISSN 0065-2598   ISSN 2214-8019 (electronic)
Advances in Experimental Medicine and Biology
ISBN 978-3-030-40405-5   ISBN 978-3-030-40406-2   (eBook)
https://doi.org/10.1007/978-3-030-40406-2

This Springer imprint is published by the registered company Springer Nature Switzerland AG.
The registered company address is: Gewerbestrasse 11, 6330 Cham, Switzerland

Preface

One of the most exciting microscopic events that can be observed in real time is probably phagocytosis. Showing it to anyone who has yet to see it (children, non-scientific adult, scientist from different disciplines, etc.) what happens during phagocytosis is always a stimulating experience. Often while watching, the observer gives encouragement to the cell ("Come on! You can do it", "Just one more") and may cheer when the phagocyte succeeds ("Good cell, you've done it", "Great job", etc.) as if the cell were responsive to the words. I guess that it is because in the non-microscopic world, trying to get a dog or a horse do something amazing is rewarded only by encouragement

But consider this. You are given a microscope to view two grey microscopic blobs, with the task of discovering which one of the two is alive. One is a grey blob of gel and the other a living cell. The school-level definitions of life include respiration, growth and excretion. This might not help to distinguish between them. Scientists setting out to discover if there is life on Mars have found this out (distinguishing life from other chemical processes is tricky). You cannot see any substructure in the grey blobs because of the poor optics available. However, you might quickly discover which is the living cell simply by poking each blob with a micropipette. If one of the grey blobs pushes out a pseudopodia at the opposite side to the one you are prodding and then moves away from the irritation, you might immediately guess that it was the living cell. A nonliving blob of gel would not and could not do this. Yet, would a small bag of chemicals do it? A phagocyte , whether a small neutrophil or a large amoeba, is not a dog or a horse to be encouraged; it is instead a bag of chemicals – admittedly, a complex bag of chemicals, with substructure (smaller bags within the larger bag), but still a bag of chemicals. When you see this bag appear to think, and decide to grab a smaller object so that you feel it is "trying" or "struggling" to grab something, it is mere anthropomorphism. It is just a bag of chemicals. But to see a bag of chemicals do something like phagocytosis, putting out pseudopodia, and holding an object and then proceeding to swallow it, is to see a miracle. *Molecular and Cell Biology of Phagocytosis* is the study of this miracle.

Somerset, UK
2019

Maurice B. Hallett

Contents

1   An Introduction to Phagocytosis   1
    Maurice B. Hallett

2   A Brief History of Phagocytosis   9
    Maurice B. Hallett

3   The Role of Membrane Surface Charge in Phagocytosis   43
    Michelle E. Maxson and Sergio Grinstein

4   Receptor Models of Phagocytosis: The Effect of Target Shape   55
    David M. Richards

5   Decision Making in Phagocytosis   71
    Jana Prassler, Florian Simon, Mary Ecke, Stephan Gruber,
    and Günther Gerisch

6   Membrane Tension and the Role of Ezrin During Phagocytosis   83
    Rhiannon E. Roberts, Sharon Dewitt, and Maurice B. Hallett

7   Molecular Mechanisms of Calcium Signaling During
    Phagocytosis   103
    Paula Nunes-Hasler, Mayis Kaba, and Nicolas Demaurex

8   Calpain Activation by Ca^{2+} and Its Role in Phagocytosis   129
    Sharon Dewitt and Maurice B. Hallett

9   The NADPH Oxidase and the Phagosome   153
    Hana Valenta, Marie Erard, Sophie Dupré-Crochet, and
    Oliver Nüße

10 Conclusions and the Futures of Phagocytosis   179
    Maurice B. Hallett

Index   183

Contributors

Nicolas Demaurex Department of Cellular Physiology and Metabolism, University of Geneva, Geneva, Switzerland

Sharon Dewitt School of Dentistry, Cardiff University, Cardiff, UK

Sophie Dupré-Crochet Université Paris-Saclay, Orsay, France
CNRS U8000, ICP, Orsay, France

Mary Ecke Max Planck Institute of Biochemistry, Martinsried, Germany

Marie Erard Université Paris-Saclay, Orsay, France
CNRS U8000, ICP, Orsay, France

Günther Gerisch Max Planck Institute of Biochemistry, Martinsried, Germany

Sergio Grinstein Program in Cell Biology, Hospital for Sick Children, Toronto, ON, Canada
Department of Biochemistry, University of Toronto, Toronto, ON, Canada
Keenan Research Centre for Biomedical Science, St. Michael's Hospital, Toronto, ON, Canada

Stephan Gruber Department of Fundamental Microbiology (DMF), Faculty of Biology and Medicine (FBM), University of Lausanne (UNIL), Lausanne, Switzerland

Maurice B. Hallett School of Medicine, Cardiff University, Cardiff, UK

Mayis Kaba Department of Cellular Physiology and Metabolism, University of Geneva, Geneva, Switzerland

Michelle E. Maxson Program in Cell Biology, Hospital for Sick Children, Toronto, ON, Canada

Paula Nunes-Hasler Department of Pathology and Immunology, University of Geneva, Geneva, Switzerland

Oliver Nüβe Université Paris-Saclay, Orsay, France
CNRS U8000, ICP, Orsay, France

Jana Prassler Max Planck Institute of Biochemistry, Martinsried, Germany

David M. Richards Living Systems Institute, University of Exeter, Exeter, UK

Rhiannon E. Roberts School of Medicine, Cardiff University, Cardiff, UK

Florian Simon Max Planck Institute of Biochemistry, Martinsried, Germany

Hana Valenta Université Paris-Saclay, Orsay, France
CNRS U8000, ICP, Orsay, France

An Introduction to Phagocytosis

1

Maurice B. Hallett

Abstract

Phagocytosis is usually defined as the cellular process by which cells internalise particulate matter larger than about 0.5 μm in diameter. It is an endocytic process, distinct from pinocytosis and macropinocytosis. These latter processes may internalise small particles suspended the extracellular fluid, but this is a by-product of internalising the fluid, and is not phagocytosis per se. In contrast, phagocytosis is targeted at solid particulates, usually microbes, which are internalised and "digested" either to provide food, or as part of the immune system of higher animals. The mechanism of phagocytosis may have, at its core, many primitive elements, but it is a highly complex and coordinated series of cell biological and molecular events which together result in the uptake of a particle. In this introduction, the basis of phagocytosis and some ideas of its origin are discussed.

Keywords

Phagocytosis · Evolution of phagocytosis · Role of phagocytosis in evolution · Phagocytic steps · Phagocytic cup formation

M. B. Hallett (✉)
School of Medicine, Cardiff University, Cardiff, UK
e-mail: hallettmb@cf.ac.uk

Introduction to Phagocytosis

Phagocytosis is usually defined as the cellular process by which cells internalise particulate matter larger than about 0.5 μm (Aderem and Underhill 1999). It is thus an endocytic process, but distinct from pinocytosis and macropinocytosis. Pinocytosis is fluid endocytosis, sometimes called "cell drinking", brought about by small invaginations of the cell membrane (Chapmanandresen 1962). This process may also internalise small particles if suspended the extracellular fluid, but this is a by-product of internalising the fluid, and is not phagocytosis per se. Macropinocytosis is a related cellular phenomenon nut uptake of fluid results in light microscopically visible fluid droplets within the cell. Classical micropinocytosis involves the formation of macropinosomes within rings of polymerised actin, formed by circular plasma membrane ruffles (Bloomfield and Kay 2016; King and Kay 2019; Lim and Gleeson 2011; Swanson 2008). In some respects, macropinocytosis thus overlaps with phagocytosis but, as it does not involve intake of large solid particles, is not strictly phagocytosis.

Phagocytosis can be seen by unicellular animals is a number of phyla and cells within multicellular organisms, such as hydra and humans. In a number of higher complex animals, phagocytosis is intimately involved with immunity and defence against infection by microbes. In humans, immune phagocytes, such as neutrophils and macrophages, are key to the innate (ie not acquired) immunity.

The Cell Biology of Phagocytosis

Phagocytosis by phagocytes in higher animals and as singled-celled protozoa, involves the actin cytoskeleton and the localised protrusion of the plasma membrane to "capture" the particulate target, internalise it by fusion of the extended pseudopodia and then the withdrawal of the captured particle within its phagosome back into the cell body. At this point phagocytosis is complete. However related phenomena involve the maturation of the phagosome within the cell. For example, once within the cell, granules (lysosomes) fuse with the phagosomal membrane, to form phagolysomes which also involves acidification of the vascular contents, and "digestion" of the internalised target; and finally (in some phagocytes) recycling of membrane components back to the plasma membrane.

Phagocytosis Dissected

Phagocytosis is a continuous process but could be divided into seven stages (Fig. 1.1). In brief outline these are:

1. *Particle recognition.* The first stage, following contact between the particle and the phagocyte, involves molecular interactions between the opposing surfaces. Receptors on the phagocyte plasma membrane "recognise" the nature of the particle by binding between the receptors and their ligand. For immune phagocytes, these receptors can be "opsonin" receptors and the particle opsonised by the complement component C3bi or specific antibodies which present the Fc portion of the molecule for binding by Fc receptors on the immune cell.

2. **Particle binding.** If the interaction between cellular receptors and the particulate ligands is sufficiently strong to hold the particle at the cell surface and overcome the Brownian motion of the particle.

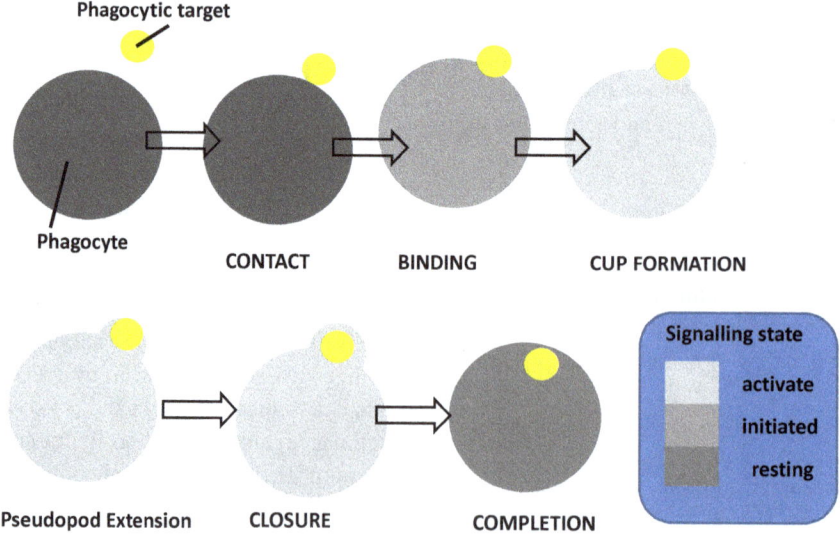

Fig. 1.1 The stages of phagocytosis. The cartoon show the stages of phagocytosis from contact with the target particle, through binding and phagocytic cup formation, to pseudopod extension and phagosome closure. The signalling activity of the cell is coloured according to the scale shown on the far right

3. **Particle signalling.** At this point, signals within the phagocyte may be generated as a result of the cellular receptors and the particulate ligands interaction

4. **Phagocytic cup formation.** The initial intracellular signals result in the formation of a phagocytic cup. The phagocytic cup holds the particle firmly at the cell surface and the target is loosely "captured". The number of cellular receptors and the particulate ligands interactions increases as the opposed surfaces increase and the possibility of inter particle-cell molecular interaction also increases.

5. **Pseudopodia extension.** Once the number of inter particle-cell molecular interactions exceeds a threshold value, a second set of cellular signals are triggered which result in the growth of the phagocytic cup around the particle by the rapid extension of phagocytic pseudopodia around the particle.

6. **Fusion of the phagocytic pseudopodia.** When the opposing phagocytic pseudopodia meet, usually at the pole of the particle opposite the phagocytic cup, the pseudopodia fuse. The target is now fully captured and cannot escape as it is within the cell.

7. **Phagosome formation and withdrawal.** Fusion of the phagocytic pseudopodia results in the outer membranes of the pseudopodia forming the cell membrane and the inner membranes of the pseudopodia forming the phagosome membrane. The pseudopodia retract and draw the particle into the cell body and phagocytosis is fully completed.

Many of the details underlying some of these steps are discussed in the subsequent chapters in this book, such as aspects of phagocyte signalling by cytosolic Ca^{2+} (Chaps. 8 and 9) and changes to lipid composition and charge (Chap. 3). Many of the details, however, are still not fully understood.

The Importance of Phagocytosis

Most of the early studies on phagocytosis were undertaken in the "animalcules" which were unknown before the invention of the microscope (see Chap. 2). As well as amoeba and Metch-nikov's starfish phagocytes (Leidy 1875; Metch-nikoff 1893; Metchnikov 1889), these include many other (some of which remain unidentified) unicellular organism from a number of different phyla. Clearly phagocytosis is a widely spread phenomena in the "animal" kingdom. Its main purpose is of course nutrition. Cells eat cells for food. The phagosome is a digestive sac and the nutrient, usually free amino acids are absorbed by the cells for use in anabolism. Immune phagocytes, of course have a different purpose. As Metchnikov showed (Metchnikov 1889), once digestion was taken over by a digestive tract, the free moving phagocytes became specialised as anti-microbial agents, and form the basis of the innate immune system. Phagocytosis was also a key step in cellular evolution.

Evolution and Phagocytosis

The step from prokaryotic cells (ie cells without intracellular organelles) to eukaryotic cells (ie cells which contain specialised intracellular organelles is thought to be of major importance (Margulis 1970; Martin et al. 2015; Sagan 1967). The highly complex interactions which occur in eukaryotic cells, usually involves interactions and signalling between organelles, as well as for basic cell function. Mitochondria and chloroplasts, contain DNA of bacterial origin, are both enclosed within double membranes, suggesting that they were internalised (by phagocytosis) at some early stage of evolution. This has led to the hypothesis that at some point in ancient evolution phagocytosis of bacteria (with fortuitously beneficial properties) by prokaryote cells lead to a symbiotic relationship that persists to this day (Margulis 1970). This simple idea now has a wealth of genetic support for this, but still faces one problem, namely that today's prokaryotes are not phagocytic. It has been suggested that the metagenomic lineage of *Lokiarchaea* have been a "phagocytosing archaeon" (Spang et al. 2015; Zaremba-Niedzwiedzka et al. 2017) having the ability to undertake "primitive phago-cytosis" (Martijn and Ettema 2013; Yutin et al. 2009). Clearly this would be the "missing link"

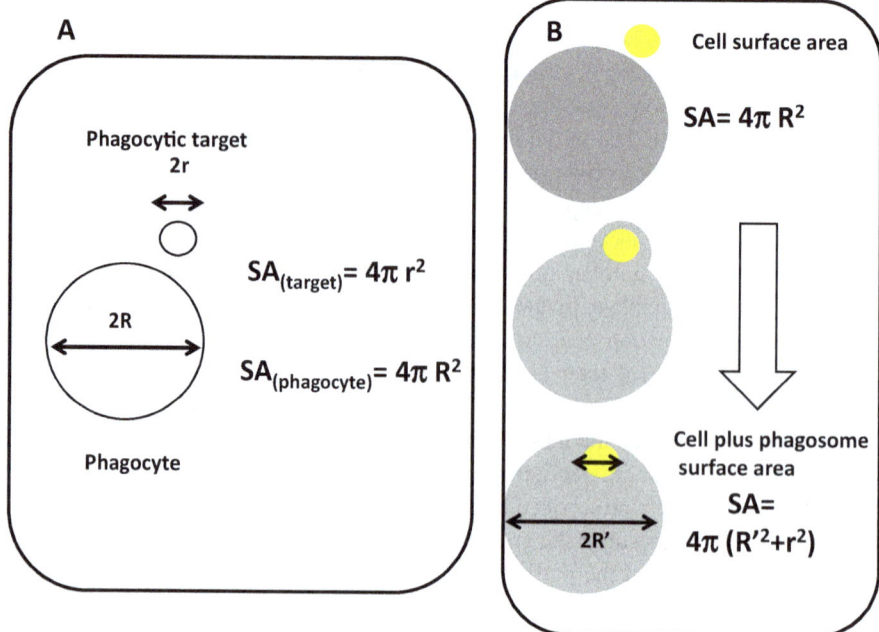

Fig. 1.2 Requirement for surface area expansion during phagocytosis. *Panel A* gives the basic and simplified geometry of phagocytosis. The surface areas of the phagocyte and the target are given, where R is the radius of the cell, and r, the radius of the target. *Panel B* shows a simple calculation of the addition cell surface are to complete phagocytosis, where R' is the radius of the cell after phagocytosis. For small targets ie r < <R, R' may be taken as R and the calculation is simple

between prokaryotes and eukaryotes. However, the idea that that phagocytosing archeon evolved phagocytosis in the same way that present day cells phagocytose with all the molecular machinery and complexes, does not see to hold true and is no longer widely supported (Martin et al. 2017). Instead, the idea that they had a "primitive phagocytic capability" of some sort seems more plausible. Presumably, it was this which evolved into phagocytosis as we know it today. Thus the "primitive phagocytic capability" may still be at the heart of the complex process of phagocytosis involving many different molecular classes and types which co-ordinate together to effect the cell phenomenon of phagocytosis.

Central to phagocytosis is the movement of the plasma membrane to enclose particles. Once phospholipid bilayer membrane touches an adjacent phospholipid bilayer membrane, fusion is a chemico-physical outcome that requires no biological control. Presumably, primitive cells were simply bags of chemicals formed by lipid membranes. As the minimum surface area required to enclose a certain volume is a sphere, presumably primitive cells were spheres. They would be incapable of phagocytosis because each internalised object requires an addition cell surface area equal or greater than the surface area of the object. Thus the internalisation of an external particle would require either that additional membrane is added to the sphere or that the membrane stretches enough to accommodate the internalised object (Fig. 1.2). The membrane is strong across the water/lipid/water sandwich but weak laterally so stretching of bilayers is unlikely. One possibility is that a primitive cell formed in low osmotic (fresh) water, ventured into higher osmotic (salt) water, the cell volume would osmotically reduce and there would be excess surface area (Fig. 1.3). The baggy plasma membrane would naturally form surface wrinkles as the cell volume reduced but the surface area remains constant (Cerda and Mahadevan 2003; Huck 2005; Pocivavsek et al. 2008; Wang et al. 2011). Were a

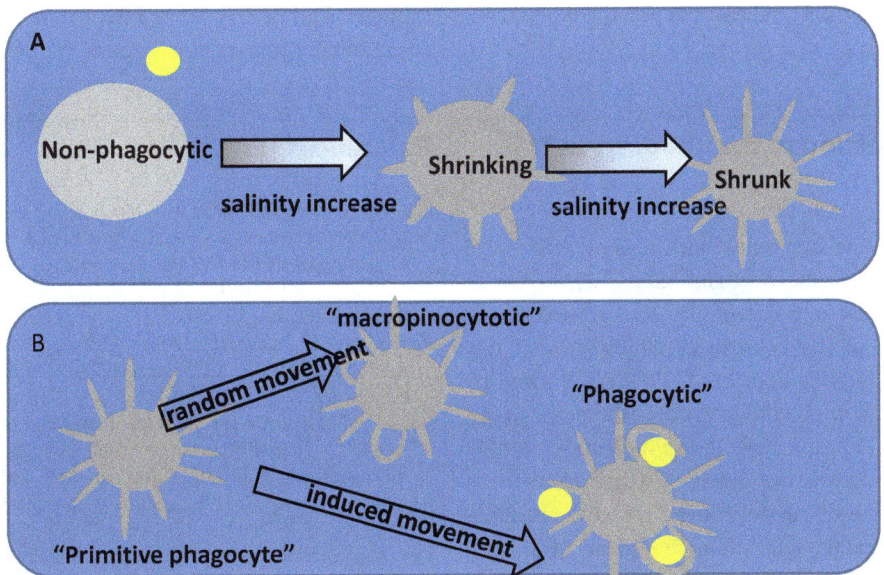

Fig. 1.3 Possible route to phagocytosis. *Panel A* shows a very primitive prokaryotic spherical non-phagocytic cell experiencing an increase in salinity (osmotic strength) of its aqueous environment in two steps. Cell surface wrinkles form as salinity increases. *Panel B* shows how the random movement of the wrinkled surface could entrap fluid as a primitive form of feeding similar to micropinocytosis; or became directed towards more efficient feeding in a manner similar to phagocytosis

passing bacterium to be caught within two membrane ridges and distort the membrane projections so that they touched each other (and fused), the bacterium would be "caught" and could be easily internalised (Fig. 1.3). However, this would be an unlikely scenario, and would occur rarely but perhaps frequently enough to provide the single mitochondrial endocytotic event that was required. Martin et al. (2017) estimate that there have been about 10^{40} prokaryotic cells on Earth in the last 2 billion years, giving a large number of opportunities for phagocytosis to create eukaryotic cells, but as they point out, "Rarity is … a desirable property of endosymbiosis, because mitochondria arose only once in 4 billion years" (Martin et al. 2017).

This primitive form of phagocytosis would require no actin, signalling molecules, clathrin or other proteins, and hence would precede genomic modification and mutation would not be necessary. However, such a "primitive" form of phagocytosis could form the start of an evolutionary movement towards improving the process which leads to the complexity seen today. The hypothetical scenario outlined above, if true, would leave a primitive imprint underlying the complex phenomenon of phagocytosis. This may be the case. Bacteria are reported to increase cell surface ruffling and so increase the change of capture by macropinocytosis (Francis et al. 1993). Phagocytic cells (especially "professional phagocytes") all have wrinkled surfaces (eg Bessis 1973; Hallett et al. 2008), which it is thought acts as a membrane reservoir for the additional surface area required for phagocytosis (Dewitt and Hallett 2007; Hallett and Dewitt 2007). The wrinkled membrane is held in place against osmotic swelling pressure by proteins which link the plasma membrane to the underlying cortical actin structure, and include ezrin. Also, the scenario outlined, has echoes in macropinocytosis, which occurs without the need for particle recognition and is simply driven by the fusion of extended sheets of plasma membrane which enclose extracellular fluid (and occasional particles). Indeed it has been suggested that micropinocytosis is probably the mechanisms of feeding used by the "earliest eukaryotic ancestral cells" (King and Kay 2019;

Yutin et al. 2009) and thus the precursor of phagocytosis.

Introduction to the Chapters in This Book

This volume continues with a short history of the study of phagocytosis (Chap. 2), in which a long view is given of the background to the subject and the context in which present day advances can be compared. This is followed by chapters in which the latest developments in our understanding of the cell and molecular biology of phagocytosis are given. In Chap. 3, there is an in-depth discussion of the roles of phospholipids in phagocytic membranes and of surface charge in phagocytosis and phagosome maturation. Chapter 4 introduces some modelling concepts of phagocytosis and especially in relation to the phagocytic target shape. Chapter 5 describes some recent discovers in amoeba about decision-making during phagocytosis where the target shape is challenging, unusually long and difficult to swallow. The outcome of the challenge is controlled by membrane tension. Chapter 6 continues to look at membrane tension during phagocytosis and focusses on the role of ezrin, a protein which links the plasma membrane to the underlying cortical actin network. In Chap. 7, the role of cytosolic Ca^{2+} in phagocytosis is introduced and followed by a detailed discussion of the mechanisms and ion channels involved in the generation of the phagocytic Ca^{2+} signal. Chapter 8 focusses on a Ca^{2+} activated protease, calpain, and presents the evidence for this being an important step in phagocytosis. Chapter 9 discusses the molecular details of the phagosomal NADPH oxidase system, and the exquisite fine control that molecular complexity can achieve. In the immune system, it could be argued that the only goal of neutrophil chemotaxis and phagocytosis is to bring the NADPH oxidase complex into close apposition with the target microbe in order to accomplish its destruction. The final chapter attempts to briefly look back at the progress in understanding phagocytosis given in the preceding chapters, and briefly looks forward to the future directions for phagocytosis research.

References

Aderem A, Underhill DM (1999) Mechanisms of phagocytosis in macrophages. Annu Rev Immunol 17:593–623. https://doi.org/10.1146/annurev.immunol.17.1.593

Bessis M (1973) Living blood cells and their ultrastructure. Springer, Berlin

Bloomfield G, Kay RR (2016) Uses and abuses of micropinocytosis. J Cell Sci 129:2697–2705. https://doi.org/10.1242/jcs.176149

Cerda E, Mahadevan L (2003) Geometry and physics of wrinkling. Phys Rev Lett 90. Article Number:074302. https://doi.org/10.1103/PhysRevLett.90.074302

Chapmanandresen C (1962) Studies on pinocytosis in amoebae. C R Trav Lab Carlsberg 33:73–264

Dewitt S, Hallett MB (2007) Leukocyte membrane "expansion": a central mechanism for leukocyte extravasation. J Leukoc Biol 81:1160–1164

Francis CL, Ryan TA, Jones BD, Smith SJ, Falkow S (1993) Ruffles induced by salmonella and other stimuli direct micropinocytosis of bacteria. Nature 364:639–642

Hallett MB, Dewitt S (2007) Ironing out the wrinkles of neutrophil phagocytosis. Trends Cell Biol 17:209–214

Hallett MB, von Ruhland CJ, Dewitt S (2008) Chemotaxis and the cell surface-area problem. Nat Rev Mol Cell Biol 9:662–662. https://doi.org/10.1038/nrm2419-c1

Huck WTS (2005) Artificial skins – hierarchical wrinkling. Nat Mater 4:271–272

King, J.S., Kay, R.R. (2019) The origins and evolution of micropinocytosis. Philos Trans R Soc B Biol Sci 374. Article Number:20180158. https://doi.org/10.1098/rstb.2018.0158

Leidy J (1875) On the mode in which Amoeba swallows its food. Proc Acad Natl Sci Phila 143. https://archive.org/details/proceedingsofaca26acad/page/142

Lim JP, Gleeson PA (2011) Macropinocytosis: an endocytic pathway for internalising large gulps. Immunol Cell Biol 89:836–843

Margulis L (1970) Origin of eukaryotic cells. Yale University Press, New Haven

Martijn J, Ettema TJG (2013) From archaeon to eukaryote: the evolutionary dark ages of the eukaryotic cell. Biochem Soc Trans 41:451–457. https://doi.org/10.1042/BST20120292

Martin WF, Garg S, Zimorski V (2015) Endosymbiotic theories for eukaryote origin. Philos Trans R Soc B Biol Sci 370:20140330. https://doi.org/10.1098/rstb.2014.0330

Martin WF, Tielens AGM, Mentel M, Garg SG, Gould SB (2017) The physiology of phagocytosis in the context of mitochondrial origin. Microbiol Mol Biol Rev 81:e00008-17. https://doi.org/10.1128/MMBR.00008-17

Metchnikoff É (1893) Lectures on the comparative pathology of inflammation, delivered at the Pasteur Institute in 1891. Kegan Paul, London. https://archive.org/details/lecturesoncompar00metcuoft/page/n67

Metchnikov E (1889) Researches sur la digestion intracellulaire (Research on intracellular digestion). Ann Inst Pasteur (J Microbiol) III:25–29. gallica.bnf.fr/ark:/12148/bpt6k6436880n/f31.image

Pocivavsek L, Dellsy R, Kern A, Johnson S, Lin B, Lee KYC, Cerda E (2008) Stress and fold localization in thin elastic membranes. Science 320:912–916

Sagan L (1967) On the origin of mitosing cells. J Theor Biol 14:255–274. https://doi.org/10.1016/0022-5193

Spang A, Saw JH, Jørgensen SL, Zaremba-Niedzwiedzka K, Martijn J, Lind AE, van Eijk R, Schleper C, Guy L, Ettema TJG (2015) Complex archaea that bridge the gap between prokaryotes and eukaryotes. Nature 521:173–179. https://doi.org/10.1038/nature14447)

Swanson JA (2008) Shaping cups into phagosomes and macropinosomes. Nat Rev Mol Cell Biol 9:639–649

Wang L, Castroac CE, Boyce MC (2011) Growth strain-induced wrinkled membrane morphology of white blood cells. Soft Matter 7:11319–11324

Yutin N, Wolf MY, Wolf YI, Koonin EV (2009) The origins of phagocytosis and eukaryogenesis. Biol Direct 4:9. https://doi.org/10.1186/1745-6150-4-9

Zaremba-Niedzwiedzka K, Caceres EF, Saw JH, Bäckström D, Juzokaite L, Vancaester E, Seitz KW, Anantharaman K, Starnawski P, Kjeldsen KU, Stott MB, Nunoura T, Banfield JF, Schramm A, Baker BJ, Spang A, Ettema TJG (2017) Asgard archaea illuminate the origin of eukaryotic cellular complexity. Nature 541:353–358. https://doi.org/10.1038/ nature21031

A Brief History of Phagocytosis

2

Maurice B. Hallett

Abstract

This chapter outlines some of the more significant steps in our understanding of the phenomenon and mechanism of phagocytosis. These are mainly historical, ranging from near the advent of microscopy in the seventeenth and eighteenth century up to the period before the Second World War (1930s). During this time, science itself moved from being the domain of the wealthy enthusiast to the professional and funded university scientist. Not surprisingly progress was slow of the first two centuries of phagocytic research, but accelerated around the late nineteenth century and the turn of the twentieth century. Since then progress has accelerated still further. This chapter however aims to put our current progress into a historical context and to explore some of the interesting personalities who have set the ground work for our current understanding of the subject of this book, namely phagocytosis.

Keywords

Animalcules · The Phagocyte theory · Early observations of phagocytosis · Early experiments on phagocytosis · Early Microinjection experiments

Introduction

In this chapter, the focus will be on the history of research into phagocytosis and some of the researchers who you may find interesting and important. As this a only "brief history", I will start with the advent of microscopy (since phagocytosis is a purely microscopic event) and will end around the inter-war period (ie 1920–1939) after which more "modern" research approaches were adopted. This end-date is of course an arbitrary decision. As is my focus on European and North America research. However, I apologise for any absence but unfortunately, I neither know of nor could comment on any Arabic, Asian or any other pre-1930 phagocytosis research. I am sure that it exists, but I have been unable to find any information on this (and would probably be unable to read it, if were pointed out to me). I therefore apologise to the descendants of those who have made significant contributions to understanding phagocytosis (from whatever continent of origin), but whom I have omitted.

M. B. Hallett (✉)
School of Medicine,, Cardiff University, Cardiff, UK
e-mail: hallettmb@cf.ac.uk

© Springer Nature Switzerland AG 2020
M. B. Hallett (ed.), *Molecular and Cellular Biology of Phagocytosis*, Advances in Experimental Medicine and Biology 1246, https://doi.org/10.1007/978-3-030-40406-2_2

Purpose of Chapter

In preparing this chapter, three things became immediately apparent.

1. The first is that collaboration and sharing of techniques and insights (sometimes unacknowledged) between seekers of true knowledge, is always a good thing. In the early stages of the history of phagocytosis there are many obvious examples where understanding leaps forward. For example, Leeuwenhoek (c 1674), rightly recognised as the Father of microbiology, discovered single cell organisms (animalcules) in "dirty" water using the simple of hand-held microscope described earlier by Robert Hooke in his book *Micrographia* (1665). Hooke himself preferred the compound (tube) microscope but when Leeuwenhoek wrote a letter to Royal Society in London describing aspects of the previously unseen microscopic world, Hooke was the one who replicated the observation (despite failure by another Royal Society Fellow, Nehemiah Grew, and by himself on two earlier occasions) and recommended its publication. What is especially remarkable was that the letters from Leeuwenhoek were written in Low Dutch which Hooke had to first to teach himself how to read (Lane 2015). All Leeuwenhoek's letters to the Royal Society were published in its journal, Philosophical Transactions (in English, having first been translated by the founding editor of the journal, Henry Oldenburg) within a few months of receipt (Lane 2015). Were Leeuwenhoek's observations not accepted and published, it is likely someone else may have done so later, but there would have been a slowing down of progress; and many others (unaware of Leeuwenhoek's findings) wasted their time by following older theories which were obviously absurd or untenable in the light of Leeuwenhoek's discovery.

2. The second is how the obvious acknowledgment of much progress in understanding cell biology has been made since the invention of the microscope: but the surprise is how little our current understanding has increased since the earliest papers. Many details of the molecular and biophysical process of phagocytosis, of course, are now understood and questions which could not have been previously asked have now been answered. These are, of course, the subject of subsequent chapters in this volume. But in the context of the big historical stories of phagocytosis research, it could be argued that these are but the necessary dotting of "i"s and the crossing of "t"s of a story already written. The earlier researchers had the fun of being constantly excited by the "remarkable" and undreamt of worlds and phenomena. It was of course much easier to discover that single cells can "eat" other smaller single cells (by simple observation) than to establish the molecular mechanism by which they do so. But it surprising to see the amazing ingenuity of the earlier researchers, who had no off-the-shelf technical kits and equipment solutions; and with these were able to discover intracellular details of the process such as the pH within the phagosome and the role of localised cytosolic Ca^{2+} changes (see later).

3. The third is that there really is no start-date for "history", especially in the history of scientific discovery. Every scientific advance depends on an earlier advance: which in turn depends on an even earlier one: and so on until we reach what is obvious or common knowledge. It is possible that somewhere in this chain, there was a "eureka" moment, on which later research was built. However, even then, I would argue that that "eureka" moment" arose to the "prepared mind" who already understood the state of knowledge to that point and so was suddenly able to see what had previously been obscured by confusion, in a sudden moment of clarity. But this clarity arose because of an observation or experiment. With Archimedes, it was a commonplace observation made by many before (ie everyone would know that a full bath would overspill as you get into it). With Metchnikov, it was an experiment that no one had done before (the thorn in the star-fish experiment – see later in this chapter) designed to test an hypothesis. However with both these

(and other) "eureka moments", the observation and the experiment were in the context of deep thought and an attempt to solve a specific puzzle. Many of the puzzles of phagocytosis have been solved, but many more remain.

The purpose of this chapter is therefore not only to set the historical background for the subsequent chapters, but also to re-tell some of the stories in the historical adventure which may inspire future imaginative researchers to see through the confusion of the thousands of papers published annually so that they may have their own "eureka" moment.

Let's Start at the Very Beginning

As Rodgers and Hammerstein wrote (1959), it is "a very good place to start". However, finding the "very beginning" of phagocytosis is not easy. It may be supposed that the "beginning" of research into phagocytosis would start with the naming of the process as being "phagocytosis". This term was first used by Ilya Metchnikov (1845–1916). Many reviews of the history of phagocytosis, not surprisingly, therefore start with Metchnikov. After all, he won the Nobel Prize in 1908 for his convincing work on his "phagocytic theory". He was, thus, undoubtedly of considerable importance to our story, as we will see later, and can be considered as the "father of phagocytosis". Olga Metchnikov, Ilya Metchnikov's second wife, gives an account of how the name phagocytosis arose in her biography of her husband, La vie d'Elie Metchnikoff (1920), later translated into English as The Life of Elie Metchnikoff (1921). She writes that:

*On the way back to Russia through Vienna (in 1882), he (Metchnikov) went to see the Professor of Zoology, (Carl Friedrich Wilhelm) Claus .... and expounded his theory (ie the unnamed phagocytic theory) to them. They were much interested, and he (Metchnikov) asked them (Claus and his colleagues) for a Greek translation of the words "devouring cells" (or "eating cells"; or, in the German language which they were presumably speaking, "Fresszell"), and that is how they (the cells) were given the name of **phagocytes**. ... It (as* the "phagocyte theory") appeared s͞c in 1883 (ie Metchnikoff 1883).

The name which Claus came up from the Greek φαγεῖν (*phagein*) "to eat or devour", appearing as the first element "phago-"; and κύτος, (*kytos*) "a container", *as the second element "cyte"* (now used as a designation for a number of cell-types eg leukocyte, lymphocyte, hepatocyte, adipocyte and many more). By extension of the "English" name for phagocyte (derived from Greek by a German for use by a Russian), the process these cells undertake is called **phagocytosis.**

But, despite having no name previously, Metchnikov was not the first to observe and describe the process of phagocytosis. In a masterly review, Thomas Stossel (1999) points to many earlier description of ingestion of particles by cells. Indeed, Metchnikov made his discovery that phagocytes in higher animals were immune cells, by first looking at the evolution of digestion, which was already known to occur in "lower animals" by phagocytosis. So it must be in "lower animals" that phagocytosis was first described. These "lower animals" are of course, single-celled animals (or animalcules, as they were called by Leeuwenhoek). There is a very accurate description of phagocytosis by amoeba published in 1875 (many years before Metchnikov's 1883 paper) by the American polymath Joseph Mellick Leidy (1823–1891; Fig. 2.9a). In 1853 (aged 30 years) he was elected Professor of Anatomy University of Pennsylvania, and later also President of the Academy of Natural Sciences of Philadelphia (where an impressive statue, 8 ft high on a 10 foot plinth still remembers him). As a palaeontologist and anatomist, he was an early supporter of Darwin's theory of evolution, ensuring Darwin's election to membership of the Academy. But Leidy was also a brilliant microscopist and in 1846 became the first person to use the microscope to solve a murder (Hare 1923) by showing that the blood on the murderers clothes was not the chicken blood that he had claimed. Under the microscope, Leidy could see that the red blood cells were not nucleated, as are avian red blood cells, and so the murder's explanation could not be true. It was also as a microscopist that Leidy reported to the Philadelphian Academy his observation of how amoeba "swallow food" ie phagocytosis (Leidy 1875). Before Leidy's report, as we shall see later, it was already known that in the microscopic world, cells could swallow other smaller cells, but the mechanism was unclear. In Amoeba, it looked as if the "food particle" becomes stuck to the viscous gel which it was thought made up the amoeba's body and simply sunk into it, just as a hazelnut in contact with a viscous sugar syrup is at first stuck on the syrup surface and then sinks into it, to finally be encased in the syrup. However Leidy reported to the Academy that "he had repeatedly observed a large Amoeba . . . . . . creep into the interstices of a mass of mud and appeared on the other side without a particle adherent". So the "sticky surface" theory did not seem to hold. He then reported what actually occurs during phagocytosis. His verbal report to the Academy, published in the third person, is as follows.

On the mode in which Amoeba swallows its food –
Prof Leidy remarked that .... on one occasion he had accidentally notice an Amoeba, with an active flagellate infusorium, a Urocentrum, included within two of its finger-like pseudopods. It so happened that the ends of these were in contact with a confervous filament, and the glass above and below between which the Amoeba was examined, effectively preventing the Urocentrum from escaping. The condition of imprisonment of the latter was so peculiar that he was led to watch it. The ends of the two pseudopods of the Amoeba gradually approached, came into contact, and then actually became fused, a thing which he had never before observed in an Amoeba. The Urocentrum continued to move actively back and forth, endeavouring to escape. At the next moment a delicate film of the ectosarc proceeded from the body of the Amoeba, above and below, and gradually extended outwardly so as to convert the circle of the pseudopods into a complete sac, enclosing the Urocentrum.

To anyone who has watched phagocytosis in real time, either by Amoeba or neutrophils, this all sounds very familiar and is a brilliantly accurate description of the event (and all from a single "accidental" observation!). The two "finger-like pseudopods" form the phagocytic cup holding the target, which gradually encroach around the particle. Then, the more rapid closure caused by the *"delicate film of the ectosarc"*, which is also often seen. Leidy's report surely confirmed the view of that phagocytosis was not simply "swallowing" by sinking into a viscous inert cell body and showed that phagocytosis was, in fact, an active and complex process undertaken by the cell.

Twenty years earlier, Claparède (Fig. 2.9d) also reported the act by which amoeba phagocytose their prey (Claparède 1854). Edouard Claparède, a Swiss anatomist working in Geneva, where there also remains an impressive statue to him, reports his impression of phagocytosis as follows.

The amebas feed in a most remarkable way. They glide slowly along, attach themselves like snakes to the prey to be swallowed and, like a soft mist moving across a landscape, completely encircle it: one has the impression that the object still lies underneath, but it has already been enclosed within the body. (translation by T.P. Stossel 1999)

Claparède use of the word "swallow" and the phrase "completely encircle it" suggests phagocytosis, but it is obviously a more romantic picture than that of Leidy. However, although it does not have modern day scientific rigour, and lacks the detail that might convince a sceptic that phagocytosis was an active process, (as does Leidy's description), it is easily recognisable as phagocytosis in action and another stepping stone in phagocytic research.

At this point in our journey into the past, it is important to point out that "the past" is truly a "foreign country" (Hartley 1953). When microscopes first revealed a world full of weird and wonderful "animalcules", there was a profusion and confusion of amazing animals. Almost anything could be believed. As well as amoeba, "dirty" water or water made "dirty" by leaving in contact with hay or soil (infusion) contained many other unicellular animals which were able to ingest food particles. One such group of animals are the heliozoans, also called sun-animalcules because they have stiff straight long projection (axopodia) made of micotubles which radiate from their spherical body (Fig. 2.1a) giving the appearance of a child-like depiction

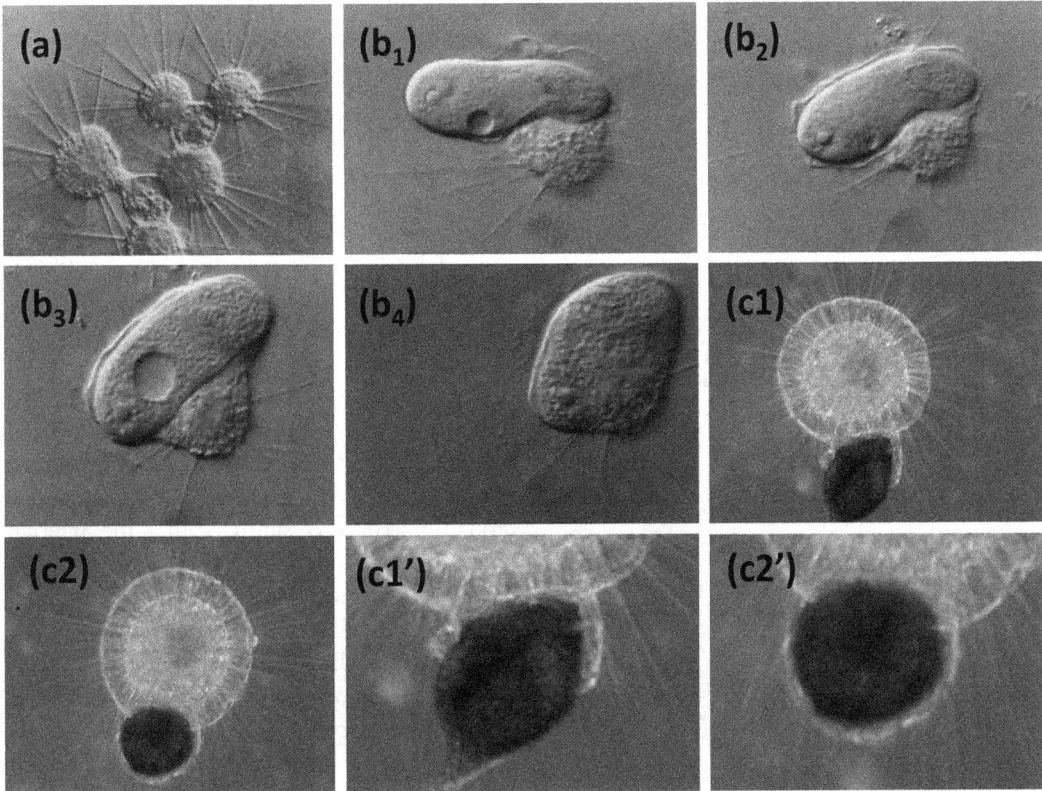

Fig. 2.1 Heliozoan Phagocytosis. (**a**) A phase contrast image showing a collection of heliozoans with typical spherical bodies and spines projecting to form "sun-like" shapes. Their size can vary from 20 μm to 1 mm in diameter. (**b1–4**) A sequence of images showing phagocytosis of a large paramecium by a heliozoan. (**c**) Images show-ing details of phagocytosis with (**c1**) and (**c1'**) showing the typical phagocytic cup and (**c2**) and (**c2'**) showing closure of the phagosome, with (**c1'**) and (**c2'**) at higher magnification. (The images are from the cell biological resource held by Institut für den Wissenschaftlichen Film Hausmann 1986)

of the sun. Surprisingly, they also undertake an amazing phagocytosis and can engulf targets of equal or greater size that themselves, once the prey has been immobilised by the spines (Fig. 2.1b). This type of phagocytosis can be slow, to complete (hence the need to catch the prey first on their spines), and can also involve a massive expansion of available cell surface area. This is achieved by exocytosis of granular membrane and, sometimes by recruiting other heliozoans, which together provide a joint phagosome which all partners can share. With smaller targets, phagocytosis procedes by the formation of a "classical" phagocytic cup (Fig. 2.1c). I have laboured this point, because it is the Heliozoan that provides the next major stepping stone. The Swiss scientist Rudolf Albert Kölliker (1817–

1905), studying Heliozoans, made perhaps the first major description of phagocytosis. Kölliker (Fig. 2.9c) held some interesting posts, including in 1844 Professor Extraordinary of Physiology and Comparative Anatomy at Zurich University and in 1847 Professor of Physiology and Micro-scopical and Comparative Anatomy at University of Würzburg. Kölliker was later elevated to the nobility in 1897 by Prince Regent Luitpold of Bavaria for his scientific contributions and in later publications, he is consequently called Albert von Kölliker (the insertion of "von" signifying ennoblement). Why Kölliker is important to our story is that in 1849, 6 years before Claparède's romantic and 26 years before Leidy's careful descriptions of phagocytosis by amoeba, he had provided a detailed and accurate description of phagocytosis by the heliozoan, Actinophrys. As this type of phagocytosis may not be as familiar as that of leucocytes or amoeba, Fig. 2.1 shows some examples, including from the movie available from the excellent archive of Institut für den Wissenschaftlichen Film (Hausmann 1986). With this in mind, we can see the accuracy of Kölliker's 1849 description (Kölliker 1849) given on page 202 of Zeitschrift für wissenschaftliche Zoologie, 1849 volume.

"The creature which is destined for food (ie trapped by the spines), gradually reaches the surface of the animal (ie the heliozoan, Actinophyrys), in particular, the thread that caught it is shortened to nothing, or, as it often happens, once trapped in the thread space, the thread unwinds from around the prey when close together and at the surface of the cell body". Here's what happens next:

The place on the cell surface where the caught animal is, gradually becomes a deeper and deeper pit (fig. 2f) into which the prey, which is attached everywhere to the cell surface, comes to rest. Now, by continuing to draw in the body wall, the pit gets deeper, and the prey which was previously on the edge of the Actinophrys, disappears completely, and at the same time the catching threads, which still lay with their points against each other, cancel each other out and extend again (fig 2g). Finally, the edges "choke" the pit, so that it is flask-shaped (flaschenformig) (fig 2g) all sides increasingly merging together, so that the pit completely closes and the prey is completely within the cortical cytoplasm (Rindensubstanz). Here it lingers for variable lengths of time (fig. 3f), but will always move towards the centre of the cell, and finally enters into the deeper part of the cell (fig. 3g), in order soon to find its finite fate within. (translated into English with the grateful assistance of *Google translate* at translate.google.com)

It is clear that this was not a single or accidental observation because Kölliker goes on to follow the fate of the ingested prey. He writes that usually the ingested infusorium (a catch-all name for the animalcules which appear after water is left in contact with hay or similar) is completely dissolved and that "the space that sheltered it (ie the phagosome) is diminished and disappears completely." However, he also reports that

On the other hand, if an indigestible remnant remains (a membrane of cellulose, a chitin skeleton,

a shell of a lynceus, or a radiolarian (raderthierche) etc.), it simply re-emerges by contraction of the homogeneous cytoplasm (leibessubstanz) (fig. 3m), in the direction the object followed on entering, until it finally leaves the whole area, while the canal and the opening which led it out, disappear without a trace.

From the drawings that Kölliker made (Fig. 2.2), it can be seen that the size of the target is smaller than the example given in Fig. 2.1b and is thus, probably, more familiar to today's phagocytologists with a phagocytic cup or pit in the cell body clearly drawn. It is also interesting that in his Fig. 2.3, the internalised material is clearly within a spherical membrane within the cell, with a small clear (water) space around it. He was surely seeing the internal phagosome. Furthermore, unlike the ciliates, such as paramecium, which have an anatomically identifiable and permanent "oral groove" (vestibulum) ending in the cytstome (cell mouth) for ingesting food, the Amoeba's "mouth" is a transient pit " opening ….(and)…. disappearing without a trace". Kölliker's published account of pseudopodia and the transient pit (phagocytic cup), and the visualisation of phagosomes is at least as accurate an account as given by Leidy 26 years later for phagocytosis by Amoeba. All the key elements of phagocytosis (as we understand it) are present in the Heliozoan. This may be, therefore, the first complete report of phagocytosis and is thus important for this alone.

But is this the "the very beginning"? Not quite.

The "swallowing" of food by single cells was known before Kölliker's report. For example, 10 years before Kölliker's paper, Andrew Pritchard (1804–1882) included some details which show that it was widely accepted that amoeba and other animalcules internalised food. Pritchard (Fig. 2.9b), an Englishman, was a member of a religious sect of Unitarianism which believed that "God and nature were one". He was also an optician/lens maker in London, UK, and had several shops in London that sold optical microscopes, microscopic accessories and microscopic preparations ready for use. He may thus have had a commercial, rather than purely academic, reason for publishing books such as

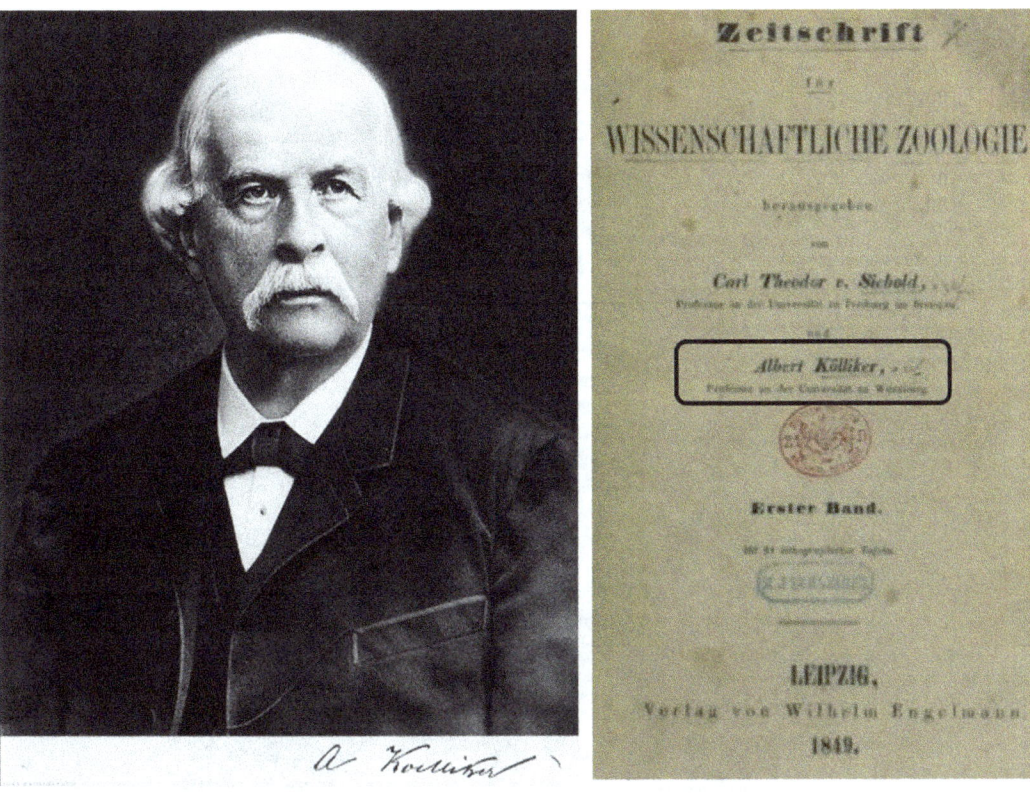

Fig. 2.2 Albert Kölliker. The image on the left shows a photograph of Albert von Kölliker and on the right the front cover of the ground–breaking report in 1849

"*List of 2000 Microscopic Objects*" (Pritchard 1835). But his book "The Natural history of animalcules" published the year before (Pritchard 1834) shows that he had a wide knowledge of the subject and is an important contribution to the history of phagocytosis. For example, he says of Vibrio punctatus (section 94): "*They are eaten by the Proteus diffluens and the large Vorticella which see*". In the same book, he earlier describes the process of "eating" by "Proteus diffluens", the melting Amoeba (Proteus section 22):

> *When in its contracted state, it (the amoeba) appears like a gelatine ball; this it readily changes, thrusting out branches of different dimensions in various directions. Some of its numerous forms are shown in the group, figures 8, 9, 10, 11 and 12. When it (the Amoeba) swallows animalcules which are covered with a crustaccous shell, as in figure 9, 10 etc, it accommodates its shape to the food. The mouth aperture is situated at the cross in figure 9.* (Pritchard 1834)

Pritchard tells us that "*the long animalcules within them (the Amoebae) are species of Bacillaria, which it has seized and eaten: they serve to exhibit the wonderful dilatation of their stomachs*". The part of the plate showing figures 8, 9, 10, 11 and 12 to which Pritchard refers in this paragraph are shown as Fig. 2.4a, b. Co-incidentally in Chap. 5 of this book, there are more modern examples of a similar phenomenon, together with an examination of the phospholipids signalling (as in Chaps. 3 and 9).

So in 1834, Pritchard, has nonchalantly stated (i) that Amoeba eat Vibrio, as if everyone already knew Amoeba ate other animalcules, and this was just a detail of its diet, (ii) the process of "eating" was by swallowing, and (iii) that the captured prey was clearly within the cytoplasm of the Amoeba. The detail that Amoebae "swallow" its food is suggestive of phagocytosis and since the objects swallowed are clearly inside the Amoeba,

Fig. 2.3 Phagocytosis by Actinophyrs (Heliozian) as reported by Kölliker in 1834. The images show figures from the original paper with their original labelling. (**a**) Kölliker 's figure 1, (**b**) Kölliker 's figure 2 and (**d**) Kölliker 's figure 3 with their original labelling referred to in the paper by Kölliker (1849). (**c**) shows an enlarged view of Kölliker 's figure 2 to more clearly show two stages of phagocytosis (original label *f*) the "pit" or phagocytic cup formation and (original label *g*) internalised prey or phagosome. There were no scale bars in the original drawings

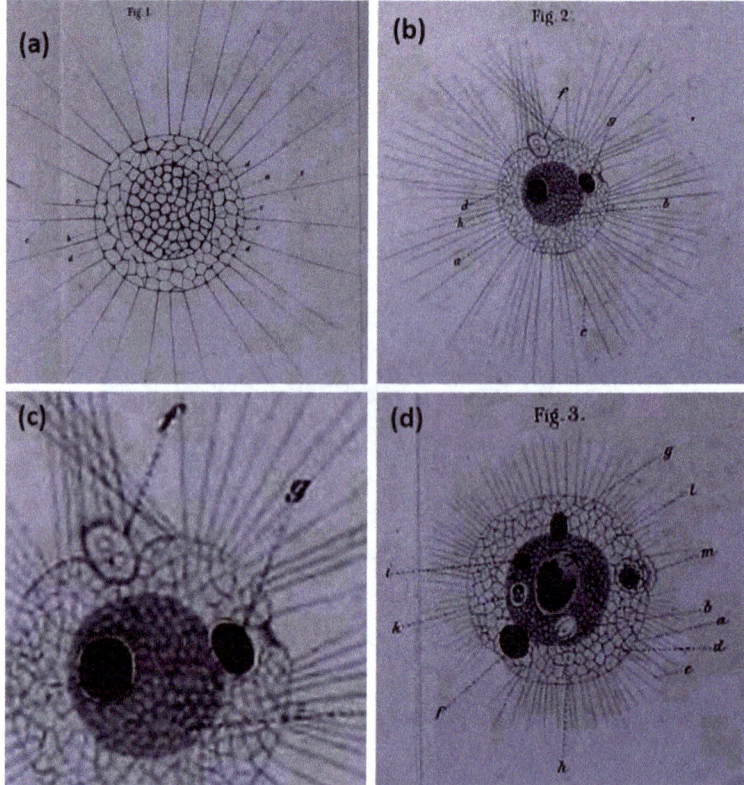

Fig. 2.4 Illustations from Andrew Pritchard's work (1834). (**a**) Shows "eating" by "Proteus diffluens" given in Proteus section 22 of The Natural history of animalcules" (1834), where the amoeba is distorted by the shape of the phagocytosed object *"covered with a crustaccous shell"*. (**b**) Shows the uptake of coloured particles (*"indigo, carmine or other minutely divided bodies"*) within phagosomes within the amoeba. (**c**) Shows heliozoan phagocytosis with the entrance and exit points on the cell marked and (**d**) shows the "proboscis", probably an extended (and stylised drawing) of the phagocytic cup

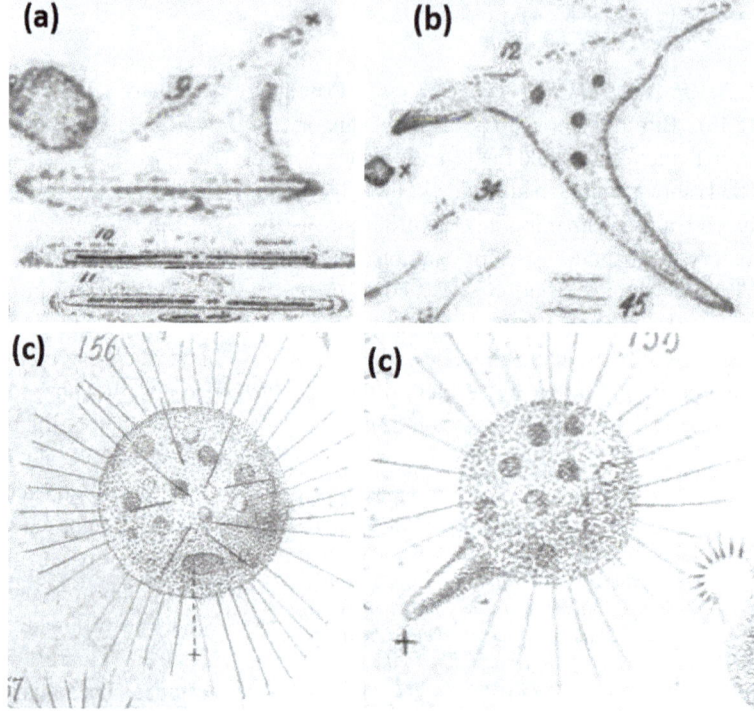

changing shape to accommodate it, may thus represent one of the first reports of phagocytosis by Amoeba. It is interesting that Pritchard must have observed the entire process of phagocytosis (without comment) as he could mark the site of the "mouth aperture" (see Pritchard's fig 9 in our Fig. 2.4a, b). This terminology may be strange to our ears, but the "mouth aperture" must be where Pritchard saw the "lips" of the Proteus open and "swallow" the object. We know that this was actually phagocytosis and we would use terms like pseudopodia, but how could Pritchard describe it other than opening a "mouth aperture"? It is interesting that he does not say that Amoeba had a mouth (an anatomically distinction mouth like he ascribes to many of the ciliates for example), but simply that its food was swallowed via an aperture. The "stomach" that he mentions is surely the phagosome, but as he has no other word for it, he used the analogy of the human stomach. When he describes the amoeba equivalent of a digestive tract, it is obvious that he is not describing a stomach. He says that the Amoeba's *"digestive organs consist of a number of sacs"* and demonstrates this by using coloured dye particles that, when ingested, ended up in these vacuoles.

When they (amoeba) are fed on indigo, carmine or other minutely divided bodies, they (the digestive sacs) remain circular: several of these are shewn in figure 12.

In Pritchard's figure 12 (our Fig. 2.4b), there are several discrete phagosomes, so presumably Pritchard was not implying a single stomach but that his coloured test objects were small enough not to distend the cell and that each ended up in its own vacuole. Pritchard also reports in section 235 that Trichoda Sol (Actinophyrs sol), the Heliozoan from which Kölliker described phagocytosis so clearly and completely, has *"as many as twenty polygastic sacs"*. He has trouble in accurately describing the way it feeds, but from his use of the word "suction", it is clearly something unusual.

This creature is an interesting object for the microscope: it preys upon other animalcules by suction and has been found attached to Kerona pulsulata. Size 1/900th of an inch (1/900th of an inch = 28 μm)

Pritchard also had trouble in describing the apparatus for this suction and gives one credible and another incredible view of the "proboscis" (see Fig. 2.4c, d). This must be the extending pseudopodia, but the description is difficult to understand.

Its mouth is elongated into a proboscis, as shewn in fig 158; this creature can contract at pleasure, and when turned towards or from the observer appears like an oval sac, as shewn by the dotted line and cross below figure 156

Presumably, the "suction" which Pritchard attributes to the Heliozoan was actually the movement of prey towards the cell by the action of the spines: or even by being attached to extending pseudopodia which, without high magnification and phase contrast, may be invisible, so that when they contract drawing the prey towards the cell body, it looks like suction. Presumably, the proboscis in fig 158 is an exaggeration, being a simple geometrical shape. However, it may be based on what was observed. The prominent phagocytic cup of the Heliozoan was shown as early as 1784 (as we shall see later). The appearance of the mouth as an "oval sac" (in figure 156) is more realistic, and it is tempting to suggest that Pritchard was seeing the "pit" or phagocytic cup described by Kölliker.

Pritchard, who was obviously well-versed in the world of microscopic animalcules, showed by his use of coloured particles to "feed" Amoeba (and other animalicules), that he must have read (or be aware of) the earlier work of Gleichen. Gleichen first used dyes particles to convince himself, and others, that animalcules really did engulf their prey.

Baron Friederich Wilhelm von Gleichen-Russworm (1717–1783), to give him his full name, began his career in the army, reaching Lieutenant Colonel by 1748 (aged 31 years), and master of stables (senior equerry) 2 years later. In 1756 he resigned from the army and moved into the "family" castle of Greifenstein (Fig. 2.5a). It was here that, fortunately for science, he put aside the militaria and "focussed" on microscopy and experimentation. The results of his work were published in 1778 in his book entitled "Treatise on seeds and infusion, and on production; with

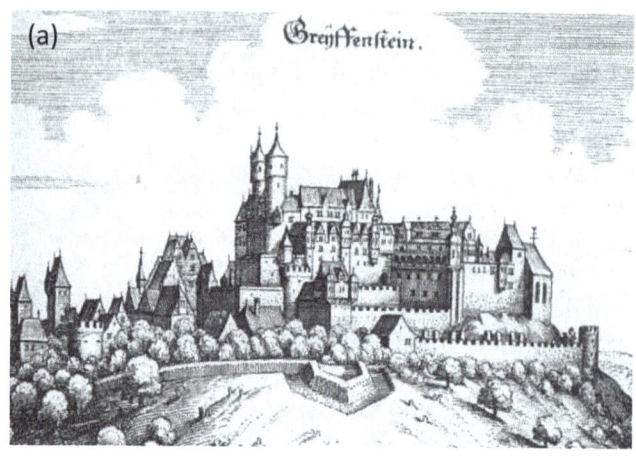

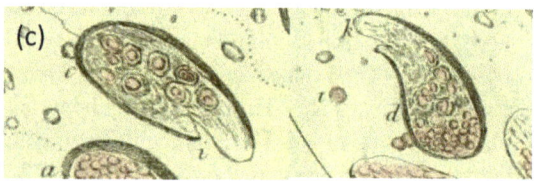

Fig. 2.5 The work of Baron Friederich Wilhelm von Gleichen-Russworm (1778). (**a**) The "family" castle of Greifenstein, where Gleichen underook his scientific studies. (**b**) the cover of his 1778 book "Treatise on seeds and infusion, and on production; with microscopic observations of the seed of animals, and various infusions" in which he reports important aspects of phagocytosis. (**c**) The uptake of red-coloured carmine particles was shown in Gleuchen's "pendeloques", which had "mouths". The clear space around the carmine is seen within the phagosomes. (**d**) Although not in colour, the drawing shows that there was no uptake of coloured particles by these animalcules which had no "mouths"

microscopic observations of the seed of animals, and various infusions" (Fig. 2.5b). In it, he explains why he is convinced that some of the objects he could see inside an animalcule which he calls "pendeloques" or pear-shaped pendants, which he thought might be unborn off-spring, were actually internalised food. He was convinced only after experimentation which he details in his explanation of Plate XXIIb entitled "Fressende Infusionsthierchen" and in the French translation "Voracious infusion animalcules" ie "Eating Infusion Animalcules". Having set up the objective of the experiment ie to test whether the vacuoles he could see in his "pendeloques" were internalised food, he continues:

> *So I coloured water with carmine, and I mixed it with an infusion of wheat, which contained many large "pear-shaped pendants" (pendeloques) and small "ovals", which lived there for some months. My expectation was fulfilled the next day; and I was not only convinced by the internal red colour of most animals, of an effective swallowing of food, but I also acquired more knowledge of their interior (4). This point, then, would be proved . . .*

Examples of the animalcules which Gleichen oberserved are shown in Fig. 2.5c, d. Now he knew the nature of the intracellular red particles, Greichen writes that

> *Henceforth I devoted all attention to the red pellets beneath my magnifier, ignoring all else*

Looking carefully at the intracellular red particles, he writes that *at first glance ... you cannot avoid the idea that the internal pellets are eggs, because they are surrounded by clear rings as seen around frog eggs.*

This is seen in his careful drawings (Fig. 2.5c) and may have been caused by the osmotic swelling of the phagosome as the insoluble dye particle begins to breakdown in the acidic environment. Interestingly, Gleischen becomes convinced that the "pendeloques" were "schluckung" or, in the French translation, "deglutition" ie *swallowing* the food. The same description (swallowing) was used earlier by Leidy in 1875, Claparède in 1854 and Pritchard in 1834 in describing phagocytosis by amoeba, and it may be tempting to think this is what Greichen also meant. However, this term was probably used by Greichen because the "pendeloques" which took up the red particles had "mouths" (Fig. 2.5c) whereas other animals which failed to take up the particles did not (Fig. 2.5d). He writes

A careful inspection of animals c and d showed in both animals an incision at their narrow side (i and k) which resembles so much a mouth in shape and position that I truly think it is one.

He also saw the indigestible dye particles being later ejected from "the *rear and once from the sides*", observations which he says "*required so much time, patience and visual concentration*" to convince himself that it really was happening. This may seem like a minor feature, but as a good scientist, he persisted with his "visual concentration" for 4 weeks and eventually saw the same event "10 to 12 times in innumerable observations" (ie n = 12 just to be sure!).

This is clearly excellent work. The drawing of the "pendolques", resemble Paramecium or similar cells which are "pear-shaped" and have a persistent oral groove. The mechanism by which Paramecium internalise particulate "food" is by drawing the particle into the mouth by beating cilia. Only once the food has been drawn down to the base of oral groove and in the vestibulum/mouth cavity, does phagocytosis occurs. Thus Greichen's observations of internalisation of dye particles were the result of true phagocytosis; but unlike the Heliozoan, phagocytosis itself was out of sight at the base of the oral groove. This would

make it difficult, without very good optics, to observe in real time. However a contemporary of Greichen, to whom he refers to several times in the book, Goeze, may have done so. The footnote to Greichen's report (possibly added after the main text was completed and containing Müller's classification name) states:

(footnote 4) Goeze was fortunate enough to have before him, in a hay infusion, a quantity of Müller infusion animalcules, a counsellor of state, and described what he named as trichorda cimex, because of hairs (silk) whose body is lined in the anterior and posterior parts. According to what he says of their voracity and their ability to seize other infusion animalcules, it is a real carnivorous animal, in the microscopic world, that can be called wolf of *the infusion.*

Johann August Ephraim Goeze (1731–1793; Fig. 2.9e) studied theology at University of Halle in Germany, becoming pastor in several places in Germany before becoming the first deacon of the seminary of Quedlinburg in 1787, where he later died. Presumably, his hobby was zoology, and he undertook microscopical research in his "spare time" during which time he published, what is recognised as the first to describe tardigrades. Was he also the first to describe phagocytosis? Here is Goeze's description of what Greichen called the "ability to seize other infusion animalcules" (Goeze 1777).

On November 8, 1776 ...
  Now I want to describe the predation scene in detail. As soon as the predator met one of the oval animals it suddenly dove at it and grasped it with the lips of its mouth which is located on its inferior side (number 8a). The captured animal defended itself as best it could. It struggled for a time in the jaws of the predator, especially if in an oblique position when seized. In that case the predator worked to turn the victim into a longitudinal position which was easier to accommodate to the tube which was its stomach. .... If the predator (number 7) seized the victim properly so that one end stuck in its throat, one might expect it to slide easily into the stomach. This is not what happens, however, in that the predator begins to choke and jerks itself backwards until the prey is fully swallowed. (Translation from Stossel 1999)

It is not clear exactly what the predator (which Greichen thought was T*richorda cimex*) was. Judging from the name given to it by Goeze,

J.A.E Goeze (1777)   Joblot (1718)

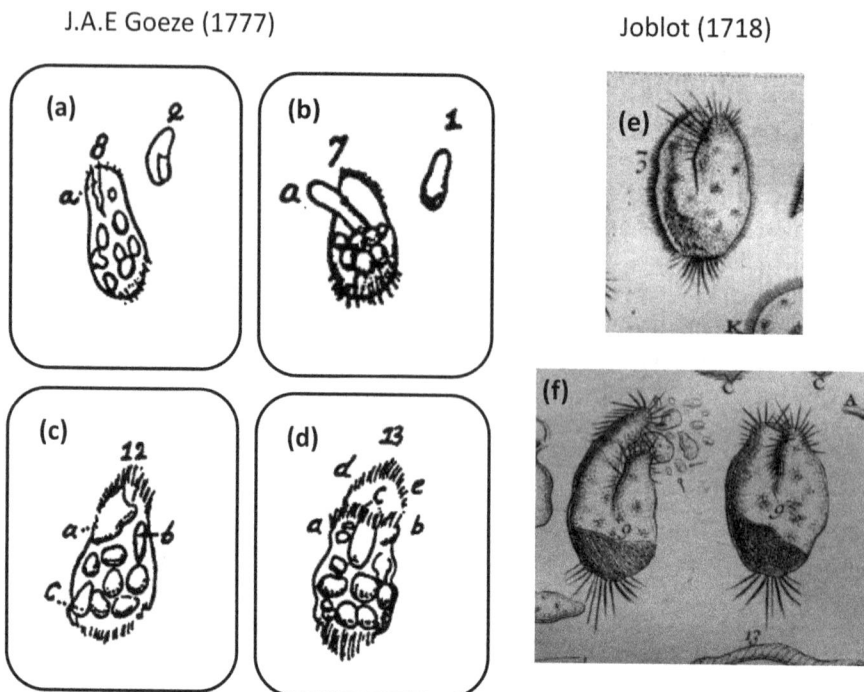

Fig. 2.6 Drawings of Devouring animalcules by Goez (1777) and Joblot (1718). (**a–d**) Drawings by Johann A.E. Goeze in his paper of 1777 showing Haarwanzen (hairy bugs) with "mouths" devouring prey. The animalcules are ciliated (hairy) and have permanent "mouths". (**e, f**)

Drawings by Louis Joblot from his book "Observations d'histoire naturelle, faites avec le microscope" showing similar cilated and "mouthed" animalcules, which he called *la grosse Araignee aquatique ("the fat water spider")*

Haarwanzen (hairy bug), and the description in Greichen's footnote as having *hairs which "lined its body in the anterior and posterior parts"* together with Goeze's drawings (Fig. 2.6a–d), it is reasonable to assume it is a ciliate and thus similar to Greichen's "pendeloque" (Fig. 2.5d). The drawings referred as numbers 7 and 8 to in the quote above are reproduced here in Fig. 2.6a, b; (as are numbers 12 and 13 in Fig. 2.6c, d, which Goeze says *"showed an astoundingly voracious predator. As I watched it, it swallowed three oval animals."*). Like Greichen's description, the mouth is an obvious feature, and as with Greichen's report, this seems to be similar to oral groove of Paramecium. If this is the case, then the description by Goeze is of the seizure of the prey animal into the mouth rather than its phagocytosis into the cell cytoplasm. Goeze's statement that after the seizure of the prey, it is "fully swallowed" seems to reflect the prey

being pulled down into the mouth (but still in the extracellular fluid) by the cilia hence the "hairy bug" appearing to *"choke and jerks itself backwards"*. Unfortunately, Goeze's description ends at this point and the phagocytotic event itself is not described. Perhaps the "the number 3 ocular and type A objective of (the) 'Composit' microscope" that he used did not have sufficient resolution or perhaps, Goeze thought it was "all over" after the "swallow". Greichen's dye particles, however, were internalised by the same or similar ciliates (see above and Fig. 2.7) showing that internalisation by phagocytosis had occurred. Goeze elegantly showed that the prey he was observing ended up within the "hairy bug" (and were still alive) by squashing the "hairy bug" to breaking point and allowing the internalised prey to escape. He reports:

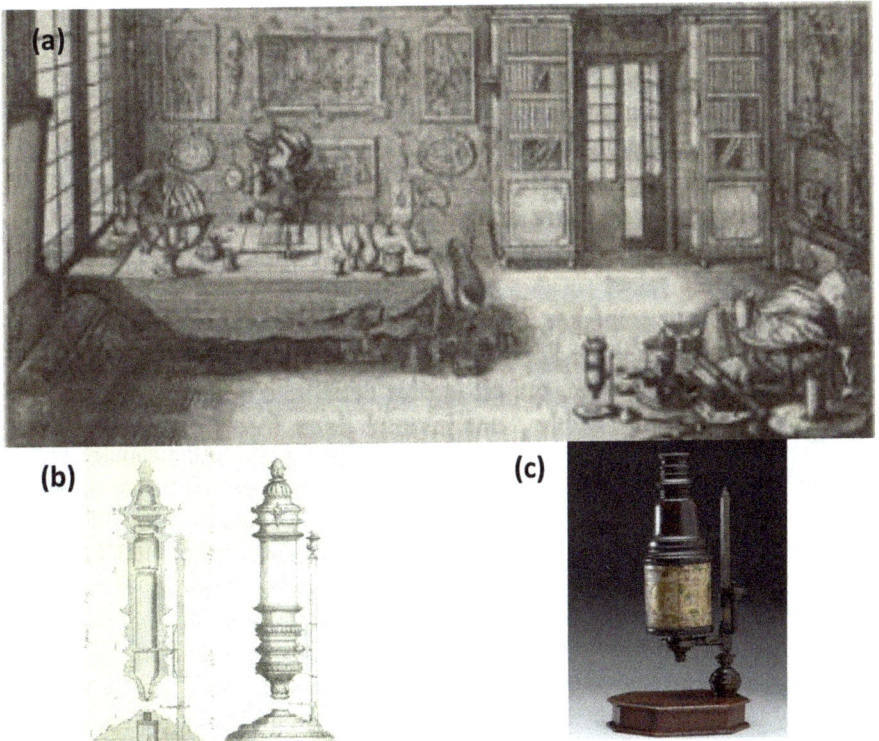

Fig. 2.7 The laboratory of Louis Joblot (1718). (**a**) shows a view of Joblot's laboratotry taken from the frontis peiec of his 1718 book "Observations d'histoire naturelle, faites avec le microscope". It shows books, microscopes and other items difficult to identify housed in a classical eighteenth century room within the Royal Academy of Painting and Sculpture in Paris. (**b**) Optical diagrams from Joblot's book showing improvements made by Joblot to the standard microscopes of his day. (**c**) A microscope of similar age and design to that shown in the frontispiece image if Joblot's lab

I must mention another experiment providing evidence that the hair bugs actually swallowed the oval animals. I applied a droplet to the lower plate of the recently devised Hoffman press with screws and small watch-springs–I will describe the device at another occasion—and superimposed the other plate, screwing it down until it touched the surface the drop. Then I gave the predators time to function and devour. One swallowed five small animals one by one, and they all were visible in its belly. Then I screwed the plate down further which caused the predator to become quiescent, whereas the swallowed animals still moved about within its body. By gentle further application of the screw, I gave them a final squeeze that caused the predator to burst, releasing the swallowed animals to freedom. The liberated animals immediately continued swimming in the liquid. I was delighted to be the deliverer of those swallowed victims, even in the microscopic world.

The last sentence seems to reflect Pastor Goeze's religious upbringing, although the zoological morality of killing one animal to save another is obviously complex. The experiment is, however, a very early example of the power of microscopic manipulation of cells to provide compelling evidence. Goeze in 1777 said of his micro-press that "*This is convincing proof of the benefits of this excellent device for microscopical experiments which otherwise would be impossible.*" His foresight was good and future "excellent devices" such as micro-manipulation, optical tweezers, microinjection and advanced microscopic imaging would indeed provide "convincing proof" of cellular events during phagocytosis "*which otherwise would be impossible.*"

For completeness, we should notice that Goeze referenced Joblot when describing the seizure of oval cells by his "peneloques". Louis Joblot (1645–1723) was French polymath and the first French microscopist. He was probable inspired by Huygens' visit to Paris in 1678 when Huygens demonstrated infusoria before the Academy of Sciences. Despite his obvious scientific interest, Joblot, from a well-off merchant family, in1680 accepted an unpaid appointment as assistant professor of mathematics at the Royal Academy of Painting and Sculpture in Paris, before becoming a full professor in 1699 (and receiving a salary). Joblot published his important book, "Observations d'histoire naturelle, faites avec le microscope" (Observations of natural history made with the microscope, 1718) while at the Academy of Painting and Sculpture (see Fig. 2.7a). In it, he described in detail some improvements to the then existing microscopes, including diaphragms in compound microscopes to correct for chromatic aberration (Fig. 2.7b). He also described and drew beautifully, but imaginatively, the animalcules he saw and named (also imaginatively).

Goeze (1777) wrote, as follows:

> The Privy Councillor did not cite an authority, but I believe it correct to see these animals in the figures of Joblot, volume 1, P. 11, tables 2.f.3 and 8.f.9.9, which he calls la grosse Araignee aquatique ("the fat water spider") on p. 78. He also mentioned that they devoured Cornemeases ("bagpipes"). The figures, however, are a bit unnatural, as is typical for Joblot's pictures.

The "fat water spider" is probably the same as Goeze's "hairy bugs" and Müller's Trichod cimex; although Müller identifies it with Joblot's "Pettit Araignee aquatique" ie small water spider). Joblot describes the encounter between his "fat water spider" and its prey in his book Observations d'histoire naturelle, faites avec le microscope (Observations of natural history made with the microscope) dated December 1714 (part 2 page 78)

> (The fat water spider) approaches the figure of an oval cell; and its slightly squeezed mouth sometimes seems split up to the middle of its body, its lips are filled with small moving hairs, whose speed seems to be communicated internally to a small body which is perhaps the heart etc … (The fat water spider) feeds on other smaller fish, which we have called Bagpipes, and which seem to move in their bodies for some time.

The drawing are, as Goeze correctly said, "a bit unnatural" but clearly show the similarity to Goeze's "hairy bugs": its mouth seemingly to split up the body, the lips with moving hairs (Fig. 2.6e, f). However, as with Goeze's description, it seems that Joblot is again seeing only the seizure of the prey and not phagocytosis itself.

We must now bring in Otto Frederik Müller, (1730–1784) a contemporary of Goeze. Goezez wrote that on seeing the carnage brought about by the "hairy bugs" on the "oval bugs" that "At first I could not believe my eyes, because my mind recalled the works of Mr. Müller in "Histor. Verm Vol. 1. p.2.p.88" declaring "Nee ullus oculatior animalcula revera ab animalculis devorari vidit"" (ie no observer really saw an animalcule devour others). It is unclear why Müller made this statement but the use of the Latin word "revera" (in reality) suggests that he was questioning what had been was observed was animalcules actually eating other animalcules. This suggests that he thought the observations of others earlier than Müller and Goeze (eg Joblot) were of prey capture rather than "devouring". What Müller will be remembered for is his book "Vermium Terrestrium et Fluviatilium, seu Animalium Infusoriorum, Helminthecorum, et Testaceorum non Marinorum, succincta Historia (1773), in which, for the first time, he arranged the "infusion animalcules" into a logical genera and species. Müller (Fig. 2.9f), a Dane, initially trained for the Church, but was never ordained. Instead, he travelled European for a few years before settling down in Copenhagen with a wealthy wife. Presumably, having a such a wife relieved him of the need to earn a living, and he took up zoology and microscopy as "hobbies" but it is obvious that they became an obsessions. His new classification made descriptive but ambiguous terms (like "pendeloques" and "hairy bugs") obsolete and enabled future progress by ensuring that similar animalcules studied by different workers were or were not referring to the same animalcule. Müller tries to back-date the new classification by giving

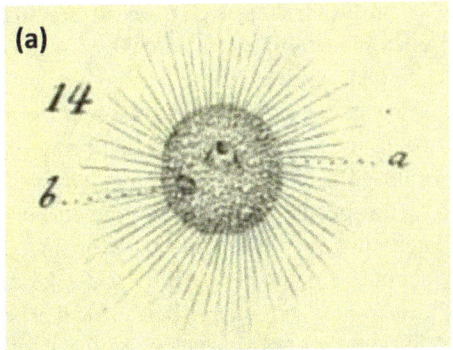

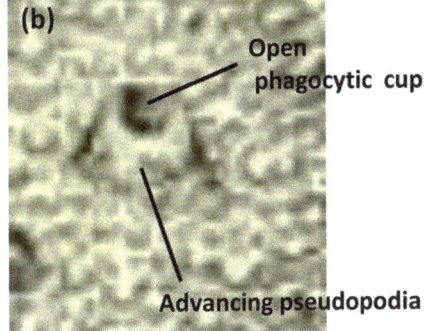

Fig. 2.8 O.F. Müller's phagocytic cup (c 1777). (**a**) Shows Müller's drawing of a heliozoan in the process of consuming the animalcule, Lynceus. The labels on the original drawing are shown and are labelled as showing "at the centre a raised nipple open, transparent (pellucentremque) that consumed Lynceus." where original label "*a*" is "papilla oris" or nipple mouth and "*b*" is "Insectum devoratum" or insect drained or devoured (presumably the shell remains of the Lynceus. (**b**) shows a closer view of the drawing of the "nipple mouth" which is probably a phagocytic cup with advancingpseudopodia, as labelled

the older names when he could. For example, the animalcule which Goeze called the "hairy bug", Greichen called by the new classification name "Trichorda cimex", which Müller tells us was called "petite araignee aquatique"("little water spider") by Joblot. So, in theory, data from different scientists could be cross referenced; eg Goeze's "hairy bug" was Joblot's "little water spider". The problem was that many animalcules look similar and, without photography, reliance must be placed on the skill of the drawing and the verbal description. In the book published in 1786, after Müller's death, "Animalcula infusoria fluviatilia et marina" useful illustrations are included and some animalcules, such as in section 177: Trichoda Sol, the Sun animalcule (Heliozoa Sol or Actinophrys Sol), which earlier featured in our story, is easily identified from the drawings in Müller book (Fig. 2.8b). Müller gives some stories and anecdotes (in difficult Latin) under each heading. He included this under Trichorda Sol the following:

Brunswigg ae aestate anni 1777. amicissimus wagner, me praesente, Lynceum ex interaneis expressit, hinc animalculum, licit maxime deses cohabitantia devorat

Google Translate says this means something like "Brunswigg the summer of the year 1777. Wagner my great friend, (said) in my presence, Lynceum from intestinal worms, on the side, that little creature, although the most lazy, dwell with them then is eaten by them", ie looking at the Heliozoan (as a lazy creature) and the Lynceum (on the side) that live with them, Wagner remarks something like, "they (lynceus) live with them (Heliozoan) until they eat them!". This makes more sense when one looks at the accompanying drawing of a Heliozoan with a *drained/devoured* body of a Lynceum indicated (Fig. 2.8a). Lynceum is a radiolarian with an indigestible "shell" as had been seen by Kölliker. There seems to be no doubt that this is a depiction of Heliozona "eating"

The description of the accompanying drawing is

Fig 13 Trich. Solem, centro clauso,
   14 eundem centro papillam elevato ac aperto,
pellucentremque devoratum Lynceum

This translates as:

Fig 13 Trich. Sun, the centre is closed,
   14 The same, with at the centre a raised nipple open, transparent (pellucentremque) that consumed Lynceus

Müller's figure 14 is shown in our Fig. 2.8a with the labelled a, and b are given as a "(a) papilla oris and (b) Insectum devoratum" translated as "(a) nipple mouth and (b) insect drained or devoured". The drained or devoured remains of the "insect" ie Lynceus is shown inside the Heliozoan. It is an amazing foreshadowing of Kolliker's report, where his Heliozoan also phagocytosed Lynceum leaving behind "the shell of a lynceus". What is more remarkable is that in fig 14, the "nipple mouth" (papilla oris) is shown in 3-dimensions (as Pritchard had attempted in 1834, shown as Fig. 2.8b). Müller shows this as not just a "pit" as described and shown in Kölliker's drawings, but has a projecting pseudopodia forming a complete phagocytic cup (Fig. 2.8b). Müller does not give a description of phagocytosis, but the drawing and its labelling is an important point on the graph tracking the history of phagocytosis as it may be the first observation of extending pseudopodia forming the phagocytic cup.

And so the journey ends, as Tolkein probably wrote. The path was winding, with many unexpected turns. But this usually is the way with scientific advance, even today. To summarise this attempt to "drill down" into the mass of old, multilingual writings, there was no single "Eureka" moment, but a gradual progress with observations being repeated and assumption bring reinforced until gradually it becomes the "obvious". Some of the people involved are shown in Fig. 2.9. Perhaps Leidy, Kölliker and (surprisingly) Müller should step forward a little and take a bow in the Awards ceremony; Leidy for the best early description of phagocytosis: Kölliker for the first description of true phagocytosis: and Müller for the first depiction of the phagocytic cup. All these important aspects of phagocytosis, were chanced upon, and the implications not really explored. But they were reported accurately and so remain as markers in our phagocytic history. Of course, this was still not "the very beginning",

Fig. 2.9 Faces of some early phagocytologists. (**a**) Joseph Leidy (1823–1891: aged about 40 years) (**b**) Andrew Pritchard (1804–1882: aged about c 46 years) (**c**) Albert Von Kölliker (1817–1905 aged about 42 years) (**d**) Édouard Claparède (1832–1871; aged 28 years) (**e**) J.A.E. Goez (1731–1793; aged 55 years); (**f**) Otto F. Müller (1730–1784: aged about 30 years)

as these pioneers would not have had that chance to observe anything had Hooke not publicised the new Microscope in 1665, and Leeunkook had not excited the world with his animalcules (1674). Perhaps the invention of the microscope marks the "very beginning'"; but that is another story.

The Eureka Moment of Metchnikoff

We must next turn to Metchnikoff (1845–1916). He took phagocytosis from being an interesting oddity of animalcules only of interests to (rich) amateurs, to an important event of the immune system in multicellular organism and especially in mammals. So much has been written about Metchnikoff (Fig. 2.10a), that I feel it is unnecessary to give only but the barest outline and so direct readers to recent (and older) reviews of Metchnikoff's work including

those by Aterman (1998), Tauber (1992; 2003), McGonagle and Georgouli (2008), Cavaillon and Legout (2016) and Korzha and Bregestovskic (2016) and Gordon (2008, 2016). It should be noted that there are many English and French variants of his Russian name Илья Méчников, including "Metchnikov/Mechnikov/Metchnikoff/Metschnikoff/Mecznikow.", with his first name "Ilya/Elie/Ellie" with a corresponding initial E. or I. (This can add confusion and lead to omissions when searching for the work of this important phagocytologist, especially if relying only on electronic searching.).

Metchnikov himself is an interesting and well-travelled European. The biography of Metchnikov given to accompany his Nobel Prize Lecture in 1908 (Nobel Media Biographical 1908) states the he was born "in a village near Kharkoff", Russia, to "an officer of the Imperial Guard", who also owned land in the Ukraine steppes. We are told "even when he was a little

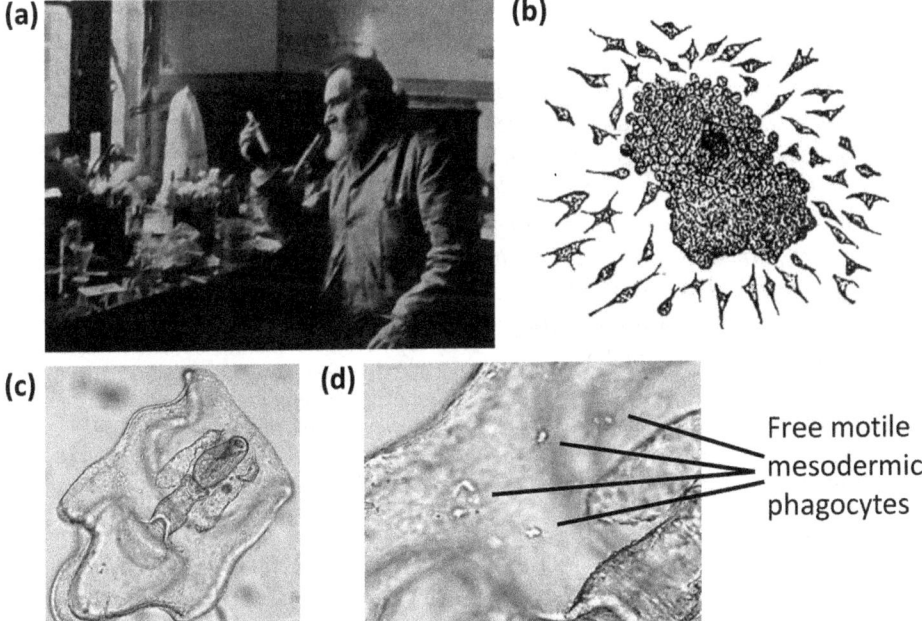

Fig. 2.10 Metchnikov. (**a**) The photograph shows Metchnikov in his laboratory at the Pasteur Institute, Paris, with his microscope behind him on the lab. bench. (**b**) The drawing (Fig. 32) in "Lectures on the Comparative Pathology of Inflammation" (Metchnikoff 1893) showing the result of the "thorn in the starfish" experiment. After injury of the starfish embryo, motile mesodermal pahgocytes were seen to accumulate at the site of injury. (**c**) The "starfish embryo", Bipinnaria asterigera, which has few internal organs and is as "clear of water". (**d**) The starfish embryo at higher magnification showing the free moving mesodermal cells clearly visible within the tissue of the embryo

boy, (he was) passionately interested in natural history, on which he used to give lectures to his small brothers and to other children". He went to the University of Kharkoff to study natural sciences, and completed the 4 year course in just 2 years. He then went to several European universities to undertake zoological studies and research. His doctoral thesis, submitted to Naples University, was on the embryonic development of the cuttle-fish and he then returned to Russia to take up a post as docent at the new University of Odessa and was duly promoted over subsequent years to the post of "Titular Professor of Zoology and Comparative Anatomy".

So far, Metchnkov's biography is a straight-forward CV of a hard-working and driven young scientist. However, in typical Russian tragic style, his life story now takes several dark turns before he finally reaches his Eureka moment.

Olga, Metchnikov's second wife tells us in her book, "The life of Elie Mechnikoff" (Metchnikoff 1921) that in St Petersburg, Metchnikov "*was devotedly fond of B's children, whom he used to take for walks on Sundays and to the theatre now and then; he was always ready to read to them and to indulge them in every possible way.*" In fact, Metchinkov's devotion to the children extended further than Sunday walks. Olga goes on to tell us that Metchnikov "*continued to entertain the dream of marrying one of them (the children) someday, and was particularly interested in the eldest, a girl of thirteen, intelligent, gifted, and lively*". Metchnikov's mother was not overjoyed by her son's choice when Metchnikov told his mother about the young girl, Ludmilla, judging by the letters which Olga published in her book. Things get even worse when it is obvious that Ludmilla is seriously unwell. Olga writes:

As Elie learnt to know his fiancée better, he became more and more attached to her. Their happiness seemed likely to be complete, but a cruel Fate had decided otherwise. The girl's health was not improving: her supposed bronchitis was assuming a chronic character (it was probably tuberculosis MBH). Yet the marriage was not postponed, and the bride had to be carried to the church in a chair for the ceremony, being too breathless and too weak to walk so far. The marriage ceremony of the bearded Metchnikov to his a young invalid bride thus took

place. It was clear however that "cruel Fate" had not finished yet.

Olga writes that after the wedding

Elie did his utmost to procure comforts for his wife, and hoped that she could still be saved by care and a rational treatment. It was the beginning of an hourly struggle against disease and poverty; his means being insufficient, he tried to eke them out by writing translations. His eyesight weakened again from overwork, and it was with atropin in his eyes that he sat up night after night, translating. There was but one well-lighted room in his flat, and he turned it into a small laboratory for the use of his pupils; his own researches he had to give up, his time being entirely taken up by teaching and translations.

Ludmilla's health fluctuated and her inevitable death hit Mechnikov hard, as Olga describes:

When Metchnikoff went back to his wife he found her with eyes wide open and so full of mortal anguish and utter despair that he could bear it no longer and went out hastily, not to show her his dismay. This was his last impression; he never saw her again … Only half conscious, he walked up and down the drawing-room, opening and closing books without seeing them, his mind full of dis-connected pictures ….. Time passed without his realising it. Then his sister-in-law came to tell him that all was over. This was on the 20th April 1873

Metchnikov did not attend his wife's funeral and sank into a dark depression.

After the catastrophe, Metchnikoff felt incapable of thinking of the future, his life seemed cut off at one blow; he destroyed his papers and reserved a phial of morphia, without any settled intention. …. He said to himself: "Why live? My private life is ended; my eyes are going; when I am blind I can no longer work, then why live?" Seeing no issue to his situation, he absorbed the morphia. He did not know that too strong a dose, by provoking vomiting, eliminates the poison.

And thus Metchnikov survived his suicide at-tempt. Olga tells us, thankfully, that eventually "*his thoughts turned towards Science; he was saved; the link with life was re-established.*"

In fact, although Metchnikov "was saved", there was still a little more tragedy yet to come. In Odessa, he lived in a flat below Olga's fam-ily; "*we were eight children, our ages ranging from one to sixteen years*" she writes. Metchnikov *"having heard that I (Olga) was interested in natural science, it occurred to him to offer to give me lessons in zoology. I was delighted. He asked and obtained permission from my parents, and we eagerly set to work.*

This was part of a plan by Metchnikov to have the"ideal scientific wife" because Olga writes that:

> *Elie, being strongly attracted by me, returned to his former idea of training a girl according to his own ideas and afterwards making her his wife.*

However, Metchnikov was forced to marry Olga before he had "fully trained her". Olga writes that Metchnikov *"might have realised his programme of completing my education first and marrying me afterwards if he had not been prevented by the complete lack of accord between his ideas and those of my father. .... Elie decided to ask for my hand without further delay"*. Olga leaves to the imagination of the reader what the *"complete lack of accord between his (Metchnikov's) ideas and those of my father"* was about, but Metchnokiv gets his way and the marriage is arranged. Olga's poignant description of the day of her wedding, highlights the lack of her maturity and the age difference between them.

"Our marriage took place in February 1875; it was a very cold winter and the ground was covered with a thick coating of glistening snow. A few hours before the ceremony my brothers came with a little hand sledge to fetch me for a last ride. "Come quick," they said, "this evening you will be a grown-up lady, and you can't play with us anymore!" I agreed, and we rushed out to the snowy carpet which covered the great yard of our house. In the midst of our mad race my mother appeared at the window; she had been looking for me everywhere and was much disturbed. "My dear child! what are you thinking of? It is late, you have hardly time to dress and to do your hair!" "One more turn, mother! It is the last time, think of it!" Other childish emotions awaited me; my wedding-dress was the first long dress I had ever worn, and I feared to stumble as I walked. Then, too, I was frightened at the idea of entering the church under the eyes of all the guests. My little brother tried to reassure me by offering to hold

my hand, and my mother made me drink some chocolate to give me courage.

Elie was awaiting us at the entrance; my shyness increased when I heard people whispering around us, "Why, she is a mere child!" The ceremony took place in the evening, after which Elie wrapped me carefully in a long warm cloak and we set off, the sledge gliding like the wind, towards our new home. In spite of the day's emotions, I rose very early the next morning in order to work at my zoology exercises and to give my husband a pleasant surprise. He was now free to superintend my education, a very difficult and delicate task when having to do with a mind as unprepared for life as mine was.

This marriage seemed to bring stability to Metchnikov's scientific life, but there were suggestions in Olga's writing that, in the early days, Olga may not have been completely happy (she met some younger perhaps) because she writes *"At a certain time, Elie, believing that happiness called me elsewhere, offered me my liberty, urging that I had a moral right to it. The nobility of his attitude was the best safeguard"*. However, she continues that *"As years went on, our lives became more and more united; we lived in deep communion of souls, for we had reached that stage of mutual comprehension when darkness flees and all is light."*

But Russia was a dangerous place to be in the 1870s and 1880 and conspiracies and reprisals were a constant worry, which combined with a clamp-down on travel trips by the University, caused Mechnikov extreme anxiety which led him, once again, to consider suicide. This time, writes Olga,

> *In order to spare his family the sorrow of an obvious suicide, he inoculated himself with relapsing fever, choosing this disease in order to ascertain at the same time whether it could be inoculated through the blood. The answer was in the affirmative.*

Fortunately, Metchnikov's suicide failed once again and he fully recovered. Perhaps "cruel fate" now relented, and from then on, things took a more positive turn. As is often the case, wealth provided the answer. Metchnikov inherited finance from Olga's parents and was thus freed from the University rules and especially its travel ban. Olga writes:

> *Thanks to my parents' inheritance, he was … to live henceforth independently. He wished to pursue researches on the shores of the Mediterranean: therefore, in the autumn of the year 1882, we went to Messina with my two sisters and my three young brothers. The children were no trouble to Elie, who loved them; on the contrary, he enjoyed organising the journey and arranging all sorts of pleasures for them.*

The importance of this release from University drudgery and its restrictions, allowing Mechnikov to go to Messina cannot be overstated. Metchnikov himself said

> *… it was in Messina that the great event of my scientific life took place. A zoologist until then, I suddenly became a pathologist. I entered into a new road in which my later activity was to be exerted.*

In order to understand this great event, we have to briefly follow Mechnikov's scientific thought processes. In his pre-doctoral days, in 1865 at the University of Giessen, he discovered intracellular digestion in one of the flatworms. As we have seen, it was well established that in single cell animals, digestion occurs after phagocytosis in "digestive vacuoles" within the cell. In higher animals, there are specialised structures, eg the gut, in which digestion occurs. In between the single cells and mammals, however, there are "intermediate examples". For example, in coelenterates there is a "gastic cavity" lined with cells that can also phagocytose. As this was the era of Darwin and the new ideas of evolution were in the air, Metchnikov's long-term project was to discover how it was possible that the gut in higher animals evolved from the unicellular animals with no gut and whether the evolution from intracellular digestion to intestinal digestion had other consequences. Olga tells us what was going on in Metchnikov's mind at that time:

> *The study of medusæ and of their mesodermic digestion confirmed him more and more in the conviction that the mesoderm was a vestige of elements with a primitive digestive function. In lower beings, such as sponges, this function takes place without being differentiated, whilst with other Cœlentera*

> *and with some Echinoderma the endoderm gives birth to a digestive cavity; yet, the mobile cells of the mesoderm preserve their faculty of intracellular digestion. As he studied these phenomena more closely, he ascertained that mesodermic cells accumulated around grains of carmine introduced into the organism.*

The last sentence about "grains of carmine" shows that Metchnokov knew of Greissen earlier work using coloured particles, and carmine in particular, being taken up into the phagosome of phagocytic animalcules (see above). Clearly Metchnikov's mind was prepared and he now had the freedom to follow a crazy idea. The time was ripe for his "Eureka moment": and here it is (in his own words):–

> *I was resting from the shock of the events which provoked my resignation from the University and indulging enthusiastically in researches in the splendid setting of the Straits of Messina.*
>
> *One day when the whole family had gone to a circus to see some extraordinary performing apes, I remained alone with my microscope, observing the life in the mobile cells of a transparent starfish larva, when a new thought suddenly flashed across my brain. It struck me that similar cells might serve in the defence of the organism against intruders. Feeling that there was in this something of surpassing interest, I felt so excited that I began striding up and down the room and even went to the seashore in order to collect my thoughts.*
>
> *I said to myself that, if my supposition was true, a splinter introduced into the body of a starfish larva, devoid of blood-vessels or of a nervous system, should soon be surrounded by mobile cells as is to be observed in a man who runs a splinter into his finger. This was no sooner said than done.*
>
> *There was a small garden to our dwelling, in which we had a few days previously organised a "Christmas tree" for the children on a little tangerine tree; I fetched from it a few rose thorns and introduced them at once under the skin of some beautiful star-fish larvæ as transparent as water.*
>
> *I was too excited to sleep that night in the expectation of the result of my experiment, and very early the next morning I ascertained that it had fully succeeded.*
>
> *That experiment formed the basis of the phagocyte theory, to the development of which I devoted the next twenty-five years of my life.*

This was a true "Eureka moment". The phagocytic cells within the starfish larvae moved to the site of injury as shown in his Eureka diagram (Fig. 2.10b). He repeated the experiment with different stimuli, and after his to visit to Claus in Vienna, he published his findings in 1883 (Metchnikov 1883: Fig. 2.10b). Metchnkov writes in "Lectures on the Comparative Pathology of Inflammation" that the same effect was observed with either a rose thorn, a sea urchin spine or a delicate glass rod; and that the mass of phagocytes was often visible to the naked eye. Furthermore, if the thorn was first soaked in carmine or indigo before insertion, these coloured particles "were eagerly devoured by the mesodermic phagocytes". Not only did Metchnikov devote the next 25 years of his life to this, by ingenious experiments, publications and arguments, but many others joined him.

Metchinov's choice of the starfish larva, which he tells us in his "Lectures on the Comparative Pathology of Inflammation" was a bipinnaria, the first stage in the larval development of the common starfish *Asterias and is as "transparent as water"* (Fig. 2.10c, d). This optical transparency was the key to the success and foreshadows, by more than a 100 years, the current use of Zebra fish as a model organism for studying inflammation and wound healing *in vivo*.

While Metchinov's starfish experiment was of obvious importance, there were a number of detractors. One of the major counter-arguments Metchnikov faced was from the "old guard" who still hung on to the theory of "spontaneous generation" of germs. The belief and evidence for this came from looking at the microscopic life which seemed to appear spontaneously in water left in contain with hay and other infusorions. When phagocytic cells were seen at sites of infection, such as pus and wounds, it was "obvious" to the old guard that these cells were "bad guys" and were simply carrying the spontaneously generated germs to the wound site. Olga Metchnikov says that the famous physician and cell biologist Rudolf Virchow warned Metchnkov of this as follows:

Metchnikoff was also greatly encouraged by Virchow, who happened to pass through Messina and came to see his preparations and his experiments, which seemed to him conclusive. However, Virchow advised him to proceed with the greatest prudence in their interpretation, as, he said, the theory of inflammation admitted in contemporary medicine was exactly contrary to Metchnikoff's. It

was believed that the leucocytes, far from destroying microbes, spread them by carrying them and by forming a medium favourable to their growth.

Metchnikov's counter argument to those who raised this was two-pronged; firstly that they should do experiments to refute his idea; Metchnikov pointed out that Pasteur's experiments had shown that germs do not "spontaneously generate"; secondly that they should consider Darwinian logic. Metchnikov (and others) could demonstrate the presence of phagocytic cells in all phyla and species, including those where they were no longer needed for digestion of food. If there were only a harmful role for phagocytyes (ie carrying germs to sites of injury), Natural Selection would exert its pressure and those animals which had no germ-carrying phagocytes would be more fit to survive, and so animals with germ-carrying phagocytes would dwindle away, to be replaced by animals with no germ-carrying phagocytes. It was easily verifiable evidence, that all animals retained motile phagocytes. This point was especially strong to the supporters of Darwin, pointing to a crucial and beneficial role for these cells. Far from phagocytes being "bad" and dangerous, these cells were clearly *essential* for health.

In 1888, Pasteur had given him a laboratory and an appointment in the newly built Pasteur Institute in Paris, and Metchnikov finally left Odessa for good. This was where he undertook many important experiments and published key papers (eg Metchnikov 1889) and books, including two volumes on the comparative pathology of inflammation (1892), and *L'Immunité dans les Maladies Infectieuses (1901)*, tranlsted into *English as* (Metchnikoff 1905). He was given many awards, most notably the Nobel Prize for Physiology or Medicine in 1908, which he shared with Paul Ehrlich. He spent the rest of his life at the Pasteur Institute in Paris and was so attached to the place that when he died in 1916, at his request, his ashes "were enclosed within an urn and placed in the library of the Pasteur Institute" (where they still remain).

The implications of Metchnikov's insight and work are more the subject of immunology than phagocytosis. However, those who are interested about the impact of Metchnikov's thorn in the starfish experiment and how it led to the understanding of the innate immune system can read, for example, (Nathan 2008, Cavaillon 2011; Merien 2016; Silverstein 2011; Teti et al. 2016; Nauseef 2014). Perhaps the profound implications of Metchnikov's simple "Eureka" experiment can be summarised best by Olga Metchnikov in her book:

> *This very simple experiment struck Metchnikoff by its intimate similarity with the phenomenon which takes place in the formation of pus, the diapedesis of inflammation in man and the higher animals. The white blood corpuscles, or leucocytes, which constitute pus, are mobile mesodermic cells. But, while with higher animals the phenomenon is complicated by the existence of blood-vessels and a nervous system, in a star-fish larva, devoid of those organs, the same phenomenon is reduced to the accumulation of mobile cells around the splinter. This proves that the essence of inflammation consists in the reaction of the mobile cells, whilst vascular and nervous intervention has but a secondary significance. Therefore, if the phenomenon is considered in its simplest expression, inflammation is merely a reaction of the mesodermic cells against an external agent. Metchnikoff then reasoned as follows: In man, microbes are usually the cause which provokes inflammation; therefore it is against those intruders that the mobile mesodermic cells have to strive. These mobile cells must destroy the microbes by digesting them and thus bring about a cure.*

Inside the Phagocyte

From the start of the twentieth century, the history of phagocytosis accelerates and becomes "modernised" and becomes more familiar to modern science quite quickly. I have therefore chosen just two discoveries, the importance of which still resonating at the start of the twenty-first century.

Phagosomal pH

The first is the pH of the phagosome which, though not strictly part of the process of phagocytosis itself, it is a post-phagocytic event clearly triggered by phagocytosis. Following Greichen in the late 1700's, who first used coloured particles

to convince himself that external particle really did end up inside living cells, this became a fairly routine approach. But it was not until 1847, that Rustizky hit on the idea that if the colour of the particle were sensitive to a chemical change, information could be gleaned. He was interested in bone resorption by "giant cells" and reasoned that, since bone is dissolved by acid, the giant cell must generate acid to absorb the bone. He therefore fed litmus particles to his cells and watched. He explains what happened in his 1847 paper,

> *In this experiment I brewed quite neutral litmus powder in the loveliest manner of Baron Dr. Baumann. The giant cells were lifted out of the bone and immediately transfer it to the object-bearer, in which was already saline solution with the litmus additive. Very soon the litmus disappeared and the giant cells became purple in colour, especially at their centres. These objects usually looked blue, but they appeared so coloured under the Gundlaeh microscope, ... especially near the nuclei of the cells, where they even had a yellowish tone.*

Rustizky may have felt that he had proved his point, as the colour change of litmus from blue to red indicated a pH change towards acidic, and yellow was even more acidic. It is not clear from Rustizky's report that he knew the litmus particles were inside vacuoles, although clearly they were. However, thirty years later, Theodor Englemann repeated this experiment on other cell types which phagocytosed particulates. Engelmann was interested in "protoplasm" and especially its contractile properties. As part of this study he used litmus particles and he reports in his paper in 1879:

> *In life, the reaction of protoplasm is generally weakly alkaline or neutral..(but)... now and then I have seen blue litmus particles change within a few minutes after being taken into the contractile endoplasm of Stylonychia mytilus and S. pustuluta, Paramcecium aurelia, and Amoeba diffluens, to a red colour and remain so.*

Engelmann makes no comment on this observation, presumably because it was peripheral to his main study. Helpfully, however, the English translator of this paper in 1884, A.G. Bourne, adds a footnote saying that *"this is possibly due to an acid secreted in an attempt, to digest the particles."*

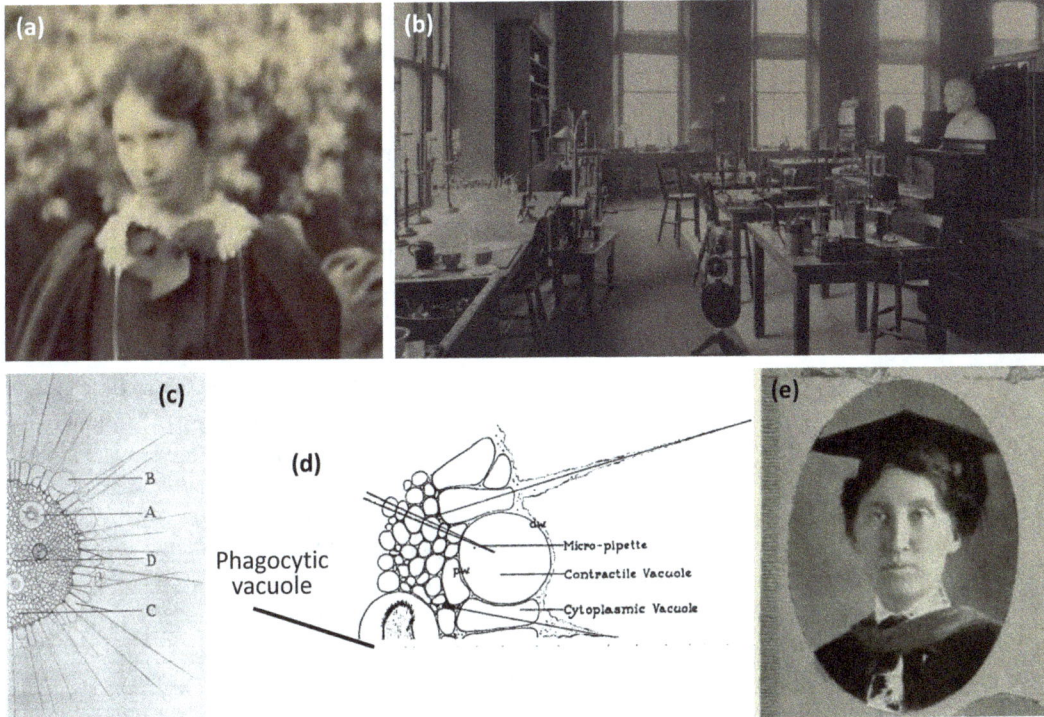

Fig. 2.11 Pioneers of intraphagosomal pH. (**a**) A photo of Marian Greenwood at Newnham College 1896 when she was Director of the Balfour Laboratory.This image was taken from a group photo of Newnham staff standing on the grass behind the Balfour Labs. (**b**) The interior of the Balfour Biological Laboratory for Women showing the simple microscopes on the benches by the windows that Greenwood may have used. The whole lab is watched over by the sightless eyes of the bust of Francis Balfour (the lab's founder). (**c**) Ruth B, Howland's drawing of the heliozoan showing the spines, and internal organelles, with the targets for microinjection indicated, A = phagocytic vacuole. (**d**) shows the micropipette in position (in this case the in contractile vacuole) taken from Howland (Howland 1930). (**e**) Ruth B. Howland at Sweet Briar College in 1919, taken from "Briar Patch", the College Year book. I am grateful to The Principal and Fellows, Newnham College, Cambridge for permission to use the rare photographs of Marion Greenwood and the Balfour Laboratory

The concept of probing the intracellular, or in this case intraphagosomal, environment with an optical probe, whose signal can be detected microscopically outside the cell is one that is still used today. But the early researchers who used the litmus test to look at phagosomal pH soon discovered some surprising things. Firstly, not every cell type or even every cell showed the acid change. An early adopter of the methodology, Greenwood (Fig. 2.11a) reports in her papers of 1886 and 1887 that the litmus indicator did not produce convincing evidence of an acid intravacuolar reaction (Fig. 2.11c). Marion Greenwood (1862–1932) was an English cytologist, who undertook her microscopical work at Newnham College Cambridge, UK, where she was one of the first women to do independent research in Cambridge University. She also directed the newly established Balfour Biological Laboratory for Women (Fig. 2.11a, b). In 1895 she was the first woman to speak about her work at a Royal Society meeting. From her equivocal work with litmus, she suggested that as litmus particles were not "food", they did not necessarily stimulate the digestive response and ultimately concluded that the cytoplasmic secretion into the vacuole was 'probably not acid'. In the following year, Meissner (1888), used a different indicator, alkanet (*Alkanna tinctoria*), which is blue in strong alkali, becoming increasingly "crimson" as acidity increases (blue at pH 10; purple at 8.8 and red at pH 6.1). Meissner showed that *"Amoeba princeps"* took up globules of olive oil dyed red with *"tincture of alkanna"* and reported that this dye gave a clear demonstration that "the digestive vacuoles of lower organisms" were acidic. A year after Meissner's paper, Metchnikoff (1889) too looked at the litmus test. He states that it is accepted that protozoa "secrete around the object they englobe, an amount of acid sufficient to turn blue litmus red". However, when looking at the important mesodermal phagocytes, he reports his own experience as follows

> *I placed a few grains of blue litmus in the water containing young spongilla … (which)… were soon englobed by the sponges and were found to be taken up chiefly by the mesodermic phagocytes. The litmus however did not change colour, even after a prolonged stay in the cells.*

As this seems to be contrary to what was reported in other cell types, Metchnikov unusually cites an earlier report to support his observation. He refers to Krukenberg (1882) who gives biochemical evidence that tryptic digestion by sponge extracts occurred without the need for acid. When Metchnokov repeats the litmus test on his phagocytic leukocytes, he also reports in his *"Lectures"* a disappointing effect

> *In a large number of experiments that I have made on the absorption of granules of blue litmus by leucocytes, I have seen the colour change to red in only a few exceptional cases*

Presumably, Metchnikov was able to distinguish between litmus particles which were adherent to the phagocyte surface and litmus particles within phagosomes. Unfortunately, he does not follow up the "exceptional cases" when he saw the litmus turn red. However, in the Annals of the Pasteur Institute, he reports one such exception:

> Although this study is still far from finished, it has already shown me the existence of facts analogous to those which have been reported for the Protozona. Thus, after having cut the end of the tail of the newt larvae, Triton taeniatus, and rubbed the wound with a blue litmus powder, I was able to observe that the incoming uninuclear leucocytes partly change litmus grains which are englobed inside them, bright red. In some of these macrophages there was, next to a red litmus grain, a vacuole filled with blue granules of the same substance, which proves that the production of intra-

cellular acidic juice can be localized in a restricted part of the cell.

The observation that the litmus particles became red in some phagosomes but not others led Metchnikov to a remarkable conclusion, namely that chemical changes can be restricted to part of the cell, a topic which (as sub-cellular localisation) is still under discussion. Netchaeff (1891) thought that in the cases when Methchnikov saw the litmus colour change, it was "simply an optical illusion". Netchaeff never saw such a colour change in his own "observations on the fate of limus granules in the interior of leukocytes". Metchnokov countered that Netcahaeff can never have looked at Protozoa, where the pH change was obvious and that Metchnikov's own research *"over a series of years, left him in no doubt as to the reality of the colour change of the litmus"*.

However Metchnikov, in his "Lectures" concludes ultimately that *"digestion is carried out in leukocytes in neutral or alkalai medium, as in the case of phagocytes from the sponges"*, leaving the confusion of whether phagosome pH changes occur or not and whether they are important or not.

The litmus colour change, as it appears to the eye, is almost a threshold effect and therefore difficult to follow dynamically. However, it was realised that the pH changes were dynamic. As early as 1891, Le Dante reported 'seeing the slow secretion of an acid … the acidity is progressive, as if it was caused by a secretion' (Le Dantec 1890) and Greenwood & Saunders, 1894 and Saint-Hilaire, 1904 also reported that the acid change was transient. A consensus was thus forming that the pH change had two phases, the first, an acid phase, the second, alkaline. However, when phenol red was used as a water phase pH indicator, its gradual colour change over the crucial range (from yellow pH 6.8 to red pH 8.2), reported that in paramecium, there was an even earlier pH phase which was alkaline (Shipley and De Garis 1925). By 1927, Shapiro had reported some exact values for the pH changes by using a range of pH indicators, neutral red, congo red and phenol red. All three indicators detected the acid phase in Paramecium, with values between pH 4.0–4.8 in the three organism tested (ie Paramecium, Vorticella and Stylonychia); with paramecium alone showing the initial small alkaline phase. Ruth B. Howland (1928) made a major step forward. After undertaking Ph.B. and Ph.M. degrees at Syracuse University and research as a graduate student at the Marine Biological Labs, Woods Hole and at Yale University, Howland (Fig. 2.11e) became Professor of Biology at Sweet Briar College, at an all-female college on the foothills of the Blue Ridge Mountains in Virginia. From here, she continued her research by making links with outstanding scientists. She decided to use a series of pH indicators, which were truly H^+ ion indicators with known pKa values, namely phenol red, bromothymol blue, bromocresol purple, bromocresol green and bromophenol blue. This in itself would be a step forward, but she wanted to microinject these indicators into phagocytic vacuoles within the cell (a heliozoan, actinosphaerium eichhorni). At this time, microinjection was in its infancy, and the success of this approach was a technical triumph. By making long and thin (and sharp) micropipettes, Howland caused the minimum of injury to the cell despite having to pass the micropipette through "the cytoplasm to reach the deep-lying vacuoles" (Fig. 2.11d). At that time, Howland was based in New York at Washington Square College and Cornell University Medical College, New York, near Chambers and his pioneering microinjection approach. Howland had published a paper with the "star pupil" of Chambers, Herbert Pollack (see more later) a year earlier (Howland and Pollack 1927a, b), so it is probable that the work on phagosomal pH could not have been done with Pollack's assistance. Without modern inverted microscopes, Howland used "micropipettes . . . bent upward at right angles and raised from below into the gastric vacuoles". She states that "*a striking feature of these injections is the complete localization of the dye in the injected vacuole. This permits remarkably accurate color determinations, since there is no loss of injection fluid by outward diffusion*". She then compared the colour of the indicator within the vacuole with that in the standard tubes (ie indicator at different pH values) by placing the standard tubes between the source of illumination and the mirror to find the pH which matched the colour. This

colour matching technique was also reported by Chambers and Pollack in 1927, which Howland acknowledges this in her paper. Using heliozoa, Howland measured the early phagosome (closed but with the prey still moving) at pH 6.6–6.9 with a decrease over the next 5–10 min reaching a final minimum of pH 4.3 ± 0.1. As the pH in the vacuole fell, the lethality increased, judging by the lack of movement of the prey. To test whether this fall in pH was a consequence of the death of the prey, she "*crushed and tore*" large ciliates and rotifers in microdroplets of bromocresol green and found that this alone caused a decrease in pH to 5.5 ± 0.1. This "usual acid of injury", as she called it, was only 1/10th the H^+ ion concentration that she found in the phagosome, and so she concluded that acid was secreted into the vacuoles by the living cytoplasm of the heliozoan.

The topic of intraphagosomal pH continues to be discussed and research undertaken 150 years after the first reports. It is surprising that similar or even the same techniques are still used today. For example, Geisow et al. (1981) produced an important paper entitled "Temporal changes of lysosome and phagosome pH during phagolysosome formation in macrophages: studies by fluorescence spectroscopy", in which the title suggests they used fluorescence intensity (rather than colour) as the indicator of phagosomal pH. The conclusions they reached for macrophage phagosomes were similar to those of the 1920s, namely that "*the pH in new phagosomes was transiently driven alkaline*", just as had been reported in paramecium (but not other cells) over 100 years before. Their re-discovery was, surprisingly, based on some familiar older indicators which they report in a way that would be familiar to Greenwood, and Howland in the 1920s (and before) reporting that "*neutral red yeasts seen entering macrophages turned from red to a pale yellow and returned to a brilliant red within 1 min*" and "*bromothymol blue yeasts (yellow-green in the BSS) turned blue after entry, then green, and then yellow*". From this they conclude "that the phagosomal pH is first increased from that of the external medium (to at least pH 7.5) and then within 5 min is reduced to a pH <6.5". Since then, advances in the design of pH fluorescent probes and the technology required to acquire ratiometric spatial data have made intraphagosomal pH measurements more secure (Nunes et al. 2015; Canton and Grinstein 2017) and the molecular details of the controlling factors for the phagosomal pH are now being established (eg Jankowski et al. 2002).

Ca²⁺ Ions and Phagocytosis

From the end of the 1800s to the first 20 years of the 1900s, there was a major increase in Universities, professional scientists. There was also an advance in technology and understanding. During this time, the study of phagocytosis began to probe the cellular mechanisms behind this event. Phagocytologists at this time wondered what factors in the extracellular environment were important for phagocytosis, including Ca^{2+}. However, there were conflicting reports that extracellular Ca^{2+} inhibited or enhanced phagocytosis, and that Mg^{2+} and other ions had similar effects. It is surprising that many of the early studies had no regard for the effects of osmolarity, and simply by adding large amounts of Ca^{2+} or Mg^{2+} or other ions (often in the 100 mM (M/8) range) phagocytosis was inhibited by osmotic shrinkage of the cells. Also no test system used by these early investigators were identical. Also there was the dawning realisation that factors in serum (when looking at blood phagocytes) opsonised some particulate stimuli, and that many reports simply reflected the requirement for divalent ions for binding of the stimulus. Thus, these studies were not really looking at phagocytosis itself, since phagocytosis was never initiated. The data at this period is therefore a muddle. However, there was a beam of light from Sidney Ringer (1835–1910) who showed that in order to maintain muscle contracts, it was crucial that Ca^{2+} ions were in the perfusion solution. Following this, Hartog Jacob Hamburger (1859 – 1924) looked a factors affecting phagocytosis, including Ca^{2+}. He clearly knew of the work of Ringer and even came up with a rival to "Ringer's solution" namely "Hamburger's solution". Hamburger (Fig. 2.12a), who was Dutch, and studied chemistry at Utrecht

University from where he received a doctorate in 1883 and in 1901 became professor of physiology at University of Groningen. It was here that he published a book entitled *Osmotischer Druck und Ionenlehre in den medecinischen Wissenschaften* ("Osmotic pressure and ion science in the medical sciences"). With this background, Hamburger would not make the osmotic mistakes of previous studies. He also hit on a robust test system for studying phagocytosis. While others were trying to reproduce in the laboratory what happen in the body (or elsewhere) even if unaware of all the complexities, Hamburger wanted an experimental system which, while artificial, was controllable and chose carbon as the phagocytic target. Hamburger could see the advantages, (i) "carbon" particles could be produced as a standard stimulus (Fenn later showed that neutrophils had a preference for carbon particles over some other particulates (Fenn 1923: Fig. 2.12b), (ii) could be visualised easily in living cells under the microscope and (iii) Hamburger thought carbon was not influenced by any unidentified factors from serum, although Fenn (Fenn 1921) later reported that in his cells, serum was required. In his first report in 1910, to the Royal Netherlands Academy of Arts and Sciences (KNAW), Hamburger reports a remarkably set of results (Hamburger 1910). He firstly showed that "small amounts of calcium" had effects on phagocytosis under normal experimental conditions. However, if the leucocytes were left in a Ca^{2+} free medium (NaCl saline) for 24 hours, phagocytosis was almost non-existent (less than 3% of cells internalised carbon particles). This then was the experimental condition that Hamburger needed to test the effect of ionic replacement on phagocytosis. He reports that adding Mg^{2+}, Sr^{2+} and Ba^{2+} to the medium had no effect on the ability of the cells to phagocytose, but that adding Ca^{2+} to the medium restored phagocytosis to its normal level (50%)

These experiments show that when the phagocytes, by being exposed a long time to NaCl 0.9 %, have almost entirely lost their power, they cannot be revived by barium (or magnesium or strontium). An isosmotic quantity of calcium however, produces this effect (ie revival of phagocytosis) in a very marked degree

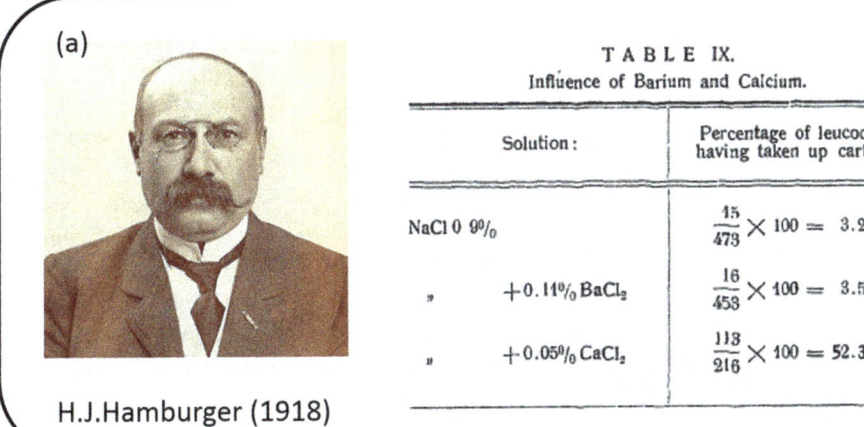

Fig. 2.12 H.J. Hamburger: Ca^{2+} ions and phagocytosis (part I). (**a**) Shows a photo of H.J. Hamburger in 1918 together with his crucial data showing the unique role of Ca^{2+} in restoring phagocytosis. This data (or slight variants of it) were shown in papers published by Hamburger in Nature, Brit Med J. and Royal Netherlands Academy of Arts and Sciences (Amsterdam) Proceedings (Hamburger 1910, 1915, 1916). (**b**) Shows some results from W. O. Fenn, showing the preferential uptake of carbon particles compared to silica and a leucocyte stretching" around a particle of silica (S) to phagocytose two particles of manganese oxide (M). The whole sequence was 120 s (Fenn 1922)

The original data for this statement in a series of tables, the key one taken from Hamburger's first paper is shown here in Fig. 2.12a. The amount of $CaCl_2$ added (0.05%) which totally restored phagocytosis (if the calcium chloride crystals were hexahydrate $CaCl_2.6H_2O$ as is usual) was 2.3 mM Ca^{2+}. In other experiments, restoration was seen with Ca^{2+} as low as 0.23 mM. These concentrations are within the usual mammalian physiological range or less than those in other environments (eg in artificial sea water $Ca^{2+} = 10$ mM). Hamburger's results were clearly physiologically significant and suddenly there was clarity. Calcium (or rather Ca^{2+} ions) was a key element in phagocytosis. Hamburger followed this paper with confirmation (Hamburger and de Haan 1910) and rightly became famous for this discovery. He reported the same observations in a number of papers, including verbatim copies in Nature (Hamburger 1915) and in the British Medical Journal (Hamburger 1916). Without knowing the importance of his finding, McJunkin (1918) also showed that simply by adding citrate (15 mg/ml ie c.70 mM), phagocytosis by neutrophils was prevented. This Ca^{2+} chelator at that concentration would have significantly depleting effect on the free Ca^{2+} in the system and seems, therefore, to be a confirmation of the reduction in phagocytosis which Hamburger saw in the absence of extracellular Ca^{2+}.

Hambuger, however, was unsure of the mechanism by which Ca^{2+} exerted this effect, but cleverly drew a conclusion which today, we now know to be true

> We might be inclined to attribute the increase (in phagocytic ability) ... as a consequence of the electric charge, caused by the entering of a number of bi-valent calcium ions. This explanation however can hardly be the correct one here, for experiments show that other bi-valent cations – namely barium, strontium, magnesium -do not augment the amoeboid motion. It must be assumed then, that the action of calcium in this case, is based upon an _unknown specific biochemical property_ of this metal

Within 20 years, a major step forward in understanding that changes cytosolic Ca^{2+} was the answer, was made by the ingenious work of the young Herbert Pollack (1906–1990). The Washington Post tells us in his obituary that during World War II, Pollack was a colonel in the Army Medical Corps, serving in Europe as the U.S. representative on the Inter-Allied Commission for the Study of Prison and Concentration Camps. He was decorated with the Bronze Star, the Purple Heart and the Army Commendation Medal. In the Korean War, Pollack visited the war zone for the surgeon general of the Army to review the medical evacuation system and that he also received an Outstanding Civilian Service Medal from the Army for work in connection with the threat of malaria in Vietnam and another for work dealing with high altitude physiology at the time of the Chinese invasion of Tibet. What the obituary does not tell us is that before he began his private medical practice in New York City in 1934, he did some amazing ground-breaking cell biological experiments.

This was when Pollack was a student of Robert Chambers, who was a pioneer of microinjection techniques which he called micrurgery (Chambers 1921, 1922). It was to Chambers' laboratory in New York that Ruth Howland came to microinject phagosomes (see above) and published a paper with Pollack (Howland and Pollack 1927a, b). She was obviously skilled at delicate microinjection because in Pollack's crucial paper, she is thanked (in a foot-note) for her "help". Herbert Pollack was only 20 years old when in Chambers laboratory (Fig. 2.13a), but he produced perhaps the key paper. Although his paper was part of a numbered series from Chambers' laboratory, Pollack is the single author. However, before we see Pollack's major contribution, it is important to see it in context. Chambers' laboratory was cutting edge having developed new microinjection technique (Fig. 2.13b–d) for investigating properties of "living protoplasm". He included line drawings of cells as had been done previously, but there are also photographs taken through the microscope of some of their experiments. This, of course, was not a first. In 1917, a "cinematographic recording" of phagocytosis was presented by Comandon (1917) to the "The Society of Biology and its subsidiaries" (La Societe de Biologie et de ses filiales). I am not sure of the impact it had, but even today an AVI.file of an experiment involving phagocytosis is often appreciated by the audience. But clearly Chambers was in the vanguard of modern science. Chambers and Reznikoff (1926) explored the effect of microinjecting ions into the cell, especially amoeba. They report that injecting Ca^{2+} caused an immediate "solidification of the cytoplasm" which results in it a pinching off. This would seem to be a pathological response, caused by extremely high cytosolic Ca^{2+} such as M/13 (77 mM Ca^{2+}), but these effect persisted at Ca^{2+} injection concentrations down to M/104 (9.6 mM Ca^{2+}). Below this concentration, they report

Fig. 2.13 (continued) alizarin and saw the first evidence of localised Ca^{2+} signalling within amoeba and its relationship to pseudopod progression. The photo was taken from the US National Library of Medicine Collection (https://collections.nlm.nih.gov/catalog/nlm:nlmuid-101439620-img) with permission of the copyright holder. (**b**) Shows the method for fabrication of the micropipette using a Bunsen burner, and the shapes they produced (from Chambers 1921, 1922). (**c**) Shows a close-up of the microscope/injector in front of Pollack and (**d**) shows a higher resolution image of the detailed assembly of pipes leading to the moveable pipette holder beneath the microscope stage as shown in Chambers (1921, 1922)

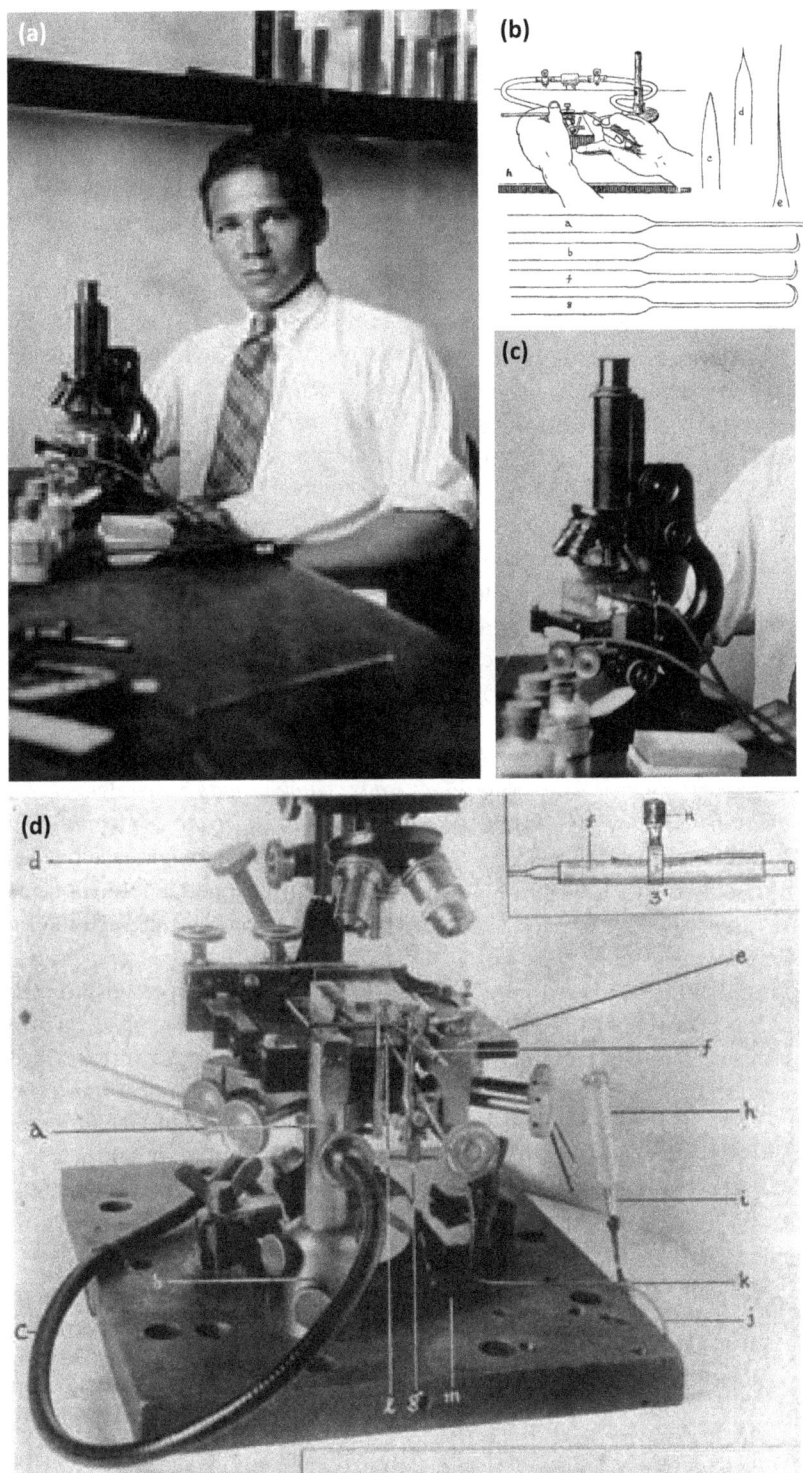

Fig. 2.13 Herbert Pollack: Ca^{2+} ions and phagocytosis (part 2). (**a**) Shows a photo of Pollack when a medical student (about 20 years old) working in Chambers' lab. In front of him is the microscope with microinjection equipment developed by Chambers' group. It was with this equipment that Pollack microinjected amoeba with

The dilution of M/208 (ie 4.8 mM Ca²⁺) appears to be the critical strength at which the pinching off is either delayed from 2 to 10 minutes or is never completed. In the latter case, the involved region is ultimately resorbed. With the dilution of M/416 (ie 2.4 mM Ca²⁺) no pinching off is even attempted and the ameba reacts as if it had been injected with water alone.

We may assume that injection of the lowest concentration of Ca^{2+}, M/416, did not elevate the level of cytosolic Ca^{2+} significantly (ie it is near the null point, now known to be about 100nM). If the injectate was approximately 0.25 the volume of the cell (as seemed to be the standard procedure for the Chambers' lab.), then the total Ca^{2+} added would elevate the cytosol by 0.6 mM (ie 2.4/4 mM). Since there was no cellular response, the free cytosolic Ca^{2+} the injection generated was at most only 100nM, allowing us to estimate the cytosolic buffering capacity for Ca^{2+} to be about 6000:1. This is not an unreasonable figure and in line with many estimates made today for the "slow Ca^{2+} buffering" component (ie from 1000 to 10,000:1). With this estimate, the M/208 Ca^{2+} injection, giving only a weak or zero response, would elevate cytosolic Ca^{2+} by 200nM; M/104, M/52 and M/26 would elevate cytosol Ca^{2+} to 400 nM, 800 nM and 1.6 μM respectively, and trigger a robust responses. These cytosol Ca^{2+} concentrations are, of course, only "guestimates" because the restricted diffusion of cytosolic Ca^{2+} would the cause a localisation of injected Ca^{2+}; and so these estimates represent minimum vales. It is, however, interesting that these are all in the range of cytosolic Ca^{2+} increases that we now know occur physiologically. It is thus possible that these represent feasibly physiological responses. This may, thus, be the first time that the concentration of cytosolic Ca^{2+} (and the first time its dynamic range for stimulation) had been estimated in phagocytes.

It is possible that Pollack was aware of the possibility that this cytosolic gelation was important in amoeba and may have already observed reversible gelation near the contractile vacuole, which was thought to be essential for vacuole contraction (Howland and Pollack 1927a, b). But this is not why phagocytologists should be interested. Instead, it is because,

pseudopodia formation, whether in amoeba or other phagocytes, is a key part of phagocytosis. Howland and Pollack (1927a) noted that cytoplasmic gelation occurred "in the greater percentage of cases . . . in the posterior portion of the ameba" and Chambers and Reznikoff (1926) reported that microinjections of "CaCl₂ . . . solidify the internal protoplasm . . . (which) . . . solidification tends to be localized . . . (and) . . . the injection of CaCl₂ accelerates movement in the regions not solidified". Pollack must also have been aware that Reznikoff and Chambers (1927) had found that the microinjection of phosphates, carbonates, and sulfates immediately but temporarily prevented amoeba from forming pseudopodia (Reznikoff and Chambers 1927). Pollack noted that these anions form insoluble salts with Ca^{2+} which raised the possibility that their effect on inhibiting pseudopodia formation was via a reduction in cytosolic free Ca^{2+}. In his paper (Pollack 1928), Pollack tested whether the effect was due to it effect of cytosolic Ca^{2+} by microinjecting *"two other organic anions whose calcium salts have relatively low solubility products, viz., tartrate and oxalate."* It must have been an exciting moment when he saw a similar effect on pseudopodia formation, and that the amoeba injected went through the same *"stages of quiescence, rounding, and pseudomembrane formation"*. Pollack reports that *"the ameba could recover from a moderate injection of M/8 solution of sodium potassium tartrate (ie 125 mM) or of M/18 solution of sodium oxalate (ie 56 mM)"* usually in a few hours. The *"peudomembrane effect"* (ie the result of quiescence and rounding) was also caused by lower amounts of the Ca^{2+} reducing agents with *"concentrations as low as M/128 of sodium potassium tartrate and M/620 of sodium oxalate (ie 1.9 mM and 400 μM respectively in the cytosol)"*. Clearly the formation of insoluble calcium tartrate ($CaC_4H_4O_6$) and calcium oxalate (CaC_2O_4) in the cytosol at these high concentrations, would have a significant and long lasting reducing effect on cytosolic free Ca^{2+} from its initial concentration of 100nM and in suppressing Ca^{2+} signals.

Pollack wrote that the "first effect of the injection of any of the calcium precipitants is absolute quiescence." This was good evidence for a role for cytosolic Ca^{2+}, but as Pollack was realising that Ca^{2+} inside the amoeba was important for pseudopodia formation, he recognised that he needed an indicator of Ca^{2+} that he could microinject into the cell. His options, at that time, were very limited. However, alizarin, a dye known from antiquity as "madder", precipitates with Ca^{2+} and had recently been reported for use as the basis of the measurement of "small quantities" of Ca^{2+} in blood (Laidlaw and Payne 1922). Pollack intended to use alizarin as an optical indicator to watch a chemical change occur within a living cell in real time. This was an ambitious and ground-breaking experiment, but having injected alizarin, he reports simply:

> The injection of a moderate quantity (1/4 the volume of the ameba) of a saturated aqueous solution of this reagent (ie alizarin reddish brown in color) causes a temporary cessation of movement.

As Pollack had expected (and hoped), alizarin acted as a simple cytosolic Ca^{2+} reducing salt, like oxalate, and pseudopodia formation was inhibited. However, unlike oxalate, Pollack could see the crystal of calcium-alizarin form within the cell. He reports:

> The ameba rounds up and the larger crystals (of undissolved alizarin) and granules may settle to the bottom. A close examination of the cytoplasm shows fine purplish red granules scattered throughout the cell, and the hyaline cytoplasm itself is diffusely colored pale red.

He noted that

> If an ameba is killed during the injections or is torn by the micro needles in a medium containing alizarin, the large crystals normally present in the ameba and some of the coagulum which is produced upon death will also take on the purplish red color characteristic of calcium alizarinate.

Thus exposure to the high Ca^{2+} of the extracellular environment was detected as *"purplish red calcium alizarinate crystals"*. Pollack then records, as follows, the key observation:

> If the ameba tries to pull forth a pseudopod as evidenced by a slight lifting of the membrane, a shower of these purplish red granules are seen to appear in this area and the pseudopod formation is immediately stopped.

The "shower of purplish red crystals" which appeared were reporting an elevation of Ca^{2+} in the cytosol. He had seen, for the first time, a localised rise in cytosolic Ca^{2+} associated with the formation of pseudopodia. This was not a chance observation, as some of the early work of phagocytosis was, but instead it was the result of a careful train of deduction and careful experiments.

It is interesting that Pollack says that the inhibited amoeba "tries to pull forth a pseudopodia" as it is difficult when observing phagocytosis or other cell movements not to feel the cell is "trying" to do something. Of course, what Pollack really witnessed was the protrusion of a pseudopodium in response to a spontaneous local elevation in cytosolic Ca^{2+}, which was then aborted by the precipitating effect of the alizarin. The localised precipitation of calcium-alizarin as a "shower of red crystals" quenched the Ca^{2+} signal with the result that that pseudopod extension stopped. The sudden appearance of the shower of red crystals, must have been the result of an elevation of cytosolic Ca^{2+} to a level above the solubility limit for calcium–alizarinate as Pollack reasoned:

> The quiescence which is induced after an injection of alizarin may be due to a removal of calcium of the protoplasm from the sphere of action.

In order to test this idea, Pollack designed an additional set of experiments aimed at reversing the effect by additional Ca^{2+}. He writes:

> When an ameba which has previously been injected with alizarin is injected with an M/208 calcium chloride solution, active flowing movements appear almost immediately which subside in a very short time.

The level of Ca^{2+} injection was estimated earlier to give a rise in cytosolic free Ca^{2+} of 200 nM at equilibrium, with higher concentrations locally and at earlier times. It is interesting that Pollack does not mention seeing red crystal forming as the Ca^{2+} injection is done, so it could be that the local Ca^{2+} change during pseudopod extension was higher than the effect of microinjection (ie 200 nM). Pollack noted that the recovery time depended on the amount of alizarin injected, and that after Ca^{2+} injection, the cells recovered the ability to form pseudopodia at a faster rate ie *"the time usually required for complete recovery after an alizarin injection is shortened from about 2 to 3 hours to ½ to 1 hour."*

Pollack's conclusion is one that has many resonances to "Ca^{2+} signallers". In his conclusion, he gives concepts with are still vitally important today eg free Ca^{2+} versus un-ionised total calcium: Ca^{2+} equilibrium in the cytosol: mobilisation of Ca^{2+} reserves (stores?). He writes:

> *The fine, purplish red granules resulting from the injection of the alizarin are, no doubt, the insoluble calcium alizarinate. Recovery of an ameba from such an injection may be explained by the postulate that the free calcium ions in the living ameba are in equilibrium with a reserve supply of unionized calcium. The equilibrium is upset when the free calcium is removed by precipitation or by other means, and the system may possibly react in such a way as to counteract the effect of the change imposed. By mobilization of the calcium from a reserve supply the ameba can therefore gradually resume its normal activity.*

These conclusions were reached by a 20 year old student nearly 100 years ago and yet are still largely accepted today. It was not until 1980 that his observations were essentially confirmed by Taylor et al. (1980) who microinjected amoeba with the chemiluminescent Ca^{2+} indicator aequorin and rediscovered the Ca^{2+} changes which accompany pseudopod formation. Pollack was clearly far advanced not only being the first to detect a change in cytosolic Ca^{2+}, but the first to relate this to a physiological event (pseudopod formation). More than this, Pollack was the first to see a dynamic change of any physiological chemistry within any living cell; and so conceptually open a whole new field of understanding.

Conclusion

As so, as this brief history of phagocytosis draws to a close, there are a few conclusions that may be drawn. The phenomenon of phagocytosis, which appeared, when it was thought that animalcules were simply very small animals, and had the same instincts and behaviours. When hungry, they eat.

When they eat, they swallow etc. There seemed no need to ask how the very small animal could swallow etc. Now, we think of cells as a well-organised collection of molecules and ions and that phagocytosis is an "emergent" phenomenon. The components of the molecular/ionic ensemble are now known and the interactions between them are increasingly understood. The more we know, the more complicated it seems. Yet paradoxically, phagocytosis must be one of the most primitive cell activities; being responsible for nutrition and probably key steps in evolution of eukaryote with the inclusion of organelles, especially mitochondria, which probably originated as symbiotic bacterium which had been phagocytosed. Primitive, often implies simple. It is possible that there is a form of simple phagocytosis and that thus has been overlain by modifiers and back-up systems which we are now faced with unravelling. In the same way the C.Elegans a simple (and primitive) organism, has led to a number of discoveries, perhaps a primitive cell displaying the "essence" of phagocytosis without the accrued overlay of complexity may be useful. The history of phagocytosis research has been in the reverse direction. So we now understand more and more of the complexities, without yet understanding the basics. Surely, since phagocytosis has been observed for hundreds of years, and with an accumulated useful knowledge base of many decades and with technology progressing exponentially over this time, fully understanding phagocytosis in the near future is an achievable objective.

References

Aterman K (1998) Medals, memoirs—and Metchnikoff. J Leuk Biol 63:515–517. https://doi.org/10.1002/jlb.63.4.515

Canton J, Grinstein S (2017) Measuring Phagosomal pH by fluorescence microscopy. Methods Mol Biol 1519:185–199

Cavaillon JM (2011) The historical milestones in the understanding of leukocyte biology initiated by Elie Metchnikoff. J Leuk Biol 90:413–424

Cavaillon JM, Legout S (2016) Centenary of the death of Elie Metchnikoff: a visionary and an outstanding team leader. Microbes Infect 18:577–594

Chambers R (1921) A simple apparatus for micro manipulation under the highest magnifications of the microscope. Science 54:411–413

Chambers R (1922) New apparatus and methods for the dissection and injection of living cells. Anat Records 24:1–19

Chambers R, Reznikoff P (1926) Micrurgical studies in cell physiology. i. the action of the chlorides of Na, K, Ca, and Mg on the protoplasm of Amoeba proteus. J Gen Physiol 8:369–401

Claparède E (1854) Ueber Actinophrys Eichhornii. Müller's Archiv:398–419. https://www.sciencedirect.com/science/article/pii/S187451729980025X

Comandan J (1917) Phagocytosis in vitro of the hematozoons of the Java sparrow (cinematographic recording). Comptes rendus des séances del Societe de Biologie et de ses filiales 80:314–316

E. R. L (1871) Edouard René Claparède. Nature 4:224–225

Engelmann TW (1879) Physiologie der Protoplasms-und Flimmerbewegung. In: Hermann L (ed) Handbuch der Physiologie. F. C. W. Vogel, Leipzig, pp 343–408

Engelmann TW (1884) Physiology of protoplasmic movement. Q J Microsc Sci 24:370–418. (Translated from Hermann's Handwörterbuch d. Physiologie by A. G. Bourne) https://jcs.biologists.org/content/joces/s2-24/95/370.full.pdf

Fenn WO (1921) The phagocytosis of solid particles. II carbon. J Gen Physiol 3:465–482

Fenn WO (1922) The adhesiveness of leucocytes to solid surfaces. J Gen Physiol 5:143–167. https://doi.org/10.1085/jgp.5.2.143

Fenn WO (1923) The phagocytosis of solid particles IV carbon and quartz. J Gen Physiol 5:169–171

Geisow MJ, D'Arcy Hart P, Young MR (1981) Temporal changes of lysosome and phagosome pH during phagolysosome formation in macrophages: studies by fluorescence spectroscopy. J Cell Biol 89:645–652

Goeze JAE (1777) Infusionstierchen die Andre fressen. Beschäftigungen der berlinischen Gesellschaft naturforschender Freunde 3:373–384. (Activities of the Berlin Society of Naturalist Companions, 3, 373–384)

Gordon S (2008) Elie Metchnikoff: father of natural immunity. Eur J Immunol 38:3257–3264

Gordon S (2016) Phagocytosis: an immunobiologic process. Immunity 44:463–475

Hamburger HJ (1910) The influence of small amounts of calcium on the motion of phagocytes. R Netherlands Acad Arts Sci (KNAW) Amsterdam Proc 13:66–79. https://www.dwc.knaw.nl/DL/publications/PU00013276.pdf

Hamburger HJ (1915) Researches on phagocytosis. Nature 96:19–23

Hamburger HJ (1916) Researches on phagocytosis. Br Med J 1916(1):37–41

Hamburger HJ, de Haan J (1910) The biology of phagocytes. VI. The effect of alkaline earth salts on phagocytosis (Ca, Ba, Sr, Mg). Biochem Z 24:470–477

Hare HA (1923) Leidy and his influence on medical science. Proc Acad Nat Sci Philadelphia 1923:73–87

Hartley, L.P. (1953) "The Go-Between" published by Hamish Hamilton

Hausmann K (1986) Prey-catching, food-intake and digestion in Actinophrys sol (Heliozoa): Beutefang, Nahrungsaufnahme und Verdauung bei Actinophrys sol (Helizoa). Institut für den Wissenschaftlichen Film (IWF). https://doi.org/10.3203/IWF/C-1533eng (https://av.tib.eu/media/9426)

Hooke R (1665) Micrographia: or some physiological descriptions of minute bodies made by magnifying glasses. With observations and inquiries thereupon. Royal Society Publication No 1

Howland RB (1928) The pH of gastric vacuoles. Protoplasma 5:127–134

Howland RB (1930) Micrurgical studies on the contractile vacuole. III. The pH of the vacuolar fluid in actinosphaerium eichhorni. J Exp Zool 55:53–62

Howland RB, Pollack H (1927a) Micrurgical studies on the contractile vacuole I. relation of the physical state of the internal protoplasm to the behavior of the vacuole II. Micro-injection of distilled water. J Exptl Zoology 48:441–458. http://jgp.rupress.org/content/8/4/369/tab-pdf

Howland RB, Pollack H (1927b) The significance of gelation in the systole of the contractile vacuole of amoeba dubia. Proc Soc Exptl Biol Med 24:377–378

Jankowski A, Scott CC, Grinstein S (2002) Determinants of the Phagosomal pH in neutrophils. J Biol Chem 277:6059–6066

Korzha V, Bregestovskic P (2016) Elie Metchnikoff: father of phagocytosis theory and pioneer of experiments in vivo. Cytol Genet 50:143–150

Krukenberg (1882) Grundzuge einer vergleichendren physiologie der verdauung. Heielberg:52

Laidlaw PP, Payne WW (1922) A method for the estimation of small quantities of calcium. Biochem J 16:494–498

Lane N (2015) The unseen world: reflections on Leeuwenhoek (1677) 'Concerning little animals'. Philos Trans R Soc B Biol Sci 370:20140344. https://doi.org/10.1098/rstb.2014.0344

Le Dantec F (1890) Recherches sur la digestion intracellulaire chez les protozoaires. Ann de l'Inst Pasteur 4:777 and 5:163

Leidy J (1875) On the mode in which Amoeba swallows its food. Proc Acad Natl Sci Philadelphia 143. https://archive.org/details/proceedingsofaca26acad/page/142

McGonagle D, Georgouli T (2008) The importance of 'Mechnikov's thorn' for an improved understanding of 21st century medicine and immunology. Scand J Immunol 68:129–139

McJunkin FA (1918) A simple tenchic for the demonstration of a phagocytic mononuclear cell in peripheral blood. Arch Intern Med (Chic) 21:59–65. https://jamanetwork.com/journals/jamainternalmedicine/article-abstract/654187

Meissner M (1888) Beitraige zur Ernaihrungsphysiologie der Protozoen. Z. Wiss Zool 46:498–516. (Contribution to the nutritional physiology of protozoa)

Merien F (2016) A journey with Elie Metchnikoff: from innate cell mechanisms in infectious diseases to quantum biology. Frontiers Pub Health 4:UNSP 125

Metchnikoff E (1883) Untersuchung uber die interrceulare Verdauung bei wirbellosen Tieren. Arbeiten des zool Inst Wien Bd v Heft ii:141

Metchnikoff É (1893) Lectures on the comparative pathology of inflammation, delivered at the Pasteur Institute in 1891. Kegan Paul, London. https://archive.org/details/lecturesoncompar00metcuoft/page/n67

Metchnikoff E (1889) Sur la propriete bactericides des humeurs. Revue critique Annals of the Inst Pasteur 3:664–671

Metchnikoff E (1905) Immunity in infective diseases (trans: Binnie FG). Cambridge University Press, Cambridge. https://archive.org/details/immunityininfec01metcgoog/page/n16

Metchnikoff O (1920) La vie d'Elie Metchnikoff. Librairie Hachette

Metchnikoff O (1921) Life of Elie Metchnikoff. English translation by Sir Ray Lancaster (Reprinted 1972). BFL Press, New York, p 119

Metchnikov E (1883) Unterchungen über die mesodermalen phagocyten einiger wirbeltieren. Biol Cent 3:560–565

Metchnikov E (1889) Researches Sur la digestion intracellulaire (research on intracellular digestion). Ann de l'Inst Pasteur (J Microbiol) III:25–29. https://gallica.bnf.fr/ark:/12148/bpt6k6436880n/f31.image

Müller OF (1786) Animalcula infusoria fluviatilia et marina, quae detexit, systematice descripsit et ad vivum delineari curavit. Copenhagen and Leipzig. https://www.biodiversitylibrary.org/item/101601#page/7/mode/1up

Nathan C (2008) Metchnikoff's legacy in 2008. Nat Immunol 9:695–698

Nauseef WM (2014) Identification and quantitation of superoxide anion: essential steps in elucidation of the phagocyte "respiratory burst". J Immunol 193:5357–5358

Netchaeff A (1891) On litmus granules in leukocytes. Virchow's Archive cxxv:448

Nobel Media Biographical (1908) Ilya Metchnikov biography. https://www.nobelprize.org/prizes/medicine/1908/mechnikov/biographical/

Nunes P, Guido D, Demaurex N (2015) Measuring phagosome pH by ratiometric fluorescence microscopy. J Vis Exp 2015(106):53402. https://doi.org/10.3791/53402

Pollack H (1928) Micrugical studies in cell physiology: VI calcium ions in living protoplasm. J Gen Physiol 11:539–545. https://doi.org/10.1085/jgp.11.5.539

Pritchard A (1834) The natural history of animalcules: containing descriptions of all the known species of Infusoria. Whittaker and co, London. https://www.biodiversitylibrary.org/bibliography/8659#/summary

Pritchard A (1835) A list of two thousand microscopic objects. Whittaker & Co., London

Reznikoff P, Chambers R (1927) Micrurgical studies in cell physiology. III. The action of CO_2 and some salts of Na, ca, and K on the protoplasm of Amoeba dubia. J Gen Physiol 10:731–738

Rodgers R, Hammerstein O (1959) Lyrics to "Do-RE-Mi": The Sound of Music

Rustizky J (1847) Untersuchunge über Knochenresorption und Riesenzellen. Virchows Arch 59:202–227

Shapiro NH (1927) The cycle of hydrogen-ion concentration in the food vacuoles of Paramecium, Vorticella and Stylonychia. Trans Am Microscop Soc XLVI:45

Shipley P, De Garis CF (1925) The third stage of digestion in Paramecia. Science 62:266

Silverstein AM (2011) Ilya Metchnikoff, the phagocytic theory, and how things often work in science. J Leuk Biol 90:409–410

Stossel TP (1999) The early history of phagocytosis. Adv Cell Mol Biol Memb Organelles 5:3–18. https://doi.org/10.1016/S1874-5172(99)80025-X

Tauber AI (1992) History of immunology-the birth of immunology- the fate of the phagocytosis theory. Cell Immunol 139:505–530

Tauber AI (2003) Metchnikoff and the phagocytosis theory. Nat Rev Mol Cell Biol 4:897–901

Taylor DL, Blinks JR, Reynolds G (1980) Aequorin luminescence during ameboid movement endocytsosis and capping. J Cell Biol 86:599–607

Teti G, Biondo C, Beninati C (2016) The phagocyte, Metchnikoff, and the Foundation of Immunology. Microbiol Spectrum 4:UNSP MCHD-0009-2015

The Role of Membrane Surface Charge in Phagocytosis

3

Michelle E. Maxson and Sergio Grinstein

Abstract

The formation and maturation of phagosomes are accompanied by acute changes in lipid metabolism. Phosphoinositides, in particular, undergo extensive modification as part of the signaling sequence that drives cytoskeletal and membrane remodeling. Because the phosphoinositides provide much of the anionic charge of the cytosolic leaflet of the plasmalemma and phagosomal membrane, the metabolic changes associated with signaling result in marked changes of the surface charge. Here we summarize the pathways involved in lipid remodeling during phagocytosis, the resultant alterations in the surface charge of the nascent and maturing phagosomes, and the consequent effects on the association of proteins attached to the membrane by electrostatic means.

Keywords

Phosphoinositides · Macrophage · Phagosome · Surface potential · Electrostatic

Introduction

Phagocytosis is central to both immunity and tissue homeostasis. Performed mainly by cells of the innate immune system, such as macrophage and dendritic cells, it is defined as the receptor-driven uptake and internalization of particles >0.5 μm. Phagocytosis plays important roles in the clearance of invading microorganisms and damaged/apoptotic cells throughout the body, placing this process at the crux of human health and disease.

Phagocytic cells continuously sample their environment via actin-rich protrusions of the plasma membrane (Flannagan et al. 2012). When such protrusions make contact with a particle, ligands exposed by the target induce the clustering of receptors on the phagocyte surface, thereby initiating intracellular signaling intended to promote engulfment (Jaumouillé and Grinstein 2016; Pauwels et al. 2017). Vast membrane and cytoskeletal remodeling drive the extension pseudopods around the target,

M. E. Maxson
Program in Cell Biology, Hospital for Sick Children, Toronto, ON, Canada

S. Grinstein (✉)
Program in Cell Biology, Hospital for Sick Children, Toronto, ON, Canada

Department of Biochemistry, University of Toronto, Toronto, ON, Canada

Keenan Research Centre for Biomedical Science, St. Michael's Hospital, Toronto, ON, Canada
e-mail: sergio.grinstein@sickkids.ca

© Springer Nature Switzerland AG 2020
M. B. Hallett (ed.), *Molecular and Cellular Biology of Phagocytosis*, Advances in Experimental Medicine and Biology 1246, https://doi.org/10.1007/978-3-030-40406-2_3

culminating in particle internalization. After sealing of the phagocytic vacuole, the nascent phagosome undergoes a "maturation" process, involving fusion events with vesicles of the endocytic pathway, while fission events recycle some of the components to the cell surface or to earlier endocytic compartments (Levin et al. 2016; Pauwels et al. 2017). Maturation of the phagosome culminates with the fusion of lysosomes, which deliver degradative enzymes to the phagosome and accentuate the acidification of its luminal fluid, promoting the breakdown of the engulfed contents for recycling or disposal.

Underlying this highly dynamic and temporally coordinated process is the marked remodeling of the plasma and phagosomal membranes. The lipid bilayers that constitute the various cellular membranes have distinct composition, and even the two monolayers constituting each of these bilayers differ markedly. These differences contribute importantly to the establishment of the individual compartment identities (Di Paolo and De Camilli 2006; Swanson 2014). It is now appreciated that specific lipids or lipid microdomains provide discrete platforms for localized, organelle-specific signaling within the cell. Such signaling is dictated by the stereospecific features of the head-groups of the lipids involved, but also by biophysical properties such as hydrophobicity, curvature and, notably, electrical charge. The purpose of this chapter is to briefly review the changes in lipid composition that accompany phagosome formation and maturation, with particular attention given to the associated alterations in the membrane surface charge.

Lipids That Contribute to Phagocytosis and Phagosome Maturation

Phagocytosis begins at the cell surface. The plasma membrane of mammalian cells, in general, is composed of cholesterol, sphingolipids, and various phospholipid species. These are largely synthesized in the endoplasmic reticulum, and reach their cellular destinations either through vesicular transport, via soluble lipid-transport proteins or by exchange proteins that tether two organelles in close proximity (Bohdanowicz and Grinstein 2013; Swanson 2014). Because cholesterol can readily flip across the membrane, it is present on both leaflets of the plasmalemmal bilayer. In contrast, sphingolipids are almost exclusively localized to the outer leaflet, where phosphatidylcholine (PtdCho) is also abundant, whereas phosphatidylserine (PtdSer), phosphatidylethanolamine (PtdEth) and phosphatidylinositides are restricted to the cytosolic leaflet. This makes the plasma membrane a paradigm of lipid asymmetry (Verkleij and Post 2014), generated and preserved through the action of energy-dependent floppases and flippases that counteract the spontaneous scrambling that tends to homogenize the lipid distribution (Holthuis and Levine 2005).

The plasma membrane contains a variety of phosphatidylinositides, generated by the phosphorylation of phosphatidylinositol (PtdIns) at positions 3, 4 and/or 5 of the inositol ring. This reversible process is temporally and spatially controlled by a variety of phosphoinositide kinases and phosphatases. The varying number and pK_a of the phosphate groups on the inositides determines the electrical charge they bear and contribute to the overall surface charge of the membrane (Holthuis and Levine 2005; McLaughlin and Murray 2005; Yeung et al. 2008). PtdSer is also an important contributor. At physiological pH, PtdSer and the phosphoinositides are deprotonated, conferring to the inner leaflet of the plasma membrane a net negative charge. The cytosolic aspect of the plasmalemma is, in general terms, thought to contain $\approx 20\%$ PtdSer, $\approx 10\%$ PtdIns, and $<5\%$ of phosphorylated PtdIns species, including PtdIns(4)P (≈ 1–2%), PtdIns(4,5)P_2 (≈ 1–2%), PtdIns(3,4)P_2 and PtdIns(3,4,5)P_3; the concentration of the inositides phosphorylated on position 3 is low, poorly defined and varies greatly with the degree of stimulation of the cells. Although not the most prevalent, the unique biochemical qualities of the charged phospholipids – especially the phosphoinositides – make them well suited as dynamic signaling platforms.

The development of lipid-specific, genetically-encoded probes has enabled the investigation of phospholipid dynamics during phagocytosis in live cells. These probes bind specifically to defined phospholipid head-groups that are cytosolically exposed. To date, probes have been developed to selectively monitor PtdIns(3)P, PtdIns(4)P, PtdIns(4,5)P_2, PtdIns(3,4)P_2, PtdIns(3,4,5)P_3, and PtdSer, as well as some of their metabolic products like diacylglycerol (DAG) and phosphatidic acid (PtdOH; Sarantis and Grinstein 2012; Wills et al. 2018). Expression of these probes in live phagocytes has allowed investigation of the distribution and metabolism of individual lipid species during particle uptake, a task that could not be accomplished previously using the conventional biochemical means of lipid detection and quantitation. The general conclusions derived from such studies are summarized in Fig. 3.1.

After particle binding, plasma membrane PtdIns(4,5)P_2 accumulates slightly in the pseudopods of the forming phagosomes (Botelho et al. 2000), presumably through the action of type I phosphatidylinositol-phosphate 5-kinase (PtdInsP5K) on PtdIns(4)P. Accordingly, PtdInsP5K and PtdIns(4)P are enriched in the extending pseudopodia (Botelho et al. 2000; Coppolino et al. 2002; Levin et al. 2017). PtdIns(4,5)P_2 is important for the recruitment of Rho-GTPases and actin nucleation-promoting factors (NPFs), and hence for the polymerization of Arp2/3-dependent branched actin networks that drive pseudopod extension during phagocytosis (Tolias et al. 2000; Botelho et al. 2000; Hoppe and Swanson 2004; Mao and Finnemann 2015). However, following this transient increase, PtdIns(4,5)P_2 rapidly disappears as the phagosome seals, followed shortly thereafter by PtdIns(4)P, by a complex process that includes hydrolysis and conversion to other lipid species (Botelho et al. 2000; Levin et al. 2017). The clearance of PtdIns(4,5)P_2, which is the result of an elaborate combination of reactions, triggers actin detachment from the late phagocytic cup (Scott et al. 2005). PtdIns(4,5)P_2 is mainly hydrolysed by phospholipase C, generating DAG and inositol *tris*phosphate (IP$_3$).

DAG and IP$_3$ stimulate PtdOH synthesis and intracellular calcium release, respectively, which in turn regulate protein effectors, the lateral spreading of phagocytic receptor signaling, and cytoskeletal remodeling (Levin et al. 2016). To a lesser extent, PtdIns(4,5)P_2 can also be hydrolyzed by inositol 5-phosphatases Inpp5b and OCRL, regenerating plasma membrane PtdIns(4)P (Schmid et al. 2004; Bohdanowicz et al. 2011). Finally, a fraction of the PtdIns(4,5)P_2 is converted to PtdIns(3,4,5)P_3 in the phagocytic cup by class I phosphatidylinositol 3-kinases (PtdIns3K), which are recruited and activated during phagocytic receptor signaling (Marshall et al. 2001; Vieira et al. 2001; Kamen et al. 2007). PtdIns(3,4,5)P_3 is an essential requirement for the phagocytosis of large, but not small particles, due to its role in the coordinated disassembly of filamentous actin (Araki et al. 1996; Schlam et al. 2015), and possibly also in membrane delivery to the nascent cup. During phagocytosis of particles ≥ 5 μm PtdIns(3,4,5)P_3 recruits GTPase-activating proteins (GAPs) that locally inactivate Rac and Cdc42, terminating the actin polymerization needed to support pseudopod extension (Schlam et al. 2015). The phagocytic cup also contains PtdIns(3,4)P_2 (Kamen et al. 2007; F. Montaño, unpublished observations), although the kinetics of its formation and role in phagocytosis remain to be determined. At the plasma membrane, and presumably the phagocytic cup, PtdIns(3,4)P_2 can be generated by the phosphorylation of PtdIns(4)P primarily by class II PtdIns3K, or by hydrolysis of PtdIns(3,4,5)P_3 by the 5-phosphatases SHIP1 and/or SHIP2 (Liu et al. 2018; Goulden et al. 2019). Subsequently, PtdIns(3,4)P_2 can be hydrolyzed by the 3-phosphatase PTEN (Malek et al. 2017; Goulden et al. 2019) to regenerate PtdIns4P. PtdIns4P, which increases transiently when phagosomes seal – ostensibly due to the action of OCRL and Inpp5b described above – is subsequently hydrolyzed by the 4-phosphatase Sac2 and possibly also by phospholipase C (Levin et al. 2017).

Interestingly, PtdSer is present at relatively constant levels in the plasma membrane, phagocytic cup, early phagosomes and even mature

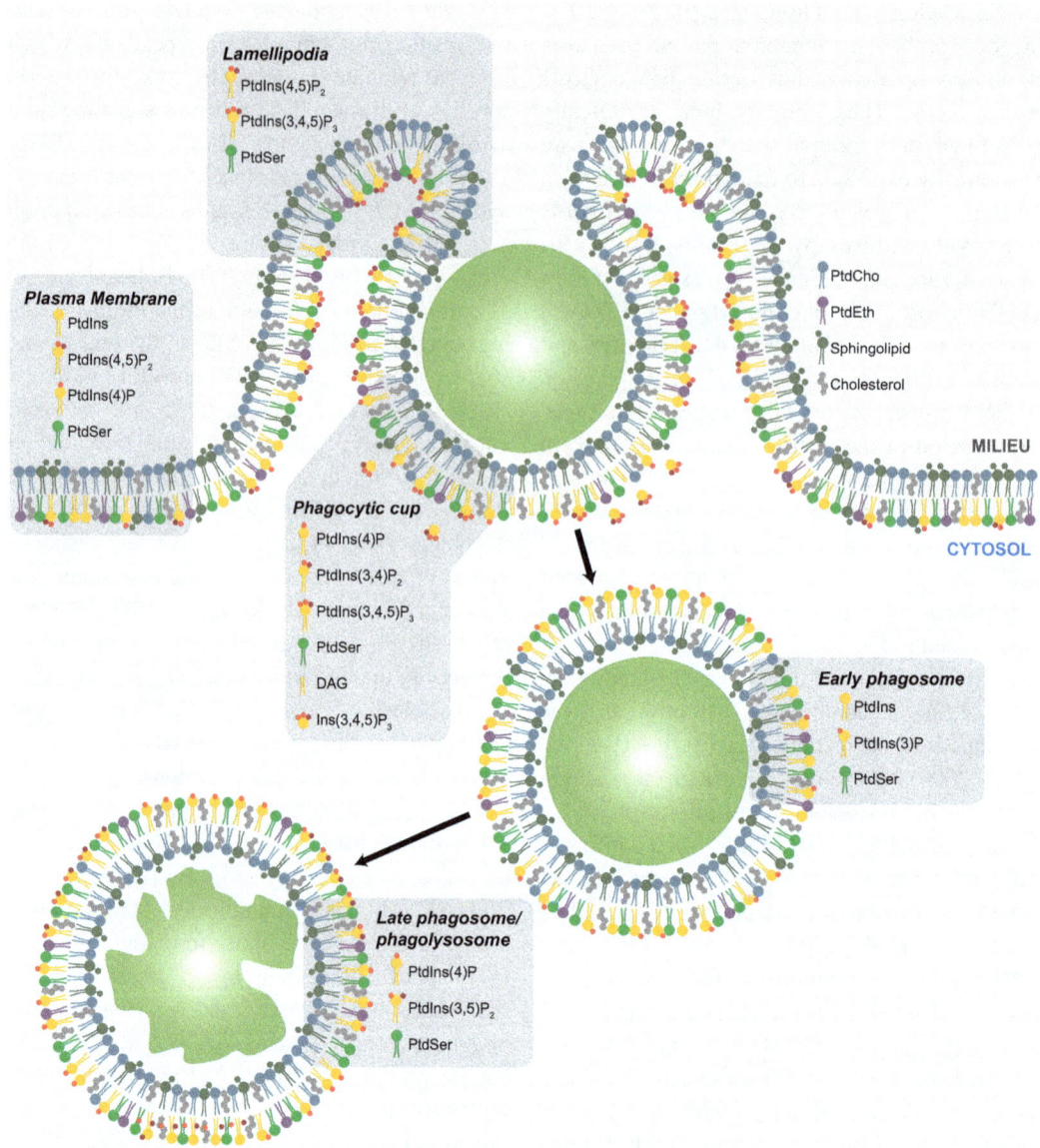

Fig. 3.1 Phospholipid distribution during phagocytosis and phagosome maturation. Localized changes to the plasma membrane inner leaflet phospholipids coordinate phagocytic signaling spatially and temporally. The inner leaflet of a resting phagocyte is rich in anionic phospholipids such as PtdIns(4)P, PtdIns(4,5)P$_2$ and PtdSer. While PtdSer levels remain relatively constant through phagosome formation and maturation, those of phosphoinositides display much more dynamic changes. PtdIns(4,5)P$_2$ and PtdIns(4)P are especially enriched in the ruffles of phagocytic cells, as well as in the lamellipodia that engage phagocytic targets. This transient increase is the result of recruited phosphatidylinositol-phosphate 5-kinase (PtdInsP5K). However, at the base of the phagocytic cup PtdIns(4,5)P$_2$ is rapidly depleted, through conversion to PtdIns(3,4,5)P$_3$ by phosphatidylinositol 3-kinase (PtdIns3K), breakdown into diacylglycerol (DAG) and inositol *tris*phosphate (IP$_3$) by phospholipase C (PLC), or hydrolysis by inositol 5-phosphatases OCRL and Inpp5b. After closure, the early phagosome acquires the phosphatidylinositol 3-kinase (PtdIns3K) Vps34, which generates PtdIns(3)P. PtdIns3P is lost during the early-to-late phagosome transition, after hydrolysis by myotubularins and conversion to PtdIns(3,5)P$_2$ by the phosphatidylinositol 3-phosphate 5-kinase, PIKfyve. Concomitant with PtdIns(3)P depletion, PtdIns(4)P kinase 2A (PtdIns4K2A) becomes enriched in late phagosomes, leading to a resurgence of PtdIns(4)P. (Refer to text for further discussion)

phagosomes (Yeung et al. 2009). Some of the PtdSer in the nascent and maturing phagosome is derived from the plasma membrane, and some is likely delivered by fusion events with compartments of the endo-lysosomal system, where PtdSer is also abundant (Yeung et al. 2008). As suggested by recent observations, phagosomal PtdIns(4)P may be exchanged for PtdSer by oxysterol-binding protein-related protein 5 (ORP5) and ORP8 (Chung et al. 2015; von Filseck et al. 2015) at sites of contact with the endoplasmic reticulum. The functional role of PtdSer on phagosomal membranes remains unclear, however preliminary data in yeast support a role for this lipid in vacuolar acidification, presumably though targeting of PtdSer-binding proteins (Yeung et al. 2009) and PtdSer may function similarly in maturing phagosomes.

Phagosome internalization is swiftly followed by the recruitment of the GTPase Rab5 and its effector, class III PtdIns3K or Vps34 (Murray et al. 2002). This kinase catalyzes the synthesis of PtdIns(3)P on the cytosolic leaflet of the early phagosome, an inositide that identifies the early phagosome and is required for its further maturation (Vieira et al. 2001; Henry et al. 2004). PtdIns(3)P recruits multiple effectors that are responsible for membrane fusion with endocytic vesicles (e.g. EEA1) that are required for the inward vesiculation of phagosomal membrane/cargo for degradation (e.g. the ESCRT complex), or for the retrograde transport of material from the phagosome to the *trans*-Golgi network for recycling (e.g. the retromer complex). These processes coincide with the transition of the phagosome from the early (Rab5-positive) to the late (Rab7-positive) maturation stage (Fairn and Grinstein 2012). PtdIns3P is lost during this transition, as a consequence of the 3-phosphatase activity of acquired myotubularins, or conversion to PtdIns(3,5)P$_2$ by the phosphatidylinositol-3-phosphate 5-kinase, PIKfyve. Remarkably, PtdIns(4)P reappears precisely as PtdIns3P disappears (Levin et al. 2017). PtdIns(4)P kinase 2A (PtdIns4K2A)

is recruited to late phagosomes (Jeschke et al. 2015; Levin et al. 2017), and is required for completion of phagosome maturation – including the full acidification that is attained via fusion with lysosomes (Levin et al. 2017) and the final resolution of the phagosomes, a poorly studied stage that entails phagolysosomal tubulation (Levin-Konigsberg et al. 2019). PtdIns(4)P recruits the retromer complex, needed for tubule scission. But perhaps more importantly, PtdIns(4)P stabilizes ARL8b-SKIP on the phagosome surface; this complex links the membrane to kinesin motor proteins to extend membrane tubules along microtubules to the periphery. Eventually, PtdIns(4)P is removed from the late phagosome via oxysterol-binding protein-related protein 1L (ORP1L), a lipid transfer protein recruited by phagosomal Rab7 that delivers PtdIns(4)P to the endoplasmic reticulum, possibly in exchange for cholesterol (Levin-Konigsberg et al. 2019). As it reaches the reticulum, PtdIns(4)P is rapidly hydrolyzed by the resident 4-phosphatase Sac1 (Zewe et al. 2018).

Phospholipid Dynamics and Inner Leaflet Charge During Phagocytosis and Phagosome Maturation

As discussed above, the plasma membrane contains about 20% negatively charged lipids that generate an electric field of approximately 10^5 V/cm (Olivotto et al. 1996) on the membrane surface. Because the bulk of the anionic lipids such as phosphoinositides and PtdSer are localized preferentially in the inner leaflet, the negative charge is exposed to the cytosol, where it can effect the recruitment of cations, notably cationic proteins, through electrostatic interactions. Phosphoinositides are especially important to dynamic changes in inner leaflet charge as they are polyvalent and are interconverted and metabolized as the plasma membrane becomes locally remodeled

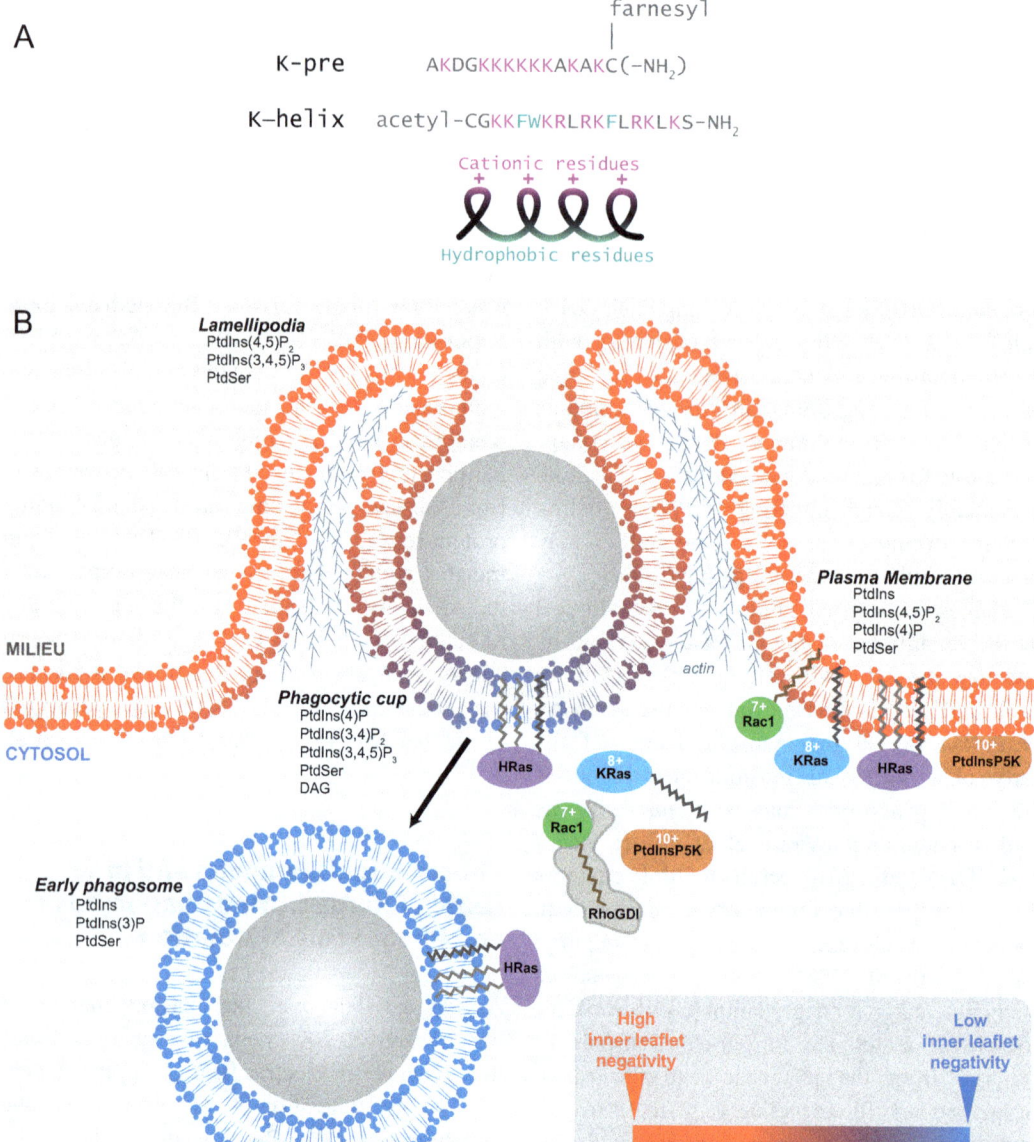

Fig. 3.2 (**a**) Structures of the genetically-encoded surface charge-sensitive probes. Probes were modeled after the C terminus of KRas, which associates with the plasma membrane in a charge-dependent manner. These probes all contain positively charged amino acid residues that allow for targeting to negatively charged membranes. For K-pre and R-Pre, association with membranes is stabilized by the presence of a farnesyl moiety. In the case of R-pre, all lysines were changed to arginines to avoid ubiquitination. K-helix has an amphiphilic α-helical structure. For all probes, acidic residues are shown in magenta, mutagenized residues blue, and hydrophobic stretches in teal. (**b**) Surface charges changes during phagocytosis. The plasma membrane carries the highest negative charge in the cell, due to the concentration of phospholipids in its inner leaflet. This favors the binding of charge-sensitive proteins such as Rac1 (stabilized by geranylgeranylation), K-Ras (stabilized by palmitoylation), and PtdInsP5K through electrostatic interactions. At the initiation of phagocytosis, the negative surface charge peaks at sites within the lamellipodia, as a result of phosphoinositide signaling initiated by particle binding. The phagocytic cup rapidly loses negative charge as phosphoinositides are hydrolyzed at its base. This change in charge coincides with dissociation of Rac1 (which binds RhoGDI in its inactive state), KRas and PtdInsP5K from

during phagocytosis (see above). Monovalent charged phospholipids like PtdSer and PtdIns, contribute the largest mole fraction of the anionic lipids; however, the polyvalent anionic phosphoinositides, PtdIns(4)P, PtdIns(4,5)P_2, and possibly PtdIns(3,4,5)P_3 (with charges of ≈ -2, -4 and -5, respectively) play a particularly important role, not only because of their higher individual charge density, but also because they can be concentrated in signaling domains by virtue of their saturated hydrophobic tails, localized generation upon receptor stimulation, or preferential binding to transmembrane proteins (Balla 2013).

Until recently, comparatively little was known about the surface charge of cellular endomembranes, which were not readily accessible to sensing devices. This predicament was overcome by the development of cytosolically-expressed fluorescent probes that can assess the charge of membranes within intact cells. Luckily, phagocytosis was one of the earliest phenomena investigated using such probes. The surface charge sensors consist of polycationic peptides of varying charge that are selectively targeted to membranes by addition of a hydrophobic moiety (Fig. 3.2a). Hydrophobicity is conferred to the peptides either by prenylation (usually near the C-terminus), or by designing amphiphilic α-helices with a cationic aspect opposed by a hydrophobic one (Silvius 1999; Roy et al. 2000; Wright and Philips 2006; Yeung et al. 2006; Quatela et al. 2008). Coincidence detection of the cationic and hydrophobic determinants minimizes confounding association with non-membranous polyanionic structures (e.g. nuclear DNA).

Initial studies revealed that the inner leaflet of the plasma membrane is the most negatively-charged surface of the cell, consistent with its unique enrichment in PtdSer and phosphoinosi-tides. Subsequently, the surface charge probes were employed to correlate changes in surface charge with phosphoinositide dynamics during phagocytosis (Fig. 3.2b). These experiments demonstrated that the charge of the inner leaflet of the phagosomal cup drops precipitously, while the bulk of the plasma membrane, not involved in the engulfment event, maintains its charge (Yeung et al. 2006). This parallels the depletion of PtdIns(4,5)P_2 and PtdIns(4)P, two highly negative phosphoinositides, that occurs as phagosomes form and seal. Because PtdSer levels remains relatively steady through formation and maturation, the phagosome surface retains an intermediate level of negativity (Yeung et al. 2009). A series of more recently developed genetically-encoded charge probes, with poly-cationic motifs of decreasing charge, confirms this notion (Yeung et al. 2006). Thus, strategic depletion of charged phosphoinositides from the maturing phagosome reduces, but does not eliminate, the negative surface charge. Moreover, the predicted appearance of PtdIns(3,5)P_2 and observed reappearance of PtdIns4P at late maturation stages (reaching levels that exceed those in the plasma membrane) are predicted to be accompanied by some restoration of the lost surface negativity. In the future, use of the available probes, perhaps in combination with newer ones, should allow for more detailed studies of membrane charge during phagocytosis, as well as comparisons to the surface charge of other cellular compartments (S. Eisenberg, unpublished observations).

The alterations of the charge of the inner leaflet can in principle have strong influence on the association and recruitment of proteins involved in the signaling, mechanical and membrane remodeling events inherent to phagocytosis. These aspects are discussed in the following section.

Fig. 3.2 (continued) However, HRas, which associates to the plasma membrane via two palmitoyl tails and a farnesyl moiety, remains membrane-bound. After internalization, lipid remodeling is expected to maintain a low negative charge on the phagosome; while it contains levels of PtdSer similar to that of the plasma membrane, it contains much lower levels of highly charged phospholipids. Approximate membrane surface charge is represented by a color gradient, ranging from high negativity (red) to low negativity (blue). (Refer to text for further discussion).

Role of Electrostatic Interactions in Protein Localization During Phagocytosis and Phagosome Maturation

Electrostatic interactions are important for the recruitment, retention, and release of soluble protein effectors from biological membranes. Additionally, the cytosolic tails of many transmembrane proteins contain polycationic stretches that are tightly associated with negative head-groups of anionic phospholipids. Conceivably, the binding of such components to membranes can be influenced by temporally-regulated changes of the membrane charge itself, or by modification of the net charge of such cationic stretches so that they no longer favor electrostatic association with negatively charged phospholipids.

This is exemplified by the KRas protein, which has a stretch of cationic residues (+8 charge) near its C terminus that is critical for targeting the protein to the plasma membrane (Wright and Philips 2006; Quatela et al. 2008). In fact, the design of the charge-sensitive probes discussed above is based on the structure of the C-terminus of KRas (S. Eisenberg, unpublished observations; Yeung et al. 2006). Like the probes introduced above, KRas is also farnesylated (Seabra 1998); however, despite the hydrophobic contribution of the farnesyl moiety, KRas detaches from the membrane when the anionic plasma membrane phospholipids (PtdIns(4,5)P_2 an PtdIns(3,4,5)P_3), are eliminated (Yeung et al. 2006; Heo et al. 2006). In contrast, the HRas isoform – which lacks the KRas polybasic region but in addition to being farnesylated is dually palmitoylated (Seabra 1998) – associates to the plasma membrane primarily via hydrophobic interactions. As such, its binding is independent of the surface charge and therefore insensitive to alterations in the composition of anionic phospholipids and the consequent changes in surface charge (Yeung et al. 2006; Heo et al. 2006). While the role of KRas in phagocytosis is poorly defined, it has been observed to detach from the membrane of the nascent phagosome, while HRas does not (Yeung et al. 2006).

Electrostatic interactions are especially compelling in the process of phagocytosis. There are several instances during the process where membrane surface charge serves as a binary switch to spatially regulate protein activity on the membrane, through changes in localization. First, the recruitment of phosphatidylinositol-4-phosphate 5-kinase (PtdInsP5K) to the plasma membrane depends on the electrostatic interaction of its cationic face with the negative surface charges (Fairn et al. 2009). The α, β and γ isoforms, which are all expressed in macrophages, catalyze the phosphorylation of PtdIns4P to PtdIns(4,5)P_2. Decreasing surface charge by pharmacological means releases PtdInsP5K from the plasma membrane (Fairn et al. 2009). Likewise, mutation of cationic residues on the face of the kinase that is involved in membrane association abrogates PtdInsP5K binding to the resting plasma membrane. In live cells, PtdInsP5K was found to detach from the forming phagosome, corresponding to the loss of negative charge seen with the surface charge probes (Fairn et al. 2009). As expected, PtdInsP5K detachment from the phagocytic cup terminates its ability to generate PtdIns(4,5)P_2, contributing to the acute depletion of the phosphoinositide from the cup that is key for the completion of phagocytosis.

Membrane surface charge also influences the localization of Rac1 to the plasma membrane. This GTPase drives at least part of the actin polymerization required for phagocyte ruffling (Cox et al. 1997) and pseudopod extension during phagocytosis (Caron and Hall 1998; Massol et al. 1998). Like KRas, Rac1 is prenylated (geranylgeranylated) near its C terminus (Wennerberg and Der 2004) and contains a polycationic motif that targets/maintains the active (GTP-bound) form at the PtdIns(4,5)P_2- and PtdIns(4)P-rich plasma membrane (Ueyama et al. 2005; Magalhaes and Glogauer 2009). Because, as discussed above, these phosphoinositides are depleted, GTP-bound Rac1 detaches from the nascent phagosome as the surface loses negative charge (Ueyama et al. 2005; Yeung et al. 2006). These interesting findings suggest that the lipid composition of the phagocytic membrane is a major determinant of the localized activity of Rac1, and possibly other GTPases with polycationic targeting motifs. Interestingly, RhoGDI contains negative charges that interact with this cationic region and a hydrophobic pocket that accommodates the prenyl tail of Rac1 when it is inactive (Ugolev et al. 2006; Ueyama et al. 2013). Electrostatic dissociation provides a unique mechanism for Rac1 regulation independent of changes in GEF and GAP activities. Instead, Rac1 activity can be terminated at the phagosome by the reduced surface charge attributed to the depletion of polyphosphoinositides.

A third example of a protein that is responsive to surface charge is myristoylated alanine-rich C kinase substrate (MARCKS), a filamentous (F) actin cross-linking protein that localizes to the plasma membrane based on its cationic effector domain (Kim et al. 1994). This cationic domain interacts with plasmalemmal PtdIns(4,5)P$_2$ (Glaser et al. 1996). Interestingly, the binding of this protein to the plasma membrane, while sensitive to the formation/disappearance of PtdIns(4,5)P$_2$, is also regulated by phosphorylation of the protein (MARCKS) itself; the accrual of negative charges by phosphorylation alters the overall charge of the protein, reducing the electrostatic attraction to the anionic membrane (Thelen et al. 1991). This causes the relocalization of the phospho-MARCKS to the cytosol, and a similar mechanism could conceivably affect the localization of many other molecules/proteins by modifying their net charge.

Relevance of Inner Leaflet Surface Charge to Phagocytes and Immunity

Because inner leaflet surface charge and its changes are indispensable for proper phagocytosis and vacuole maturation, it is not surprising that several microbial pathogens manipulate plasmalemmal phospholipids during their interactions with the host (Walpole et al. 2018). These virulence mechanisms can allow microbes to evade innate immunity through the invasion of non-phagocytic cells, or directly interfere with the maturation of the microbial vacuole after internalization to evade killing. There exist several described microbial mechanisms targeting negatively charged phospholipids of the host cell.

Shigella flexneri translocates a potent inositol 4-phosphatase, IpgD, into the host. This virulence factor specifically hydrolyzes plasma membrane PtdIns(4,5)P$_2$ into PtdIns(5)P, to promote cell entry (Niebuhr et al. 2000, 2002). Similarly, *Salmonella enterica* injects SopB, which hydrolyzes PtdIns(4,5)P$_2$ and measurably decreases the negative charge of the plasma membrane, to promote invagination and invasion, and to block lysosomal targeting of the *Salmonella*-containing vacuole (Terebiznik et al. 2002; Bakowski et al. 2010). *Burkholderia pseudomallei* also translocates an inositol-4 phosphatase, BopB, into host cells (Stevens et al. 2004; Ungewickell et al. 2005). In contrast, *Yersinia pseudotuberculosis* maintains PtdIns(4,5)P$_2$ on its prevacuole, essentially stalling vacuole scission, and hindering internalization (Wong and Isberg 2003; Sarantis et al. 2012).

Legionella pneumophila and *Chlamydia trachomatis* both deplete PtdSer from their intracellular vacuoles and thereby manipulate the vacuolar surface charge, diverting maturation away from the endolysosomal pathway and evading microbial killing (Yeung et al. 2009). Interestingly, *L. pneumophila* additionally targets negatively-charged PtdIns4P by means of the Icm/Dot effector SidC to facilitate the recruitment of endoplasmic reticulum proteins to the *Legionella*-containing vacuole. Considering the numerous examples described here, it is possible that manipulation of host membrane charge represents a universal strategy for diverse intracellular pathogens to evade host-cell innate and adaptive defense mechanisms.

Concluding Remarks

Stereospecific and hydrophobic interactions have been studied extensively as determinants of the association of proteins with biological membranes. By comparison, electrostatic attraction has been largely neglected. This oversight is particularly shortsighted in the case of the plasma membrane, which is endowed with a sizable electronegative surface charge density, capable of attracting cationic proteins and peptides, and of distorting the structure of cationic cytosolic domains of transmembrane proteins, which abound. As a consequence, functional changes resulting from diminished electrostatic attraction may have been overlooked. As illustrated in this review, rather severe changes in charge have been recorded during phagocytosis and are most likely to accompany also macropinocytosis and other types of endocytic processes. Perhaps these and other processes will similarly be appreciated to undergo charge alterations that can modify their overall composition and function to a significant extent.

Acknowledgements Supported by grant FDN-143202 from the Canadian Institutes of Health Research.

References

Araki N, Johnson MT, Swanson JA (1996) A role for phosphoinositide 3-kinase in the completion of macropinocytosis and phagocytosis by macrophages. J Cell Biol 135:1249–1260

Bakowski MA, Braun V, Lam GY et al (2010) The phosphoinositide phosphatase SopB manipulates membrane surface charge and trafficking of the Salmonella-containing vacuole. Cell Host Microbe 7:453–462. https://doi.org/10.1016/j.chom.2010.05.011

Balla T (2013) Phosphoinositides: tiny lipids with giant impact on cell regulation. Physiol Rev 93:1019–1137. https://doi.org/10.1152/physrev.00028.2012

Bohdanowicz M, Grinstein S (2013) Role of phospholipids in endocytosis, phagocytosis, and macropinocytosis. Physiol Rev 93:69–106. https://doi.org/10.1152/physrev.00002.2012

Bohdanowicz M, Balkin DM, De Camilli P, Grinstein S (2011) Recruitment of OCRL and Inpp5B to phagosomes by Rab5 and APPL1 depletes phosphoinositides and attenuates Akt signaling. Mol Biol Cell 23:176–187

Botelho RJ, Teruel M, Dierckman R et al (2000) Localized biphasic changes in phosphatidylinositol-4,5-bisphosphate at sites of phagocytosis. J Cell Biol 151:1353–1368

Caron E, Hall A (1998) Identification of two distinct mechanisms of phagocytosis controlled by different Rho GTPases. Science 282:1717–1721. https://doi.org/10.1126/science.282.5394.1717

Chung J, Torta F, Masai K et al (2015) INTRACELLULAR TRANSPORT. PI4P/phosphatidylserine countertransport at ORP5- and ORP8-mediated ER-plasma membrane contacts. Science 349:428–432. https://doi.org/10.1126/science.aab1370

Coppolino MG, Dierckman R, Loijens J et al (2002) Inhibition of phosphatidylinositol-4-phosphate 5-kinase Ialpha impairs localized actin remodeling and suppresses phagocytosis. J Biol Chem 277:43849–43857. https://doi.org/10.1074/jbc.M209046200

Cox D, Chang P, Zhang Q et al (1997) Requirements for both Rac1 and Cdc42 in membrane ruffling and phagocytosis in leukocytes. J Exp Med 186:1487–1494

Di Paolo G, De Camilli P (2006) Phosphoinositides in cell regulation and membrane dynamics. Nature 443:651–657. https://doi.org/10.1038/nature05185

Fairn GD, Grinstein S (2012) How nascent phagosomes mature to become phagolysosomes. Trends Immunol 33:397–405

Fairn GD, Ogata K, Botelho RJ et al (2009) An electrostatic switch displaces phosphatidylinositol phosphate kinases from the membrane during phagocytosis. J Cell Biol 187:701–714. https://doi.org/10.1083/jcb.200909025

Flannagan RS, Jaumouillé V, Grinstein S (2012) The cell biology of phagocytosis. Annu Rev Pathol 7:61–98. https://doi.org/10.1146/annurev-pathol-011811-132445

Glaser M, Wanaski S, Buser CA et al (1996) Myristoylated alanine-rich C kinase substrate (MARCKS) produces reversible inhibition of phospholipase C by sequestering phosphatidylinositol 4,5-bisphosphate in lateral domains. J Biol Chem 271:26187–26193. https://doi.org/10.1074/jbc.271.42.26187

Goulden BD, Pacheco J, Dull A et al (2019) A high-avidity biosensor reveals plasma membrane PI(3,4)P2 is predominantly a class I PI3K signaling product. J Cell Biol 218:1066–1079. https://doi.org/10.1083/jcb.201809026

Henry RM, Hoppe AD, Joshi N, Swanson JA (2004) The uniformity of phagosome maturation in macrophages. J Cell Biol 164:185–194. https://doi.org/10.1083/jcb.200307080

Heo WD, Inoue T, Park WS et al (2006) PI(3,4,5)P3 and PI(4,5)P2 lipids target proteins with polybasic clusters to the plasma membrane. Science 314:1458–1461. https://doi.org/10.1126/science.1134389

Holthuis JCM, Levine TP (2005) Lipid traffic: floppy drives and a superhighway. Nat Rev Mol Cell Biol 6:209–220. https://doi.org/10.1038/nrm1591

Hoppe AD, Swanson JA (2004) Cdc42, Rac1, and Rac2 display distinct patterns of activation during phagocytosis. Mol Biol Cell 15:3509–3519

Jaumouillé V, Grinstein S (2016) Molecular mechanisms of phagosome formation. Microbiol Spectr 4:1–19. https://doi.org/10.1128/microbiolspec.MCHD-0013-2015

Jeschke A, Zehethofer N, Lindner B et al (2015) Phosphatidylinositol 4-phosphate and phosphatidylinositol 3-phosphate regulate phagolysosome biogenesis. Proc

Natl Acad Sci USA 112:4636–4641. https://doi.org/10.1073/pnas.1423456112

Kamen LA, Levinsohn J, Swanson JA (2007) Differential association of phosphatidylinositol 3-kinase, SHIP-1, and PTEN with forming phagosomes. Mol Biol Cell 18:2463–2472. https://doi.org/10.1091/mbc.e07-01-0061

Kim J, Shishido T, Jiang X et al (1994) Phosphorylation, high ionic strength, and calmodulin reverse the binding of MARCKS to phospholipid vesicles. J Biol Chem 269:28214–28219

Levin R, Grinstein S, Canton J (2016) The life cycle of phagosomes: formation, maturation, and resolution. Immunol Rev 273:156–179. https://doi.org/10.1111/imr.12439

Levin R, Hammond GRV, Balla T et al (2017) Multiphasic dynamics of phosphatidylinositol 4-phosphate during phagocytosis. Mol Biol Cell 28:128–140. https://doi.org/10.1091/mbc.E16-06-0451

Levin-Konigsberg R, Montaño-Rendón F, Keren-Kaplan T et al (2019) Phagolysosome resolution requires contacts with the endoplasmic reticulum and phosphatidylinositol-4-phosphate signalling. Nat Cell Biol 21:1234–1247. https://doi.org/10.1038/s41556-019-0394-2

Liu S-L, Wang Z-G, Hu Y et al (2018) Quantitative lipid imaging reveals a new signaling function of phosphatidylinositol-3,4-bisphophate: isoform- and site-specific activation of Akt. Mol Cell 71:1092–1104.e5. https://doi.org/10.1016/j.molcel.2018.07.035

Magalhaes MAO, Glogauer M (2009) Pivotal advance: phospholipids determine net membrane surface charge resulting in differential localization of active Rac1 and Rac2. J Leukoc Biol 87:545–555. https://doi.org/10.1189/jlb.0609390

Malek M, Kielkowska A, Chessa T et al (2017) PTEN regulates PI(3,4)P2 signaling downstream of class I PI3K. Mol Cell 68:566–580.e10. https://doi.org/10.1016/j.molcel.2017.09.024

Mao Y, Finnemann SC (2015) Regulation of phagocytosis by Rho GTPases. Small GTPases 6:89–99. https://doi.org/10.4161/21541248.2014.989785

Marshall JG, Booth JW, Stambolic V et al (2001) Restricted accumulation of phosphatidylinositol 3-kinase products in a plasmalemmal subdomain during Fc gamma receptor-mediated phagocytosis. J Cell Biol 153:1369–1380. https://doi.org/10.1083/jcb.153.7.1369

Massol P, Montcourrier P, Guillemot JC, Chavrier P (1998) Fc receptor-mediated phagocytosis requires CDC42 and Rac1. EMBO J 17:6219–6229. https://doi.org/10.1093/emboj/17.21.6219

McLaughlin S, Murray D (2005) Plasma membrane phosphoinositide organization by protein electrostatics. Nature 438:605–611. https://doi.org/10.1038/nature04398

Murray JT, Panaretou C, Stenmark H et al (2002) Role of Rab5 in the recruitment of hVps34/p150 to the early endosome. Traffic (Copenhagen, Denmark) 3:416–427

Niebuhr K, Jouihri N, Allaoui A et al (2000) IpgD, a protein secreted by the type III secretion machinery of Shigella flexneri, is chaperoned by IpgE and implicated in entry focus formation. Mol Microbiol 38:8–19. https://doi.org/10.1046/j.1365-2958.2000.02041.x

Niebuhr K, Giuriato S, Pedron T et al (2002) Conversion of PtdIns(4,5)P(2) into PtdIns(5)P by the S.flexneri effector IpgD reorganizes host cell morphology. EMBO J 21:5069–5078. https://doi.org/10.1093/emboj/cdf522

Olivotto M, Arcangeli A, Carlà M, Wanke E (1996) Electric fields at the plasma membrane level: a neglected element in the mechanisms of cell signalling. BioEssays 18:495–504. https://doi.org/10.1002/bies.950180612

Pauwels A-M, Trost M, Beyaert R, Hoffmann E (2017) Patterns, receptors, and signals: regulation of phagosome maturation. Trends Immunol 38:407–422. https://doi.org/10.1016/j.it.2017.03.006

Quatela SE, Sung PJ, Ahearn IM et al (2008) Analysis of K-Ras phosphorylation, translocation, and induction of apoptosis. Methods Enzymol 439:87–102. https://doi.org/10.1016/S0076-6879(07)00407-7

Roy MO, Leventis R, Silvius JR (2000) Mutational and biochemical analysis of plasma membrane targeting mediated by the farnesylated, polybasic carboxy terminus of K-ras4B. Biochemistry 39:8298–8307. https://doi.org/10.1021/bi000512q

Sarantis H, Grinstein S (2012) Monitoring phospholipid dynamics during phagocytosis: application of genetically-encoded fluorescent probes. Methods Cell Biol 108:429–444. https://doi.org/10.1016/B978-0-12-386487-1.00019-5

Sarantis H, Balkin DM, De Camilli P et al (2012) Yersinia entry into host cells requires Rab5-dependent dephosphorylation of PI(4,5)P$_2$ and membrane scission. Cell Host Microbe 11:117–128. https://doi.org/10.1016/j.chom.2012.01.010

Schlam D, Bagshaw RD, Freeman SA et al (2015) Phosphoinositide 3-kinase enables phagocytosis of large particles by terminating actin assembly through Rac/Cdc42 GTPase-activating proteins. Nat Commun 6:8623

Schmid AC, Wise HM, Mitchell CA et al (2004) Type II phosphoinositide 5-phosphatases have unique sensitivities towards fatty acid composition and head group phosphorylation. FEBS Lett 576:9–13. https://doi.org/10.1016/j.febslet.2004.08.052

Scott CC, Dobson W, Botelho RJ et al (2005) Phosphatidylinositol-4,5-bisphosphate hydrolysis directs actin remodeling during phagocytosis. J Cell Biol 169:139–149

Seabra MC (1998) Membrane association and targeting of prenylated Ras-like GTPases. Cell Signal 10:167–172. https://doi.org/10.1016/s0898-6568(97)00120-4

Silvius JR (1999) Fluorescence measurement of lipid-binding affinity and interlayer transfer of bimane-labeled lipidated peptides. Methods Mol Biol (Clifton, NJ) 116:177–186. https://doi.org/10.1385/1-59259-264-3:177

Stevens MP, Haque A, Atkins T et al (2004) Attenuated virulence and protective efficacy of a Burkholderia pseudomallei bsa type III secretion mutant in murine models of melioidosis. Microbiology 150:2669–2676. https://doi.org/10.1099/mic.0.27146-0

Swanson JA (2014) Phosphoinositides and engulfment. Cell Microbiol 16:1473–1483. https://doi.org/10.1111/cmi.12334

Terebiznik MR, Vieira OV, Marcus SL et al (2002) Elimination of host cell PtdIns(4,5)P2 by bacterial SigD promotes membrane fission during invasion by Salmonella. Nat Cell Biol 4:766–773. https://doi.org/10.1038/ncb854

Thelen M, Rosen A, Nairn AC, Aderem A (1991) Regulation by phosphorylation of reversible association of a myristoylated protein kinase C substrate with the plasma membrane. Nature 351:320–322. https://doi.org/10.1038/351320a0

Tolias KF, Hartwig JH, Ishihara H et al (2000) Type Iα phosphatidylinositol-4-phosphate 5-kinase mediates Rac-dependent actin assembly. Curr Biol 10:153–156. https://doi.org/10.1016/S0960-9822(00)00315-8

Ueyama T, Eto M, Kami K et al (2005) Isoform-specific membrane targeting mechanism of Rac during FcγR-mediated phagocytosis: positive charge-dependent and independent targeting mechanism of Rac to the phagosome. J Immunol 175:2381–2390. https://doi.org/10.4049/jimmunol.175.4.2381

Ueyama T, Son J, Kobayashi T et al (2013) Negative charges in the flexible N-terminal domain of rho GDP-dissociation inhibitors (RhoGDIs) regulate the targeting of the RhoGDI–Rac1 complex to membranes. J Immunol 191:2560–2569. https://doi.org/10.4049/jimmunol.1300209

Ugolev Y, Molshanski-Mor S, Weinbaum C, Pick E (2006) Liposomes comprising anionic but not neutral phospholipids cause dissociation of Rac(1 or 2) x RhoGDI complexes and support amphiphile-independent NADPH oxidase activation by such complexes. J Biol Chem 281:19204–19219. https://doi.org/10.1074/jbc.M600042200

Ungewickell A, Hugge C, Kisseleva M et al (2005) The identification and characterization of two phosphatidylinositol-4,5-bisphosphate 4-phosphatases. Proc Natl Acad Sci USA 102:18854–18859. https://doi.org/10.1073/pnas.0509740102

Verkleij AJ, Post JA (2014) Membrane phospholipid asymmetry and signal transduction. J Membr Biol 178:1–10. https://doi.org/10.1007/s002320010009

Vieira OV, Botelho RJ, Rameh L et al (2001) Distinct roles of class I and class III phosphatidylinositol 3-kinases in phagosome formation and maturation. J Cell Biol 155:19–25

von Filseck JM, Čopič A, Delfosse V et al (2015) Phosphatidylserine transport by ORP/Osh proteins is driven by phosphatidylinositol 4-phosphate. Science 349:432–436. https://doi.org/10.1126/science.aab1346

Walpole GFW, Grinstein S, Westman J (2018) The role of lipids in host-pathogen interactions. IUBMB Life 70:384–392. https://doi.org/10.1002/iub.1737

Wennerberg K, Der CJ (2004) Rho-family GTPases: it's not only Rac and Rho (and I like it). J Cell Sci 117:1301–1312. https://doi.org/10.1242/jcs.01118

Wills RC, Goulden BD, Hammond GRV (2018) Genetically encoded lipid biosensors. Mol Biol Cell 29:1526–1532. https://doi.org/10.1091/mbc.E17-12-0738

Wong K-W, Isberg RR (2003) Arf6 and phosphoinositol-4-phosphate-5-kinase activities permit bypass of the Rac1 requirement for beta1 integrin-mediated bacterial uptake. J Exp Med 198:603–614. https://doi.org/10.1084/jem.20021363

Wright LP, Philips MR (2006) Thematic review series: lipid posttranslational modifications. CAAX modification and membrane targeting of Ras. J Lipid Res 47:883–891. https://doi.org/10.1194/jlr.R600004-JLR200

Yeung T, Terebiznik M, Yu L et al (2006) Receptor activation alters inner surface potential during phagocytosis. Science 313:347–351. https://doi.org/10.1126/science.1129551

Yeung T, Gilbert GE, Shi J et al (2008) Membrane phosphatidylserine regulates surface charge and protein localization. Science 319:210–213. https://doi.org/10.1126/science.1152066

Yeung T, Heit B, Dubuisson J-F et al (2009) Contribution of phosphatidylserine to membrane surface charge and protein targeting during phagosome maturation. J Cell Biol 185:917–928. https://doi.org/10.1083/jcb.200903020

Zewe JP, Wills RC, Sangappa S et al (2018) SAC1 degrades its lipid substrate PtdIns4P in the endoplasmic reticulum to maintain a steep chemical gradient with donor membranes. elife 7:1019. https://doi.org/10.7554/eLife.35588

Receptor Models of Phagocytosis: The Effect of Target Shape

4

David M. Richards

Abstract

Phagocytosis is a remarkably complex process, requiring simultaneous organisation of the cell membrane, the cytoskeleton, receptors and various signalling molecules. As can often be the case, mathematical modelling is able to penetrate some of this complexity, identifying the key biophysical components and generating understanding that would take far longer with a purely experimental approach. This chapter will review a particularly important class of phagocytosis model, championed in recent years, that primarily focuses on the role of receptors during the engulfment process. These models are pertinent to a host of unsolved questions in the subject, including the rate of cup growth during uptake, the role of both intra- and extracellular noise, and the precise differences between phagocytosis and other forms of endocytosis. In particular, this chapter will focus on the effect of target shape and orientation, including how these influence the rate and final outcome of phagocytic engulfment.

Keywords

Mathematical modelling · Computer simulation · Receptors · Target shape dependence

Introduction

The complexity of the process of phagocytosis, which requires substantial changes in cell shape, membrane reorganisation and cytoskeletal remodelling, is often seen as a hurdle to mechanistic understanding. Naïvely, this complexity might also be thought to hinder attempts to understand phagocytosis from a theoretical standpoint. However, this is often precisely when mathematical and computational modelling approaches come into their own, allowing progress in novel, unanticipated directions. In fact, it could be argued that it is precisely the complicated nature of phagocytosis that makes it ideally suited for a multidisciplinary approach that combines biology, biophysics, mathematics and computing.

Some of the following will no doubt appear quite mathematically-heavy to many readers. However, this does not mean that the underlying ideas are particularly complicated or difficult to understand. In fact, many of these underlying concepts are remarkably simple. I urge the non-

D. M. Richards (✉)
Living Systems Institute, University of Exeter, Exeter, UK
e-mail: david.richards@exeter.ac.uk

© Springer Nature Switzerland AG 2020
M. B. Hallett (ed.), *Molecular and Cellular Biology of Phagocytosis*, Advances in Experimental Medicine and Biology 1246, https://doi.org/10.1007/978-3-030-40406-2_4

specialised reader to not be intimidated by the mathematical details. Most of the equations can be skipped over with few problems. As often with any modelling approach, it is the fundamental ideas that are far more important than the technical details.

This chapter will be organised as follows. After a brief introduction to mathematical modelling in general, including what models are and what they can achieve, I will describe a type of phagocytosis model that focuses on the motion of membrane-bound receptors. These models can address many issues such as how the target size affects the rate of engulfment, the role of actin and signalling molecules, and the fundamental differences between phagocytosis and other types of endocytosis. The next section will cover how modelling is starting to shed light on the uptake of non-spherical target particles, particularly shapes such as capped cylinders and tubes, which are far closer to the actual particles that phagocytes encounter in the real world. Finally, I will explore the possible future of theoretical approaches to phagocytosis, including currently unexplored avenues, potentially deep connections to other immune processes, and how continued progress in this area is likely to require multidisciplinary research that intimately combines traditional wet-lab experiments with mathematical modelling.

Why Modelling?

Since at least the middle of the last century, mathematical and computational modelling in all areas of biology and medicine has become increasingly popular, important and useful (Mackey and Maini 2015; Reed 2004). This includes areas as diverse as action potentials in neurons (Hodgkin and Huxley 1952), bacterial chemotaxis (Tindall et al. 2008a,b), protein structure (Dorn et al. 2014) and evolution (Nowak 2006). Further, the whole gamut of theoretical approaches has been called upon, from those in mathematics and physics to those in chemistry and computing.

At times, the whole area of modelling in biology is confused by a number of other meanings of the word "model", most of which are entirely un-

connected to the mathematical and computational models considered here. For example, model organisms, disease models or mechanistic-sketches are clearly unrelated. Similarly, statistical models (that summarise underlying trends in data) and network models (of, for example, metabolic or protein interaction networks) are distinct to the type of modelling used below. In addition, other quantitative areas of biology, including bioinformatics and image analysis, although often drawing heavily on similar mathematics, statistics and computer science, are again of a different nature to traditional mathematical and computational models.

Here, by model, I always refer to mathematical and computational approaches to describing real biological (or biophysical) systems. The distinction is often fuzzy, but mathematical modelling tends to include the use of difference equations or differential equations (e.g. ordinary, partial, delay, stochastic and integro-differential equations), whereas computational modelling includes agent-based models, molecular dynamics and numerical simulation. Of course, these examples are not meant to be exhaustive: many type of modelling do not fall directly into these well-defined categories and often straddle disciplines.

There are two broad classes of mathematical/computational model: "all-inclusive" models and "abstracted" models. "All-inclusive" models attempt to include every component, process and interaction. Examples include whole-cell models (of the full proteome, genome, transcriptome and metabolome) (Carrera and Covert 2015; Feig and Sugita 2019; Goldberg et al. 2018), simulations of the human brain (such as the Human Brain Project (The Human Brain Project 2013)), and molecular dynamics (of, for example, protein folding or virus assembly) (Hollingsworth and Dror 2018; Hospital et al. 2015). Potential issues with this type of modelling is the sheer number of (typically not fully known) ingredients and parameters, often leading to a substantial drop in predictive power and the risk of "putting more in than we get out."

The other broad type of model—"abstracted" models—involve a simplified or approximate description of a system. The aim is not to include every detail, but rather to focus on the key components and processes that govern the variables of interest. Examples of this kind of approach include using a random walk to describe the position of a particular organelle (Berg 1993; Codling et al. 2008), modelling reaction kinetics using the law of mass action (Voit et al. 2015), and simulating flocking behaviour with a simple agent-based model (An et al. 2017). The potential issues now include the difficulty of knowing which components to include and exclude, and the risk that an overly-simplified system may in no way describe reality.

Since "abstracted" models are, by design, only an approximation to the real system, it is reasonable to ask what their use is. There are at least four possible answers. First, models can tell us how things cannot work or can tell us that something is wrong or missing. They can sometimes even predict what it is that is wrong or missing. Second, modelling can often speed-up research that would take many times longer with an experimental approach. Similarly, modelling can also be substantially cheaper, less technically demanding and involve fewer ethical issues. For example, consider trying to determine which of a hundred particle shapes is the easiest to phagocytose. A lab-based approach may well need several tens of years and require extensive funding to fabricate all hundred shapes and test each separately, whereas a computational model could simulate all the shapes in a fraction of the time. The third use of modelling is to provide or simplify understanding by, for example, determining the key basic mechanisms and processes that give rise to a particular behaviour. The final utility of models is to make predictions, suggest new experiments or motivate novel research avenues. Predictions can then be tested experimentally, which in turn can inform and suggest modifications to the model, so leading to further (hopefully more refined) testable predictions.

Although there is sometimes a role for "all-inclusive" models, it is in my opinion often the more focussed "abstracted" type of modelling that leads to the best and quickest progress. One reason for this is that "abstracted" models normally have only a handful of parameters, compared to tens of thousands (or more) for "all-inclusive" models. Fewer-parameter models are easier to study, easy to fit, and typically result in more testable predictions. Of course, it is possible to have a model that is too simplified or abstract and so fails to adequately capture the real system. Finding the happy medium is part of the art of model design and is invariably facilitated by working as closely as possible with real data and experimentalists: a multidisciplinary approach that carefully combines ideas from both experiments and modelling is almost always preferable to single-discipline work.

Receptor Models of Phagocytosis

There is little truth to the notion that the complexity of phagocytosis limits the scope of mathematical and/or computational modelling in this field. In fact, on the contrary, this is a case where simple "abstracted" models can be tremendously valuable. Such models can potentially identify the key processes underlying phagocytosis, leading to the fundamental biophysical mechanisms that govern engulfment. These mechanisms could include, for example, membrane curvature, the density of actin, or the availability of receptors. Once such a model is in place it becomes possible to address questions that might otherwise appear intractable, such as the effect of target shape and orientation, the role the actin network plays, and deep (perhaps evolutionary) links with other immune processes.

The following will focus on one particular type of phagocytosis modelling, arguably the most developed, which has led to a number of important results in recent years. These models take the view that the key to progression of phagocytosis lies with the receptors, particularly their spatial location. At all stages of engulfment, the phagocytic cup can only extend further if receptors are present at the edge of the cup, where they can bind some ligand on the target particle (see Fig. 4.1). Thus, it is natural to enquire how receptors find their way to the edge of the cup. In particular, is receptor motion a passive or active process?

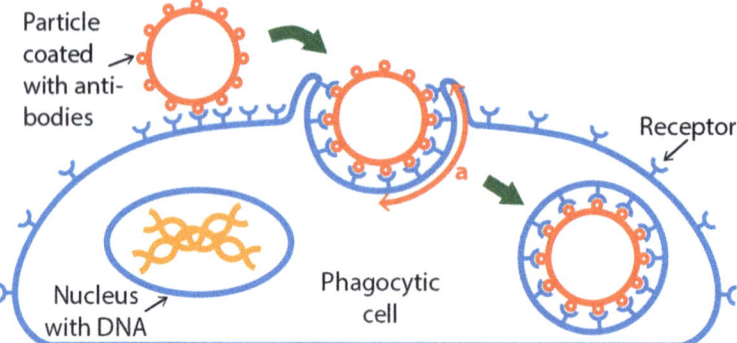

Fig. 4.1 Cartoon of the progression of phagocytosis, showing a target particle (red) engulfed by a phagocyte (blue). The key variable is the cup size a, which gradually increases during phagocytosis. This happens because receptors in the cell membrane irreversibly bind ligands on the target particle, causing the cell membrane to track the shape of the target. Questions about the ultimate fate of phagocytosis and the rate of engulfment thus reduce to understanding how receptors move to the edge of the phagocytic cup. (Image used with permission from Richards and Endres 2014)

Although driven by the motion of receptors, it is important to point out that these models do not solely focus on receptors at the expense of other processes and components. These other ingredients (such as the cytoskeleton, signalling molecules, membrane shape and target shape) typically also appear in these models, but their influence is felt indirectly, perhaps via certain model parameters or via the background geometry within which the receptors move. These are essentially models of receptors, but models that are also guided by membrane physics, understanding of actin networks, and the behaviour of signalling cascades.

The Basic Receptor Model

The basic version of the phagocytosis receptor model is built upon similar models for other types of endocytosis (Freund and Lin 2004; Gao et al. 2005). The membrane is two-dimensional and so can be described by two coordinates, typically taken to be the polar coordinates r (the distance from the point where the target first touches the membrane) and θ (the angle around the target as measured from directly above). Crucially, many models neglect stochastic effects and consider only targets that have circular symmetry, so that all the dynamics take place only in the radial direction. This collapses the system to one dimension, parameterised by the single coordinate r.

The simplest models are concerned with the dynamics of just two variables: the density of receptors $\rho(r, t)$ and the phagocytic cup size $a(t)$, where t is the time. The receptor density depends on both membrane position and time, whereas the cup size is only a function of time. The cup size is typically measured along the circumference of the target, so that full engulfment for a spherical particle of radius R corresponds to $a = \pi R$.

For simplicity, the density of ligand on the target is typically taken to be constant, given by ρ_L. With the assumption that all ligands within the cup region are bound to receptors, this means that the receptor density is always ρ_L within the cup, i.e. $\rho = \rho_L$ for $r < a$. As a consequence, all the interesting behaviour of ρ, which will be governed by the equations below, takes place outside the cup (i.e. where $r > a$). Initially, before engulfment has started (so that $a = 0$), it is assumed that there is a constant (independent of r) receptor density ρ_0 over the entire membrane. Figure 4.2a shows the general model set-up.

The heart of this model (as with most models) is the dynamics, i.e. the rules that determine how the variables change with time. First, consider the receptor density, ρ. The simplest assumption is that receptor motion outside the cup is entirely controlled by diffusion, with no active type

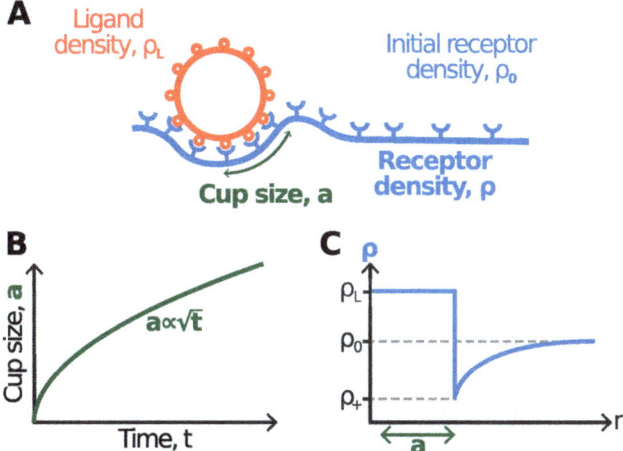

Fig. 4.2 The basic receptor model. (**a**) The simplest models consist of two variables: the cup size a and the receptor density ρ. The target ligand density, ρ_L, determines ρ within the cup, and the constant density ρ_0 gives both the initial receptor density (over the whole membrane) and the density at $r = \infty$ (at all times). (**b**) Characteristic of many diffusion-controlled processes, the cup size grows like the square-root of time. (**c**) The receptor density profile splits into two distinct regions corresponding to the parts of the membrane inside and outside the phagocytic cup. The density at the cup edge, ρ_+, is often found from free energy considerations. The positive gradient just outside the cup ($\partial\rho/\partial r > 0$) shows there is an inward receptor flux at the cup edge, causing a gradual increase in cup size. (Image taken with permission from Richards and Endres 2016)

of motion. Then the dynamics of ρ (for $r > a(t)$) are controlled by the radial part of the two-dimensional diffusion equation (Berg 1993)

$$\frac{\partial\rho}{\partial t} = \frac{D}{r}\frac{\partial}{\partial r}\left(r\frac{\partial\rho}{\partial r}\right),\qquad(4.1)$$

where D is the receptor diffusion constant. The angular part of this equation will only become important when lower-symmetry target particles are considered.

Now, consider the cup size, a. Cup growth (i.e. increase of a) is associated with new ligand-receptor bonds, which tie together the cell membrane and the target. This means that cup growth is related to inward flow of receptors at the cup edge. With the assumption that all receptors within the cup are bound to ligands, the equation for $a(t)$ is found to be (Richards and Endres 2017)

$$\frac{da}{dt} = \frac{D\rho_+'}{\rho_L - \rho_+},\qquad(4.2)$$

where ρ_+ is the receptor density at the cup edge, i.e. at $r = a$.

Interestingly, although this is a model of receptor dynamics during phagocytosis, it is identical to a famous problem in physics—the Stefan problem—which describes phase transitions such as the melting of ice (Gupta 2003; Meirmanov 1992; Stefan 1891). In fact, more precisely, this model is equivalent to the supercooled Stefan problem, which describes, for example, the freezing of supercooled water. Because of this, it is tempting to interpret phagocytic engulfment as a phase transition between unbound and bound receptors.

As it stands, the model is still not fully specified. One extra piece of information is needed to give a system with a unique solution. Starting with work in endocytosis (Gao et al. 2005), this extra condition has often been motivated by free energy considerations. In short, by requiring that there is no jump in free energy at the cup edge, it is possible to find an equation for ρ_+, the receptor density at the edge of the cup (Freund and Lin 2004):

$$\frac{\rho_+}{\rho_L} - \ln\left(\frac{\rho_+}{\rho_L}\right) = \mathcal{E} - \frac{2\mathcal{B}}{\rho_L R^2} + 1.\qquad(4.3)$$

This introduces three extra physical parameters: the binding energy per receptor-ligand bond ($\mathcal{E}$), the membrane bending modulus ($\mathcal{B}$), and the target particle radius (R). Of course, since the cup keeps growing, the point where this boundary condition is imposed continually changes.

Solving this model gives the behaviour of $a(t)$ and $\rho(r, t)$ as functions of time (Fig. 4.2b, c). This shows that, at least in a model where receptor motion is dominated by diffusion, the cup size grows with the square-root of time ($a(t) \propto \sqrt{t}$). Such behaviour is often found in diffusion processes and is characteristic of Stefan-like problems in general.

Phagocytosis Versus Other Types of Endocytosis

The model given above was originally designed to describe endocytosis in general, particularly processes like clathrin-mediated endocytosis and pinocytosis that deal with internalising relatively small particles and volumes. Phagocytosis, on the other hand, must deal with substantially larger particles, sometimes even larger than $20\,\mu m$ in diameter (Cannon and Swanson 1992).

Because of this, phagocytosis exhibits substantial differences compared to these other endocytic processes. These include more dramatic changes in membrane shape (with phagocytes often extending outward to wrap targets rather than the target sinking into the cell (Richards and Endres 2014)), a more important role for the cytoskeleton (Axline and Reaven 1974; Zigmond and Hirsch 1972) and more complex receptor-signalling cascades (Swanson 2008; Underhill and Ozinsky 2002). However, perhaps surprisingly, the above model can relatively easily be adapted or extended to apply also to phagocytosis. This is because, despite the differences, the fundamental behaviour of receptors (such as the type of motion they undergo and when they reach the cup edge) applies equally well to phagocytosis as to other types of endocytosis.

One of the first attempts to model phagocytosis in this way was by van Zon, Tzircotis, Caron and Howard, who were interested in understanding why engulfment sometimes succeeds and sometimes fails (van Zon et al. 2009). Crucially, along with some other less dramatic changes, this work included the spatial distribution of actin, with bound receptors recruiting actin to the site of engulfment. Actin within the cup region then provides an outward force that helps wrap membrane around the target. This force, along with a membrane curvature force (that acts to flatten curved membranes), determines the shape of the phagocytic cup.

This model is naturally able to explain the bimodal distribution of cup sizes, with one peak around half engulfment and one around full engulfment. This is the case even when only one shape and size of target particle is considered. The authors argue this is due to a "mechanical bottleneck" with the cell needing to generate the greatest force at around half engulfment when the cup is widest. This means that any cell that is able to overcome the "bottleneck" at half engulfment is then likely to proceed to full engulfment. Further, this model then predicts that reducing the level of actin does not always lead to stalled phagocytosis (van Zon et al. 2009).

Another possible extension of the basic endocytosis model involves considering more carefully how receptors move. Although diffusion (perhaps in the form of confined diffusion or subdiffusion) is undoubtedly involved, there may also be more active receptor motion. For example, this would be the case if receptors were transported along some element of the cytoskeleton. If receptors were moved (along the membrane) directly towards the phagocytic cup then this would result in very different motion to diffusion. In particular, it would result in a type of "drift" behaviour, with Eq. (4.1) replaced by

$$\frac{\partial \rho}{\partial t} = \frac{v}{r} \frac{\partial}{\partial r} (r\rho), \qquad (4.4)$$

where v is the drift velocity (Richards and Endres 2014). This would fundamentally change the rate of cup growth, which would now be linear in time. Of course, it is unlikely that the motion of real receptors is either pure-diffusion or pure-drift. Rather, receptors probably move under a combination of diffusion and drift, so that the growth behaviour of the cup lies somewhere between linear and square-root in time.

The consequences of simple signalling by bound receptors can also be included in this model. The is based on the idea that, once receptors start to bind ligands on the target, they initiate a signalling cascade that ultimately changes the way that the remaining unbound receptors move towards the cup edge (Flannagan et al. 2012; García-García and Rosales 2005; Greenberg 2001; Kwiatkowska and Sobota 1999). Although there are likely to be many different proteins in this signalling cascade, it is sufficient to consider just one effective signalling molecule that acts as a proxy for the entire cascade. This makes it unimportant (and fairly meaningless) to ask for the identity of this signalling molecule. In the simplest models, the signalling molecule is activated by bound receptors, which then increases the drift velocity of the unbound receptors. By having an initial low level of receptor drift, this means that active receptor drift is effectively turned on once a sufficient number of ligands have been bound.

The model with receptor diffusion, receptor drift and a proxy signalling molecule (see Fig. 4.3a) makes several interesting predictions about the nature of phagocytosis (Richards and Endres 2014). First, consider how the total engulfment time (the time between $a = 0$

and $a = \pi R$) depends on the target radius R (Champion et al. 2008; Koval et al. 1998; Pratten and Lloyd 1986; Tabata and Ikada 1988). There are two competing effects: membrane curvature and target surface area. Smaller particles are more tightly curved meaning that the wrapped membrane must also be more highly curved (leading to slower engulfment), whereas larger particles have greater surface area and so need more bound membrane to be fully wrapped (again leading to slower engulfment). This means there is some intermediate particle radius corresponding to the shortest total engulfment time (Fig. 4.3b) (Gao et al. 2005; Richards and Endres 2014). For sufficiently small particles (with correspondingly sufficiently high curvature), the model predicts that phagocytosis will never finish, stalling when the target is only partially engulfed. Of course, for very small particles, types of endocytosis other than phagocytosis will become relevant.

As well as the target radius, the model can also make predictions about how engulfment outcome and rate depend on the other model parameters. These include properties of the target (such as the ligand density ρ_L), properties of the membrane (such as the membrane bending modulus $\mathcal{B}$) and properties of the receptors (such as the diffusion constant D, the initial receptor density ρ_0 and the binding energy per receptor-ligand bond $\mathcal{E}$). For example, the model predicts that the total

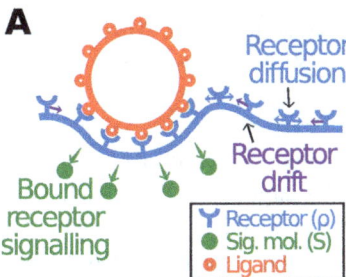
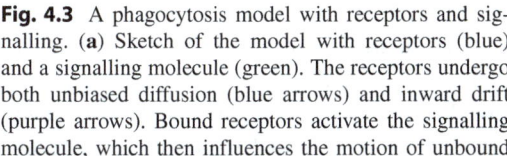
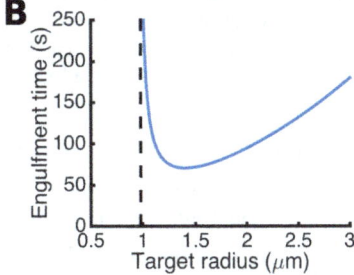

Fig. 4.3 A phagocytosis model with receptors and signalling. (**a**) Sketch of the model with receptors (blue) and a signalling molecule (green). The receptors undergo both unbiased diffusion (blue arrows) and inward drift (purple arrows). Bound receptors activate the signalling molecule, which then influences the motion of unbound

receptors (by, for example, increasing the drift velocity). (**b**) The predicted total engulfment time as a function of the target particle radius shows a minimum at an intermediate radius. Due to the high curvature, sufficiently small particles never reach complete engulfment. (Image taken with permission from Richards and Endres 2017)

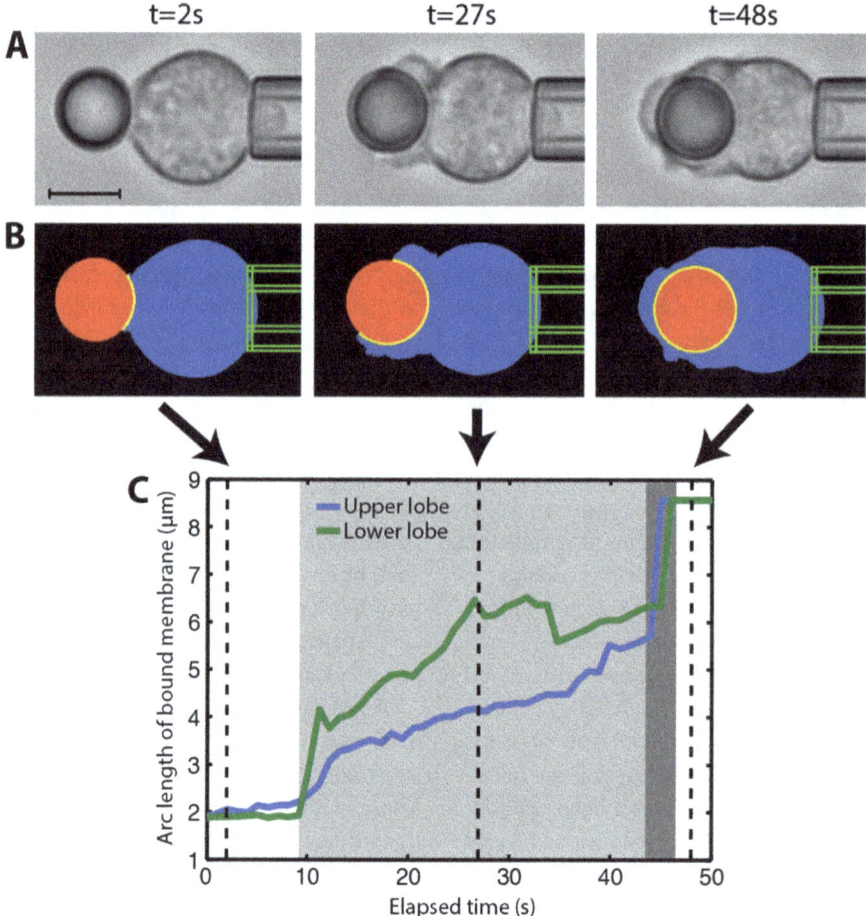

Fig. 4.4 Phagocytosis of a polystyrene bead. (**a**) Three snapshots from a time-lapse movie of a neutrophil (held by a micropipette) engulfing a 4.6 μm IgG-coated polystyrene bead. The first image, at t = 2s, shows the bead adhered to the cell but before engulfment has begun. The second image is taken around the point of half engulfment. The final image shows the bead completely inside the cell, after the engulfment stage of phagocytosis has finished. Scale bar: 5 μm. Data from Herant et al. (2006). (**b**) The same frames as in (**a**) with the cell (blue), bead (red), pipette (green) and phagocytic cup (yellow) identified. (**c**) Cup growth against time showing four distinct steps: initial adherence ($t < 10$ s), slow engulfment (light grey; 10 s $< t < 44$ s), fast engulfment (dark grey; 44 s $< t < 46$ s) and complete engulfment ($t > 46$ s). Although the two lobes are part of the same phagocytic cup (and therefore linked), engulfment proceeds (at least to some extent) independently for each. (Image used with permission from Richards and Endres 2014)

engulfment time as a function of the target ligand density ρ_L (Pacheco et al. 2013; Zhang et al. 2010) shows similar behaviour to that for the target radius, with an optimal density leading to the quickest engulfment (Richards and Endres 2014).

One of the main features of this model is that it describes the rate of cup growth. This can be directly compared to experimental data from dual-micropipette experiments, where micropipettes are used to control accurately both a phagocyte and a target (see Fig. 4.4). In fact, fitting to this type of data is how some of the model parameters were determined (Richards and Endres 2014). Importantly, the two distinct stages of engulfment (an initial slow rate of cup growth followed by a substantially quicker second stage) can be captured in the full diffusion-drift-and-signalling model. The sharp switch between the two stages corresponds, in the model, to the number of bound receptors reaching the threshold for activation of receptor drift. It is tempting to speculate that this second, quicker, more active stage may be an evolutionary add-on to ensure that phagocytosis completes in reasonable time.

The Effect of Target Shape

Up until recently, most studies of phagocytosis have only considered spherical target particles. This applies to both experimental and theoretical work. This has several advantages including easier particle fabrication, quicker image analysis and simpler modelling. However, targets in the real world are only rarely spherical. Instead they take a variety of shapes including capped cylinders, spheroids, tubes, filamentous networks and even hourglasses. Further, some organisms can deliberately change their shape to avoid (or perhaps sometimes to even encourage) phagocytosis. Although in some cases assuming a spherical target will not hamper progress, many of the most pressing questions in the subject require a proper treatment of the full non-spherical target shape. Such questions include the ultimate fate of phagocytosis (i.e. whether a particle can be fully ingested), the rate of engulfment, the optimal shape for microparticle drug carriers, and how the immune system can be supplemented to aid with the uptake of particularly difficult pathogens.

Known Experimental Results

The role of target shape in phagocytosis has been studied for over 50 years (Lengerová et al. 1957). Perhaps intuitively, during engulfment of a given particle, continued progress at any moment depends on the part of the target that is just about to be engulfed. For example, it has been argued that the outcome of phagocytosis is controlled not by the overall target shape, but by the local curvature at the edge of the cup (Champion and Mitragotri 2006). In particular, whether engulfment can even begin is determined by the target curvature at the point where the target first makes contact with the cell.

Perhaps the best studied non-spherical shapes are spheroids, which can be thought of as spheres that are squashed or stretched in a single direction. The squashed case resembles the shape of a Smartie and is called an oblate spheroid, whereas the stretched case looks like a rugby football and is termed a prolate spheroid. More precisely, a spheroid is an ellipsoid of revolution, i.e. an ellipsoid with two axes the same length. Sharma et al. investigated how uptake differs between spheres and spheroids, finding that oblate spheroids are typically easier to engulf than spheres, and that spheres are easier to engulf than prolate spheroids (Sharma et al. 2010). However, this result was strongest for smaller ($\leq 1\,\mu$m) particles; larger $3.6\,\mu$m particles showed much smaller differences. Similarly, Paul et al. determined that a prolate spheroid with surface area $\approx 75\,\mu$m^2 (and eccentricity over 0.95) took more than five times longer to engulf than a sphere of surface area $\approx 110\,\mu$m^2 (Paul et al. 2013).

Other target shapes, such as needles and rods, have also been examined, although most work has focused on nanometre-sized particles, which are less relevant to phagocytosis than micron-sized particles. It is not uncommon for phagocytes to encounter these kinds of shape either in living organisms (such as filamentous bacteria) or in inorganic structures (such as asbestos fibres). Although these particles, characterised by high aspect ratios, can sometimes be considered as extreme types of spheroids, it is important to appreciate that they can also show substantial differences. For example, Lu et al. compared spheres, rods and needles and found that spheres were the easiest to phagocytose and needles the hardest (Lu et al. 2010). Work by Champion et al. showed that long, thin, flexible particles (a good model for filamentous *E. coli*) are hardly ever engulfed (Champion and Mitragotri 2009). It is likely that some bacteria such as *E. coli* assume such elongated shapes precisely to inhibit removal by phagocytes (Möller et al. 2012). In a similar vein, shorter micelles have been shown to have substantially higher uptake rates than equivalent longer shapes (Geng et al. 2007). The phagocytosis of extended, thin shapes is made even more fascinating by suggestions of a distinct form of uptake for some bacteria (such as *Borrelia burgdorferi* and *Legionella pneumophila*), so-called coiling phagocytosis, which involves pseudopods progressively coiling around the bacteria (Horwitz 1984; Rittig et al. 1992).

Once non-spherical targets are examined, not just shape but also orientation must be considered: the same shape presented to the cell in different orientations may lead to very different outcomes. It seems plausible that bacteria use this to deliberately align themselves in order to hinder phagocytosis. This orientation dependence is naturally explained by the idea that it is the local curvature at the point of first contact that governs the ultimate fate of engulfment (Champion and Mitragotri 2006). Most work in this area has found that particles are easier to phagocytose when the more highly-curved regions are engulfed first. For example, both prolate spheroids and rods enter cells much more efficiently when presented to the cell pointed-tip first (Champion and Mitragotri 2006, 2009). Interestingly, it seems that needle-like particles with very large aspect ratios cannot be phagocytosed in any orientation (Lu et al. 2010).

Modelling Shape Dependence

One of the advantages of the basic receptor model described above is that it can easily be adapted to deal with non-spherical targets. There are two main differences that must be considered. First, the curvature around the target (and so the curvature of the wrapped membrane) is no longer constant. This alters the equation for ρ_+ (Eq. (4.3)), where the radius R must be replaced by the local radius of curvature at the edge of the cup. Since, for a general target particle, the curvature at the cup edge changes during engulfment, the ρ_+ boundary condition also changes and becomes a function of the cup size a.

The second change involves the model dimension. This is a conceptually small difference, but leads to models that are substantially more difficult to solve (both analytically and numerically). For a spherical target (at least when stochastic effects are ignored), the angular coordinate θ

plays no role. Whether this is still the case for other targets depends on both the target shape and the exact orientation that the target meets the cell. If, when looking directly down on the membrane from above, the target particle has circular symmetry then the θ-coordinate can still be ignored and the one-dimensional model remains sufficient. Such shapes include prolate spheroids and capped cylinders that are standing directly on their pointed ends. However, for general target shapes and general orientations, there is no such circular symmetry. It is then necessary to consider the full two-dimensional model described by both the r and θ coordinates.

First, consider shapes and orientations that can be studied with the one-dimensional model. Most previous modelling work of non-spherical particles has focused on the endocytosis of relatively small particles that are not relevant to phagocytosis. In such cases, engulfment is essentially passive membrane wrapping and so can be modelled simply by minimising the membrane energy (Bahrami et al. 2014, 2016). However, recent work has started to address the shape dependence of larger target particles finding that, even in the case where receptors only diffuse, the phagocytic cup no longer grows with the square-root of time (cf. Fig. 4.2b). Further, since the rate of engulfment is influenced by the local curvature at the cup edge, higher curved regions typically take longer to engulf than flatter regions, with regions of sufficiently high curvature unable to ever be engulfed and so giving rise to stalled or aborted phagocytosis (Richards and Endres 2014).

Because of the link between local curvature and engulfment rate, prolate spheroids (when standing upright) tend to demonstrate slow-fast-slow engulfment, with an initial slow phase of wrapping around the bottom followed by quicker engulfment near the mid-point and then slower engulfment again when the final, top region is reached. Oblate spheroids (when lying flat) display the opposite behaviour, with a fast-slow-fast form of engulfment (Richards and Endres 2016). The distribution of high curvature (which occurs at the tips of prolate spheroids and the middle of oblate spheroids) likely also explains why phagocytic stalling (when it occurs) takes place near full engulfment for prolate spheroids and around half engulfment for oblate spheroids (van Zon et al. 2009).

Shapes other than spheroids can also be studied using the one-dimensional model. These include capped cylinders and hourglasses (but, again, only when standing upright). Capped cylinders, also called spherocylinders, are formed of cylinders (of radius R and length L) capped at each end by hemispheres (of radius R; see cross section in Fig. 4.5a). They are typically a much better model for bacillus-shaped bacteria and archaea (such as *B. subtilis* and *E. coli*) than spheres. Modelling predicts they exhibit a slow-fast-stall-slow form of engulfment (Fig. 4.5b) (Richards and Endres 2016). As with spheroids, this is again caused by the distribution of

curvatures, with the fast stage corresponding to the lower-curvature cylindrical part and the slower stages to the higher-curved caps. The stalled region occurs at the top of the cylinder region as the cell starts to wrap the upper cap; there is a time delay whilst the receptor density at the cup edge increases enough to continue engulfment. These models also predict that all capped cylinders take longer to engulf than the equivalent sphere of the same surface area (Fig. 4.5c).

Hourglass-shaped particles may at first seem exotic, but are relevant to, for example, budding yeast in the process of budding (Clarke et al. 2010; Dieckmann et al. 2010). There is a degree of freedom in exactly how such a shape is specified (Fig. 4.5d), but they are often associated with

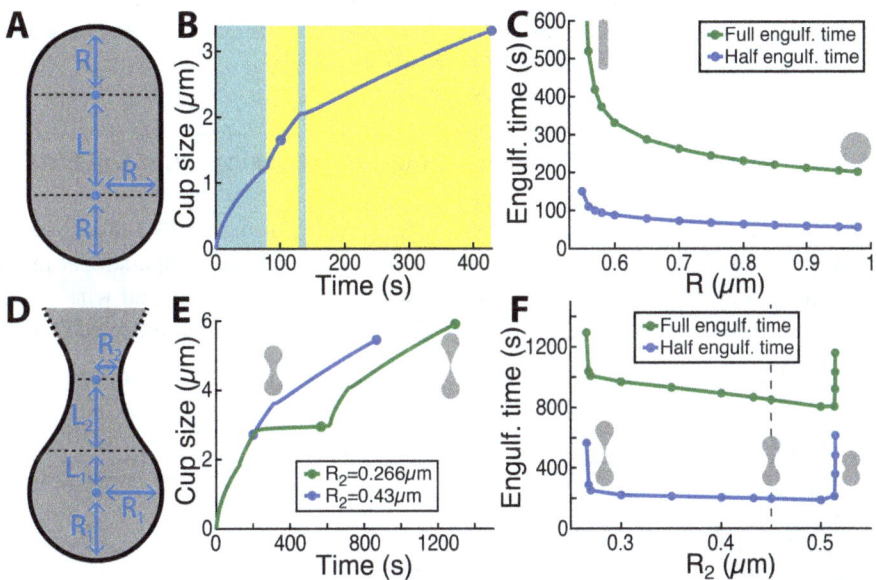

Fig. 4.5 Phagocytosis of capped cylinders (**a–c**) and hourglasses (**d–f**). (**a**) An idealised capped cylinder consists of a cylindrical body (of radius R and length L) with each end capped by a hemisphere (of radius R). (**b**) The predicted cup growth for a capped cylinder with $R = 0.8\,\mu m$ and $L = 0.8\,\mu m$ shows four distinct stages characterised in turn by slow, fast, stalled and slow engulfment. The solid circles show the points of half and full engulfment. (**c**) Both the half- and full-engulfment times decrease with increasing radius R, with the quickest engulfment corresponding to a sphere (when $L = 0$). Here, for a given value of the radius R, the cylinder length L is always chosen so that the total surface area is fixed. (**d**) An hourglass can be defined by two radii (R_1 and

R_2) and two lengths (L_1 and L_2). There are a variety of functional forms that can be taken for the neck region. (**e**) Two examples of cup growth for a narrow-neck hourglass ($R_1 = 0.8\,\mu m$, $L_1 = 0.53\,\mu m$, $R_2 = 0.266\,\mu m$, $L_2 = 0.754\,\mu m$) and a wide-neck hourglass ($R_1 = 0.8\,\mu m$, $L_1 = 0.53\,\mu m$, $R_2 = 0.43\,\mu m$, $L_2 = 0.383\,\mu m$). The solid circles show the points of half and full engulfment. (**f**) The half- and full-engulfment times as a function of the neck radius show the fastest engulfment at an intermediate R_2. The dashed line denotes the shape with the minimum neck curvature. In (**c**), (**e**) and (**f**), grey shapes show sketches of the particle cross section. (Image taken with permission from Richards and Endres 2016) stalled phagocytosis around the middle of the neck region (Richards and Endres 2016). Further, modelling suggests there are a wide range of engulfment behaviours depending on the precise shape (Fig. 4.5e). By varying the neck radius, there is seen to be an optimal intermediate shape that is engulfed quicker than all other shapes; this shape does not correspond to the hourglass with the smallest neck curvature (Fig. 4.5f) (Richards and Endres 2016).

Now, consider situations that require the full two-dimensional model. These include all but the most symmetric shapes presented to the cell in the most symmetric orientations. For example, a general ellipsoid (with three different axis lengths; see Fig. 4.6a) or a spheroid in a general orientation fall into this category. In addition to time, the cup size $a(\theta, t)$ now also depends on the angle coordinate θ. This also applies to the receptor density $\rho(r, \theta, t)$. The effect of this is that engulfment can now proceed at different rates in different directions, i.e. at different θ (Fig. 4.6b). The two-dimensional model predicts profound differences between phagocytosis of oblate and prolate spheroids: prolate spheroids engulf quickest when the highly-curved tip is presented to the cell first, whereas oblate spheroids are quicker when lying down (Fig. 4.6c). As the spheroid eccentricity is increased, complete engulfment only becomes possible in some orientations. For sufficiently eccentric cases, complete engulfment is not possible in any orientation (Richards and Endres 2016). Agreeing with the results of Sharma et al., the shape that is engulfed quickest (for a fixed surface area) is not a sphere, but a slightly oblate spheroid (Sharma et al. 2010).

Finally, models of shape dependence are helping shed light on the differences between phagocytosis and other forms of endocytosis. The engulfment of prolate spheroids is similar in both cases, but oblate spheroids are easier to engulf in phagocytosis. This is related to the extra push that phagocytosis seems to be able to provide at around half engulfment, when the more active, inward drift of receptors is triggered. Although there are always shapes that cannot be engulfed by any form of endocytosis, modelling predicts that phagocytosis is able to engulf the greatest range of target shapes (Richards and Endres 2016).

Model Extensions and the Future

As explained at the start of this chapter, the art of mathematical and computational modelling is knowing what to include and what to leave out. Simple models are often the most predictive and

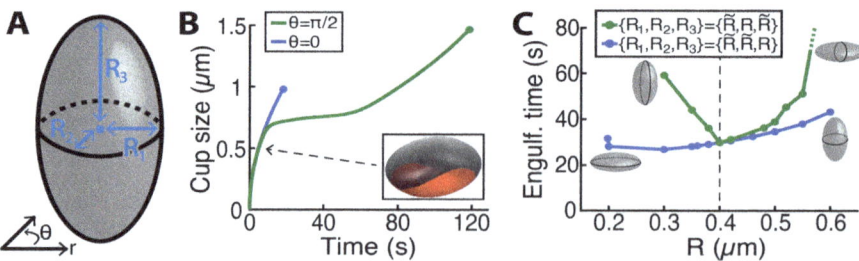

Fig. 4.6 The phagocytosis of ellipsoids. (**a**) A general ellipsoid is characterised by its three semi-principal axis lengths R_1, R_2 and R_3. Due to the reduced symmetry, understanding such a shape requires a two-dimensional model. (**b**) Cup growth at two perpendicular angles ($\theta = 0$ and $\theta = \pi/2$) for a lying-down prolate spheroid with $R_1 = R_3 = 0.3\,\mu$m and $R_2 = 0.6\,\mu$m. Importantly, different angles around the target now engulf at different rates. (Insert) The cup shape at around half engulfment, with the target in grey and the bound cell membrane in red. (**c**) The total engulfment time for a range of spheroids. In each case $\tilde{R}$ is chosen so that the total surface area is the same as that of a sphere of radius 0.4 μm. The green and blue lines show engulfment of the same shape but presented to the cell at perpendicular orientations. The dashed line indicates the spherical case when $R_1 = R_2 = R_3 = 0.4\,\mu$m; spheroids to the left/right of this line are oblate/prolate. Grey figures show sketches of the particle shape. (Image taken with permission from Richards and Endres 2016)

lead to the greatest insight, but they must constantly be checked to ensure they capture the true nature of the system in question. Existing models of phagocytosis focus on understanding the basic biophysical and biological processes, including membrane bending, receptor dynamics, the nature of signalling and the role of actin. However, there is plenty of scope for extending these models as understanding develops and novel experiments are performed.

First, consider the cell. Current models do not distinguish between different phagocytes despite their substantial differences. Similarly, there is little work trying to understand the role of different receptors. Models that address these issues, which could simply mean adjusting the current model parameters (such as the membrane bending modulus, the receptor-ligand binding energy, the receptor diffusion constant and the drift velocity), would shed light in this area. Similarly, the addition of multiple receptor types (Flannagan et al. 2012) and non-permanent receptor-ligand binding is likely to be important in future work. It is also typically assumed that there are enough receptors to make a density description ($\rho(r, \theta, t)$) appropriate. This may not always be the case.

Second, there are a number of possible extensions relating to the target particle. At present, almost all theoretical work focuses on rigid particles, which are far removed from the typical targets that phagocytes encounter. It is known that particle stiffness directly affects phagocytosis (Beningo and Wang 2002) and so it will be critical in the coming years to model the consequences of non-rigid targets. In a similar vein, it will be important to include the possibility that target surfaces may be modified (such as with various coatings, flagella or cilia) and to allow for scenarios where only a fraction of ligands need to be bound to receptors. Further, targets are usually taken as inert and non-rotating. Real targets are likely to rotate during engulfment both in a passive manner (due to the pulling forces generated by the cell) and an active manner (such as a bacterium swimming to try to avoid being phagocytosed).

Third, there are various possible ways that the current models themselves could be developed.

These include allowing the drift velocity to depend on r (the distance from the cup), considering the effect of non-zero spontaneous membrane curvature, modifying the form of the ρ_+ boundary condition (Eq. (4.3)), explicitly modelling actin as a branched network, and adding a role for stochasticity. In addition, signalling is at present included in a fairly ad hoc manner. The possibility of multiple, interacting signalling molecules could also be introduced (although only if this leads to a more useful model). Furthermore, signalling could have a variety of effects other than to increase the drift velocity, such as to alter the receptor diffusion constant (via post-translational modification), change the membrane bending modulus, produce new unbound receptors or increase the availability of spare membrane. These could all easily be investigated within existing models.

It is worth noting that we have focussed here on models that give primacy to the spatio-temporal dynamics of receptors. There are other options, such as models that are based on cellular forces (Heinrich 2015; Herant et al. 2005, 2006, 2011) and models that concentrate on minimisation of an energy functional (Deuling and Helfrich 1976; Helfrich 1973; Tollis et al. 2010). Perhaps the most successful future models will be hybrids that utilise a combination of these concepts.

We are still a substantial way from comprehensively understanding even the basics of phagocytosis. As such, there are numerous open questions, many of which modelling will be able to help address. These can be grouped into three categories: the cell itself, the phagocytic cup, and the target particle. For the cell itself, important questions include the role of filopodia in searching and grabbing potential targets (Kress et al. 2007; Vonna et al. 2007), how force is generated to wrap the cell membrane around the target, similarities and differences between receptor types (Allen and Aderem 1996; Kaplan 1977), and how phagocytosis proceeds even in the presence of stochasticity (Tollis et al. 2010).

In terms of the phagocytic cup, open issues include the cup thickness and the sharp angle at the cup base (Herant et al. 2006). It is also known that targets are initially pushed away from the cell before gradually being pulled inwards. How the push-out distance depends on properties of the cell and target is still not well understood and is an area that naturally lends itself to a modelling approach (Herant et al. 2011). The multi-stage nature of engulfment (Richards and Endres 2014) also prompts a host of interesting questions such as whether the switch between stages is always sharp, whether there are only ever two stages, whether the switch always occurs near half engulfment, and what triggers the switch.

Although work is now starting to address target size and shape dependence, there are still questions concerning particle composition, coating and stiffness. These lead into the possibility that the target shape and orientation may change during engulfment. Further, most previous work has assumed an even distribution of ligand around the target. This need not, of course, be the case and is another area where modelling should have direct relevance in the near future.

Finally, often the most fascinating discoveries in biology (if not all sciences) involve unsuspected connections between different areas. Phagocytosis is no exception. There are tantalising suggestions that phagocytosis may have deep (potentially evolutionary) links not just to other types of endocytosis, but also to areas as diverse as chemotaxis (Heinrich and Lee 2011), formation of *E. coli* pedestals (Goosney et al. 1999; Kaper et al. 2004), *B. subtilis* sporulation (Ojkic et al. 2014, 2016) and immunological synapse formation (Niedergang et al. 2016). Modelling is likely to be an excellent method of discovering these deep connections.

Summary

Although mathematical and computational models of phagocytosis are still in their infancy, they are already making striking predictions about the nature of phagocytic engulfment. The most useful modelling almost always concentrates on only a small number of key ingredients that are sufficient to capture the system behaviour. For phagocytosis, these key ingredients include the motion of receptors, the shape of the membrane, the actin cytoskeleton, and the size and shape of the target. In particular, models that focus on the dynamics of receptors, whilst also including a role for the cell membrane and the physical properties of the target, have made important advances in recent years.

It is likely that research in phagocytosis is on the verge of entering a period of substantial progress, with the potential for significant impact in numerous areas from autoimmune diseases to the design of micro/nanoparticle drug delivery systems. Continued success will require close interaction between more traditional experimental approaches and the type of focussed modelling described here.

References

Allen LA, Aderem A (1996) Molecular definition of distinct cytoskeletal structures involved in complement- and Fc receptor-mediated phagocytosis in macrophages. J Exp Med 184:627–637

An G, Fitzpatrick BG, Christley S, Federico P, Kanarek A, Neilan RM, Oremland M, Salinas R, Laubenbacher R, Lenhart S (2017) Optimization and control of agent-based models in biology: a perspective. Bull Math Biol 79(1):63–87

Axline SG, Reaven EP (1974) Inhibition of phagocytosis and plasma membrane mobility of the cultivated macrophage by cytochalasin B. Role of subplasmalemmal microfilaments. J Cell Biol 62:647–659

Bahrami AH, Raatz M, Agudo-Canalejo J, Michel R, Curtis EM, Hall CK, Gradzielski M, Lipowsky R, Weikl TR (2014) Wrapping of nanoparticles by membranes. Adv Colloid Interface Sci 208:214–224

Bahrami AH, Lipowsky R, Weikl TR (2016) The role of membrane curvature for the wrapping of nanoparticles. Soft Matter 12(2):581–7

Beningo KA, Wang Y-L (2002) Fc-receptor-mediated phagocytosis is regulated by mechanical properties of the target. J Cell Sci 115:849–856

Berg HC (1993) Random walks in biology. Princeton University Press, Princeton. ISBN: 978-0-69100-064-0

Cannon GJ, Swanson JA (1992) The macrophage capacity for phagocytosis. J Cell Sci 101(Pt 4):907–913

Carrera J, Covert MW (2015) Why build whole-cell models? Trends Cell Biol 25(12):719–722

Champion JA, Mitragotri S (2006) Role of target geometry in phagocytosis. Proc Natl Acad Sci USA 103(13):4930–4934

Champion JA, Mitragotri S (2009) Shape induced inhibition of phagocytosis of polymer particles. Pharm Res 26(1):244–249

Champion JA, Walker A, Mitragotri S (2008) Role of particle size in phagocytosis of polymeric microspheres. Pharm Res 25(8):1815–1821

Clarke M, Engel U, Giorgione J, Müller-Taubenberger A, Prassler J, Veltman D, Gerisch G (2010) Curvature recognition and force generation in phagocytosis. BMC Biol 8:154

Codling EA, Plank MJ, Benhamou S (2008) Random walk models in biology. J R Soc Interface 5(25):813–834

Deuling H, Helfrich W (1976) The curvature elasticity of fluid membranes: a catalogue of vesicle shapes. Journal de Physique 37(11):1335–1345

Dieckmann R, von Heyden Y, Kistler C, Gopaldass N, Hausherr S, Crawley SW, Schwarz EC, Diensthuber RP, Côté GP, Tsiavaliaris G, Soldati T (2010) A myosin IK-Abp1-PakB circuit acts as a switch to regulate phagocytosis efficiency. Mol Biol Cell 21:1505–18

Dorn M, E Silva MB, Buriol LS, Lamb LC (2014) Three-dimensional protein structure prediction: Methods and computational strategies. Comput Biol Chem 53:251–276

Feig M, Sugita Y (2019) Whole-cell models and simulations in molecular detail. Annu Rev Cell Dev Biol 35:191–211

Flannagan RS, Jaumouillé V, Grinstein S (2012) The cell biology of phagocytosis. Annu Rev Pathol Mech Dis 7:61–98

Freund LB, Lin Y (2004) The role of binder mobility in spontaneous adhesive contact and implications for cell adhesion. J Mech Phys Solids 52:2455–2472

Gao H, Shi W, Freund L (2005) Mechanics of receptor-mediated endocytosis. Proc Natl Acad Sci USA 102(27):9469–9474

García-García E, Rosales C (2005) Adding complexity to phagocytic signaling: phagocytosis-associated cell responses and phagocytic efficiency. In: Molecular mechanisms of phagocytosis. Springer, pp 58–71, ISBN 978-0-387-25419-7

Geng Y, Dalhaimer P, Cai S, Tsai R, Tewari M, Minko T, Discher DE (2007) Shape effects of filaments versus spherical particles in flow and drug delivery. Nat Nanotechnol 2(4):249–255

Goldberg AP, Szigeti B, Chew YH, Sekar JA, Roth YD, Karr JR (2018) Emerging whole-cell modeling principles and methods. Curr Opin Biotechnol 51:97–102

Goosney DL, de Grado M, Finlay BB (1999) Putting E. coli on a pedestal: a unique system to study signal transduction and the actin cytoskeleton. Trends Cell Biol 9(1):11–14

Greenberg S (2001) Diversity in phagocytic signalling. J Cell Sci 114:1039–1040

Gupta SC (2003) The classical Stefan problem: basic concepts, modelling and analysis. North-Holland series in Applied mathematics and mechanics (Book 45). JAI Press. ISBN: 978-0-44451-086-0

Heinrich V (2015) Controlled one-on-one encounters between immune cells and microbes reveal mechanisms of phagocytosis. Biophys J 109:469–476

Heinrich V, Lee C-Y (2011) Blurred line between chemotactic chase and phagocytic consumption: an immunophysical single-cell perspective. J Cell Sci 124:3041–3051

Helfrich W (1973) Elastic properties of lipid bilayers: theory and possible experiments. Z Naturforsch 28:693–703

Herant M, Heinrich V, Dembo M (2005) Mechanics of neutrophil phagocytosis: behavior of the cortical tension. J Cell Sci 118:1789–1797

Herant M, Heinrich V, Dembo M (2006) Mechanics of neutrophil phagocytosis: experiments and quantitative models. J Cell Sci 119:1903–1913

Herant M, Lee C-Y, Dembo M, Heinrich V (2011) Protrusive push versus enveloping embrace: computational model of phagocytosis predicts key regulatory role of cytoskeletal membrane anchors. PLoS Comput Biol 7:e1001068

Hodgkin AL, Huxley AF (1952) A quantitative description of membrane current and its applications to conduction and excitation in nerve. J Physiol (Lond) 117:500–544

Hollingsworth SA, Dror RO (2018) Molecular dynamics simulation for all. Neuron 99(6):1129–1143

Horwitz MA (1984) Phagocytosis of the Legionnaires' disease bacterium (Legionella pneumophila) occurs by a novel mechanism: engulfment within a pseudopod coil. Cell 36(1):27–33

Hospital A, Goñi JM, Orozco M Gelpí JL (2015) Molecular dynamics simulations: advances and applications. Adv Appl Bioinforma Chem 8:37–47

Kaper JB, Nataro JP, Mobley HL (2004) Pathogenic Escherichia coli. Nat Rev Microbiol 2(2):123–140

Kaplan G (1977) Differences in the mode of phagocytosis with Fc and C3 receptors in macrophages. Scand J Immunol 6:797–807

Koval M, Preiter K, Adles C, Stahl PD, Steinberg TH (1998) Size of IgG-opsonized particles determines macrophage response during internalization. Exp Cell Res 242(1):265–273

Kress H, Stelzer EHK, Holzer D, Buss F, Griffiths G, Rohrbach A (2007) Filopodia act as phagocytic tentacles and pull with discrete steps and a load-dependent velocity. Proc Natl Acad Sci USA 104:11633–11638

Kwiatkowska K, Sobota A (1999) Signaling pathways in phagocytosis. Bioessays 21(5):422–431

Lengerová A, Lenger VJ, Esslová M, Tuscany R, Volfová M (1957) The influence of the shape of dust particles on the rate of phagocytosis in vitro. Br J Ind Med 14(1):43–46

Lu Z, Qiao Y, Zheng XT, Chan-Park MB, Li CM (2010) Effect of particle shape on phagocytosis of CdTe quantum dot-cystine composites. MedChemComm 1:84–86

Möller J, Luehmann T, Hall H, Vogel V (2012) The race to the pole: how high-aspect ratio shape and heterogeneous environments limit phagocytosis of filamentous

Escherichia coli bacteria by macrophages. Nano Lett 12(6):2901–2905

Mackey MC, Maini PK (2015) What has mathematics done for biology? Bull Math Biol 77(5):735–738

Meirmanov AM (1992) The Stefan problem. De Gruyter expositions in mathematics. Walter de Gruyter, Berlin. ISBN: 3-11-011479-8

Niedergang F, Di Bartolo V, Alcover A (2016) Comparative anatomy of phagocytic and immunological synapses. Front Immunol 7:18

Nowak MA (2006) Evolutionary dynamics: exploring the equations of life. Harvard University Press, Cambridge. ISBN: 978-0-67402-338-3

Ojkic N, López-Garrido J, Pogliano K, Endres RG (2014) Bistable forespore engulfment in *Bacillus subtilis* by a zipper mechanism in absence of the cell wall. PLoS Comput Biol 10(10):e1003912

Ojkic N, López-Garrido J, Pogliano K, Endres RG (2016) Cell-wall remodeling drives engulfment during *Bacillus subtilis* sporulation. Elife 5:e18657

Pacheco P, White D, Sulchek T (2013) Effects of microparticle size and Fc density on macrophage phagocytosis. PLoS One 8:e60989

Paul D, Achouri S, Yoon Y-Z, Herre J, Bryant CE, Cicuta P (2013) Phagocytosis dynamics depends on target shape. Biophys J 105(5):1143–1150

Pratten MK, Lloyd JB (1986) Pinocytosis and phagocytosis: the effect of size of a particulate substrate on its mode of capture by rat peritoneal macrophages cultured in vitro. Biochim Biophys Acta 881(3):307–313

Reed MC (2004) Why is mathematical biology so hard? Notices Am Math Soc 51(3):338–342

Richards DM, Endres RG (2014) The mechanism of phagocytosis: two stages of engulfment. Biophys J 107:1542–1553

Richards DM, Endres RG (2016) Target shape dependence in a simple model of receptor-mediated endocytosis and phagocytosis. Proc Natl Acad Sci USA 113:6113–6118

Richards DM, Endres RG (2017) How cells engulf: a review of theoretical approaches to phagocytosis. Rep Prog Phys 80(12):126601

Rittig MG, Krause A, Häupl T, Schaible UE, Modolell M, Kramer MD, Lütjen-Drecoll E, Simon MM, Burmester GR (1992) Coiling phagocytosis is the preferential phagocytic mechanism for *Borrelia burgdorferi*. Infect Immun 60:4205–4212

Sharma G, Valenta DT, Altman Y, Harvey S, Xie H, Mitragotri S, Smith JW (2010) Polymer particle shape independently influences binding and internalization by macrophages. J Control Release 147(3):408–412

Stefan J (1891) On the theory of ice formation, particularly in the polar seas (Über die Theorie der Eisbildung, insbesondere über die Eisbildung im Polarmeere). Annalen der Physik und Chemie 42:269–286

Swanson JA (2008) Shaping cups into phagosomes and macropinosomes. Nat Rev Mol Cell Biol 9(8):639–649

Tabata Y, Ikada Y (1988) Effect of the size and surface charge of polymer microspheres on their phagocytosis by macrophage. Biomaterials 9(4):356–362

The Human Brain Project (2013) https://www.humanbrainproject.eu/en/

Tindall MJ, Porter SL, Maini PK, Gaglia G, Armitage JP (2008) Overview of mathematical approaches used to model bacterial chemotaxis I: the single cell. Bull Math Biol 70(6):1525–1569

Tindall MJ, Maini PK, Porter SL, Armitage JP (2008) Overview of mathematical approaches used to model bacterial chemotaxis II: bacterial populations. Bull Math Biol 70(6):1570–1607

Tollis S, Dart AE, Tzircotis G, Endres RG (2010) The zipper mechanism in phagocytosis: energetic requirements and variability in phagocytic cup shape. BMC Syst Biol 4:149–165

Underhill DM, Ozinsky A (2002) Phagocytosis of microbes: Complexity in action. Annu Rev Immunol 20:825–852

van Zon JS, Tzircotis G, Caron E, Howard M (2009) A mechanical bottleneck explains the variation in cup growth during FcγR phagocytosis. Mol Syst Biol 5:298–309

Voit EO, Martens HA, Omholt SW (2015) 150 years of the mass action law. PLoS Comput Biol 11(1):e1004012

Vonna L, Wiedemann A, Aepfelbacher M, Sackmann E (2007) Micromechanics of filopodia mediated capture of pathogens by macrophages. Eur Biophys J 36:145–151

Zhang Y, Hoppe AD, Swanson JA (2010) Coordination of Fc receptor signaling regulates cellular commitment to phagocytosis. Proc Natl Acad Sci USA 107(45):19332–19337

Zigmond SH, Hirsch JG (1972) Effects of cytochalasin B on polymorphonuclear leucocyte locomotion, phagocytosis and glycolysis. Exp Cell Res 73:383–393

Decision Making in Phagocytosis

5

Jana Prassler, Florian Simon, Mary Ecke, Stephan Gruber, and Günther Gerisch

Abstract

Dictyostelium cells are professional phagocytes that are capable of handling particles of variable shapes and sizes. Here we offer long bacteria that challenge the uptake mechanism to its limits and report on the responses of the phagocytes if they are unable to engulf the particle by closing the phagocytic cup. Reasons for failure may be a length of the particle much larger than the phagocyte's diameter, or competition with another phagocyte. A cell may simultaneously release a particle and engulf another one. The final phase of release can be fast, causing the phagosome membrane to turn inside-out and to form a bleb. Myosin-II may be involved in the release by generating tension at the plasma membrane, it does however not accumulate on the phagosome to act there directly in expelling the particle. Labeling with GFP-2FYVE indicates that processing of the phagosome with phosphatidylinositol 3-phosphate begins at the base of a long phagosome already before closure of the cup. The decision of releasing the particle can be made even at the stage of the processed phagosome.

Keywords

actin · *Dictyostelium* · membrane blebbing · myosin-II · phosphatidylinositol 3-phosphate · phosphatidylinositol (3,4,5)-trisphosphate · tubular phagocytic cup

Introduction

The amoeboid cells of *Dictyostelium discoideum* are professional phagocytes, of about the size of a neutrophil, which under exponential growth conditions take up 5 *E.coli* cells per minute and grow with a doubling time of 3 h by digesting about 1100 normal-sized bacteria for the generation of one *Dictyostelium* cell (Gerisch 1960). *Dictyostelium* has been used as a model to study phagocytosis mechanisms (Bozzaro 2013; Dunn et al. 2018), in particular host responses to *Legionella* (Bozzaro et al. 2013; Swart et al. 2018) and to mycobacterial infections (Cardenal-Muñoz et al. 2018).

Electronic supplementary material The online version of this chapter (https://doi.org/10.1007/978-3-030-40406-2_5) contains supplementary material, which is available to authorized users.

J. Prassler · F. Simon · M. Ecke · G. Gerisch (✉)
Max Planck Institute of Biochemistry, Martinsried, Germany
e-mail: gerisch@biochem.mpg.de

S. Gruber
Department of Fundamental Microbiology (DMF), Faculty of Biology and Medicine (FBM), University of Lausanne (UNIL), Lausanne, Switzerland

To challenge the uptake mechanism of *Dictyostelium* cells, we exposed these phagocytes to extremely long *E.coli* bacteria. Uptake of cylindrical particles with a high aspect ratio means that the phagocyte has to supply a large membrane area for a given particle volume to be incorporated. The length of the bacteria used often exceeded the diameter of the phagocytosing cell. In the following, we describe the engulfment of long bacteria and the reversal of this process if uptake cannot be completed.

There are two situations in which a phagocyte runs into problems that can only be solved by ejecting a particle: (1) when it does not find the end of the particle because the particle is too long or firmly attached to another structure, and (2) when two phagocytes are competing for the same particle. In these cases, a phagocyte has to decide whether or not to continue uptake. To discontinue and to get rid of the particle, *Dictyostelium* cells activate a release mechanism: membrane of the phagocytic cup is converted into plasma membrane by turning inside out, often forming a bleb after the particle is spilled out. This bleb indicates internal pressure due to membrane tension and de-stabilization of the cell cortex at the opening of the cup. The release of a particle as a result of membrane tension implies that the actin system responsible for uptake must act against that tension.

The methods used for the uptake and release of long bacteria by *D. discoideum* AX2 cells were essentially as described previously (Clarke et al. 2010a, b). The bacterial strain was an *E. coli* Re mutant, strain F 515 (Zähringer et al. 1985), transfected with the plasmid pWM1736 (Thanedar and Margolin 2004), which contains an arabinose inducible promoter. The addition of the monosaccharide arabinose induces the expression of SulA, which in turn prevents FtsZ polymerization and thereby inhibits cytokinesis. Bacteria from an overnight culture were incubated for 3–4 h with 0.2% arabinose at 37 °C, pelleted by centrifugation, resuspended in 1 ml 17 mM K/Na-phosphate buffer, pH 6.0 (PB), supplemented with 150 mM sorbitol to avoid osmotic stress, and were then slowly further diluted into PB.

Uptake, Processing, and Release of Long Particles

Uptake of a Long Particle

Figure 5.1 illustrates features characteristic of the uptake of long bacteria: the formation of tubular phagocytic cups, the sticking out of a membrane protrusion at the end of the bacterium opposite to the site of uptake, and the bending of the bacterium when its length exceeds the diameter of the phagocyte. Typically, the phagosomal membrane is not uniformly coated with an actin network during the uptake of a long particle, but the coat is locally strengthened by prominent actin rings. The movie sequence from which the images of Fig. 5.1 were taken, is shown in full as Movie 5.1.

Phagosome Processing

As an early endosomal marker, we visualized phosphatidylinositol 3-phosphate PI(3)P by labeling with GFP-2FYVE (Ellson et al. 2001). Previously we found during the uptake of yeast particles or normal-sized bacteria by *Dictyostelium* cells that PI(3)P appeared at phagosomal membranes only after the cup had closed (Clarke et al. 2010a, b). During the uptake of long bacteria, however, we observe 2FYVE decoration already before the cup is closed.

The uptake shown in Fig. 5.2 (and in full in Movie 5.2) illustrates details of the endosomal interplay with a long phagocytic cup from 2FYVE decoration to its depletion. Decoration begins, with the cup still open, at the basal region where the actin coat has disappeared (116 s frame) and, after closure of the cup, extends along the phagosome. The early-phase 2FYVE decoration is linked to intense interaction of highly mobile, heavily 2FYVE-decorated vesicles with the phagosome membrane. Subsequently, the phagosome becomes uniformly decorated with 2FYVE on its entire length (466 s frame).

Remarkable features, seen in the 417 s and 434 s frames of Fig. 5.2 to initiate the decline of 2FYVE decoration, are long flexible extensions that eventually vesiculate. Subsequently, dispatch

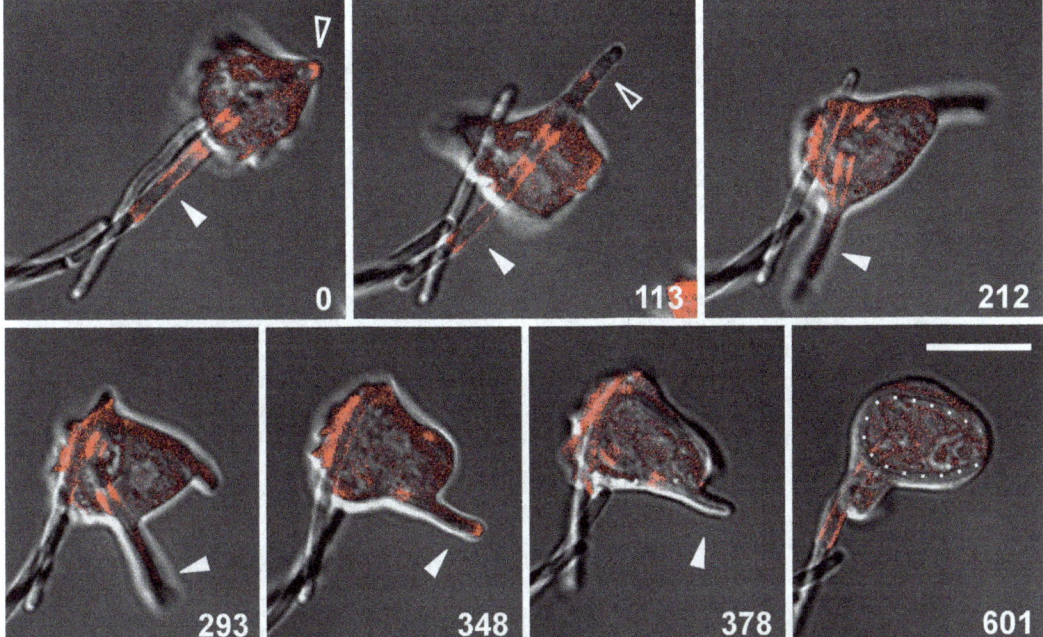

Fig. 5.1 Uptake of a long bacterium into a *Dictyostelium* cell. The cell expresses mRFP-LimEΔ as a label for filamentous actin (Fischer et al. 2004). The fluorescence images are superimposed in red on DIC bright-field images. Closed arrowheads indicate the bacterium taken up. In the final frame, the internalized bacterium has assumed a horseshoe-shape (dotted line) and the phagosome is no longer surrounded by actin. Before the bacterium coils within the cell, it pushes on the cell membrane opposite of the site of uptake, inducing an accumulation of actin and protrusion of the cell membrane (open arrows within the 0-s and the 113-s frames, respectively). Time is indicated in seconds. Bar, 10 μm. The imaging sequence from which these frames were taken is shown as Movie 5.1

of the label indicating decline of PI(3)P is associated with buckling of the phagosome membrane and its connection to a vesicular network (589 s and 599 s frames), resulting in highly 2FYVE-decorated vesicles distributed throughout the cell (last frame). To give a rough time scale, the 2FYVE decoration began about 4 min before cup closure, became more intense within the first minute after closure, decayed about 10 min after its beginning, and had nearly disappeared 5 min later.

Release of a Particle

If the *Dictyostelium* cell does not find the end of a particle to close the phagocytic cup there, it has the choice of rejection. This process is illustrated in Fig. 5.3. To specify the stage at which the particle is released, we used again cells expressing GFP-2FYVE to label PI(3)P. Release of the particle at the stage of 2FYVE decoration underlines that PI(3)P accumulation in the phagosome membrane does not require closure of the cup.

One possibility of turning uptake into release would be disassembly of the actin network surrounding the phagosomal membrane. As seen for uptake, this network is strengthened by actin rings. We explored whether these rings disintegrate prior to particle release and found that this is not always the case. The 100–112 s frames of Fig. 5.3 point to one detail of interest. During release, the phagosome membrane shortens, as indicated by the 2FYVE label. At the final stage of release, an actin ring is persisting. When the particle is released, this actin ring remains in place and the phagosomal membrane tube slips through it, indicating that the actin network is anchored within the cell but is disconnected from the phagosomal membrane at the stage of release.

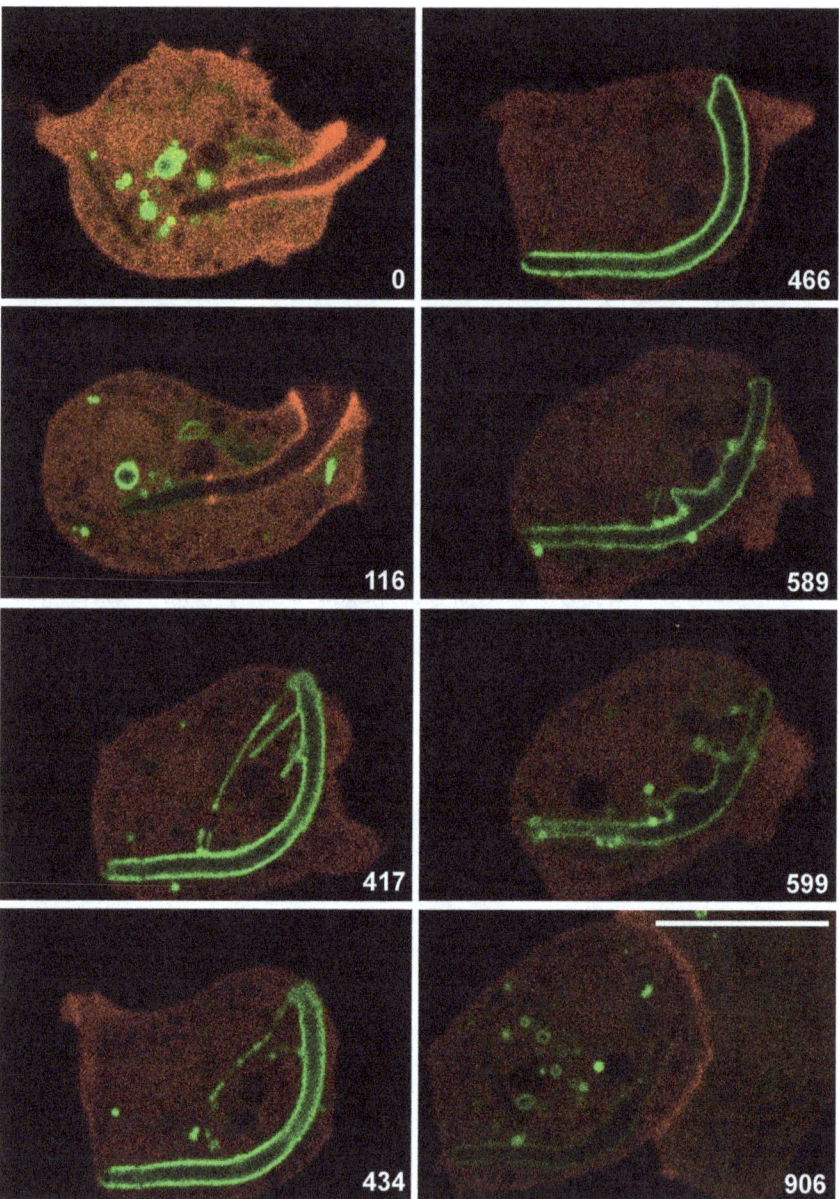

Fig. 5.2 Phagosome processing visualized in a cell expressing GFP-2FYVE as a label for the early endosomal marker PI(3)P (green) and mRFP-LimEΔ for actin (red). At the beginning of the time series, vesicles of the early endosomal system interact with the nascent phagosome, which at its base becomes PI(3)P-decorated before the phagocytic cup is closing (0-s to 116-s frames). Features not always observed are long extensions from the phagosome (417-s frame) that decay by dispersal. From the fully PI(3)P-decorated, closed phagosome (466-s frame) PI(3)P is retrieved under membrane buckling and vesiculation (589-s to 599-s frames). Time is indicated in seconds. Bar, 10 μm. The imaging sequence from which these frames were taken is shown as Movie 5.2

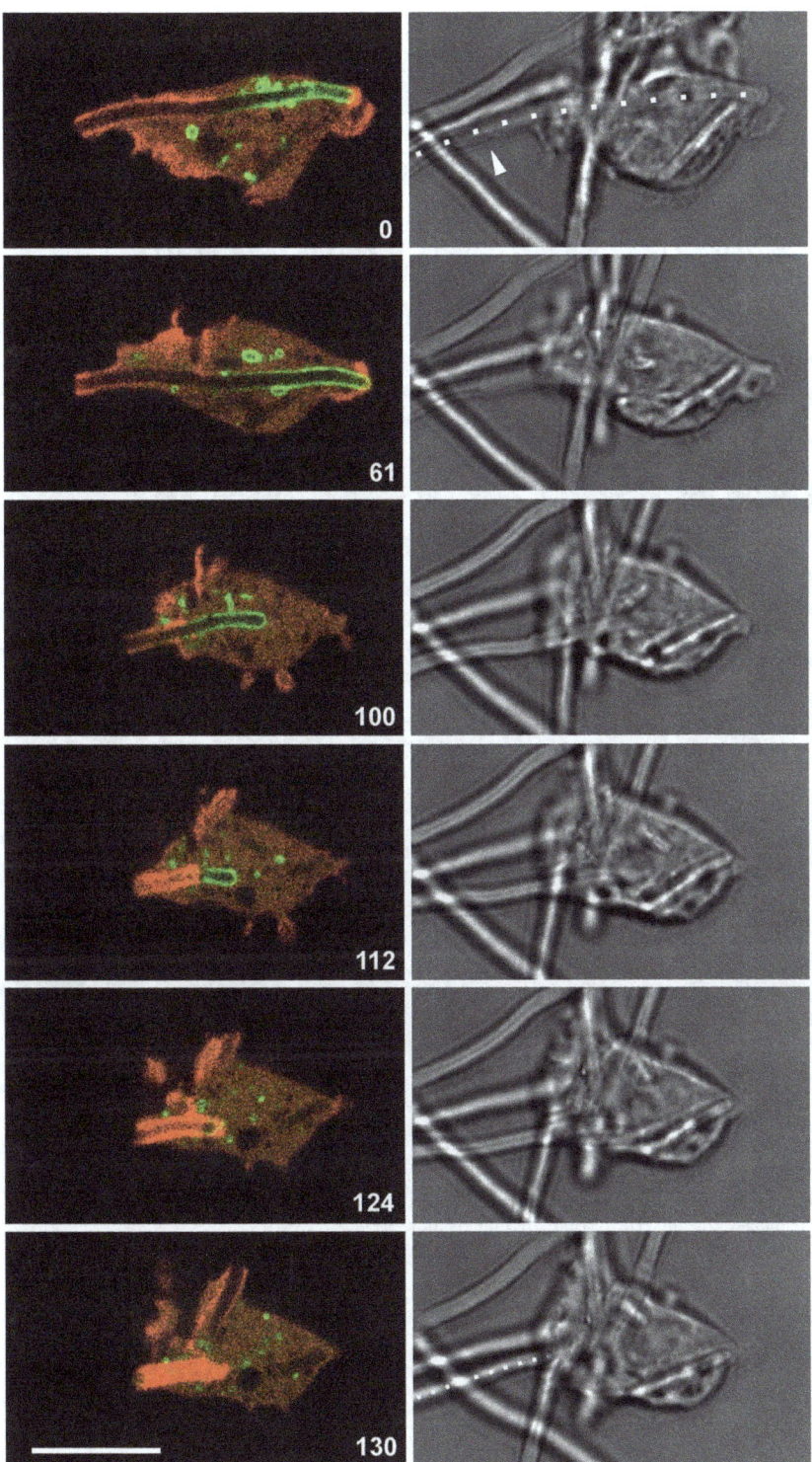

Fig. 5.3 Release of a long bacterium by a cell expressing GFP-2FYVE for PI(3)P (green) and mRFP-LimEΔ for actin (red). Left panels: Fluorescence images. The released bacterium is pushed with its surrounding PI(3)P-decorated bacterium is indicated by a dotted line. Time is indicated in seconds. Bar, 10 μm phagosomal membrane through the persisting actin tube. Right panels: Bright-field images. In the first and last frame the bacterium is indicated by a dotted line. Time is indicated in seconds. Bar, 10 μm

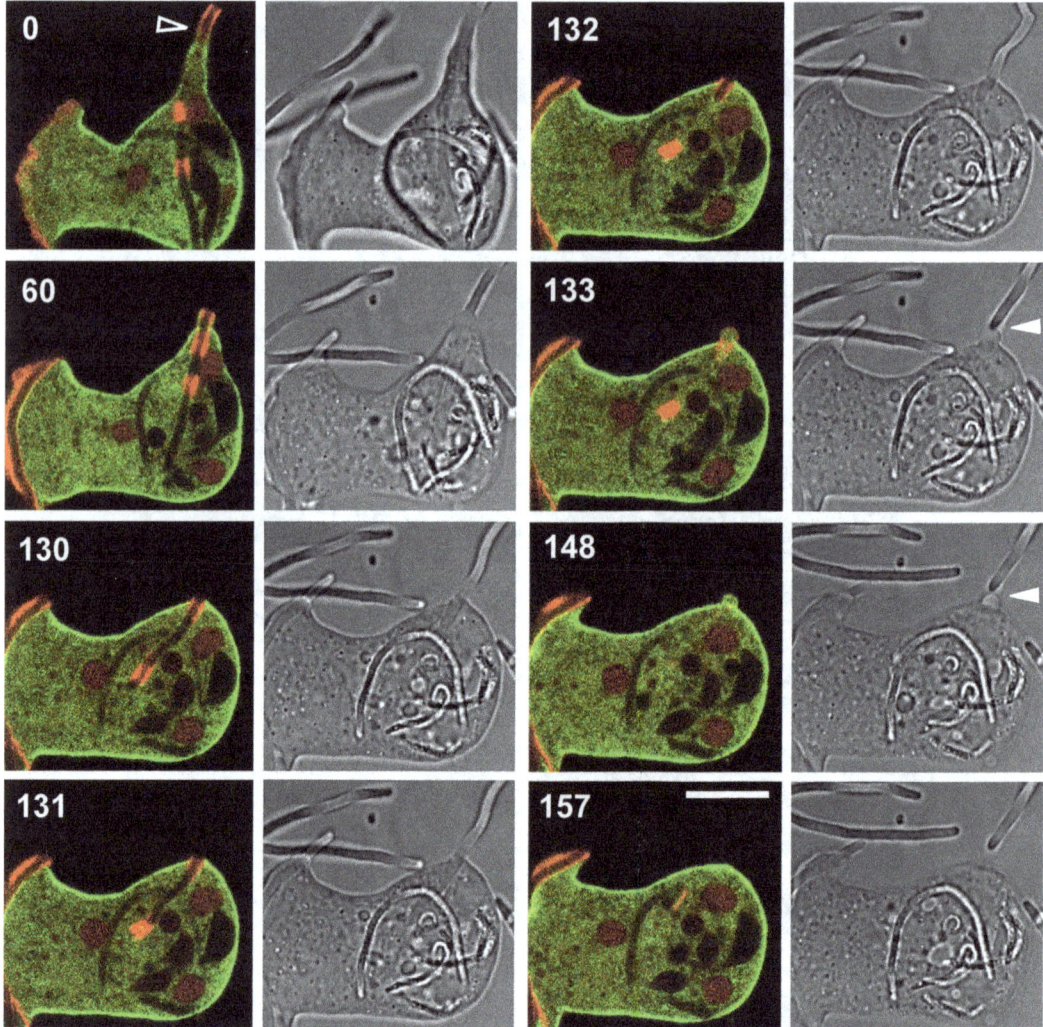

Fig. 5.4 Partial uptake and release of a long bacterium with no detectable myosin-II accumulation at the phagosome. Left panels: The cell expresses GFP-myosin-II heavy chain (green) and mRFP-LimEΔ to label actin (red). Right panels: Bright field images. The open arrowhead in the 0-s fluorescence image points to the tubular extension of the phagocytic cup. The closed arrowheads in the 133-s and 148-s bright-field images indicate the bleb formed after release of the bacterium. The distorted appearance of the bacterium in the 132-s frame is a scanning artifact due to rapid expulsion of the bacterium. Time is indicated in seconds. Bar, 10 μm. The imaging sequence from which these frames were taken is shown as Movie 5.3

Myosin-II Does Not Accumulate on the Incipient Phagosome

The question of how forces that result in uptake of a particle are replaced by forces that lead to its ejection, implies the role of the conventional two-headed myosin-II, a candidate motor protein that is known to be responsible for contraction of the cell cortex (Moores et al. 1996).

Accordingly, cells expressing GFP-myosin-II heavy chains showed a distinct labeling of the cell cortex. Does it associate with the phagocytic cup to shorten it? At the final stage of release, the filamentous bacterium is moved out with a velocity of up to 7 μm s^{-1} (Fig. 5.4). This high speed suggests that the ejection of a particle is not controlled by actin-dependent movement. In accord with this notion, we did not find myosin-

Fig. 5.5 Conversion of a tubular phagosome into a bleb during the release of a long bacterium. Left panels: The cell on the left boarder expresses superfolder (sf) GFP-PHcrac (Müller-Taubenberger and Ishikawa-Ankerhold 2013) to label phosphatidylinositol (3,4,5)-trisphosphate (PIP3) in the phagosomal membrane (green) and mRFP-LimEΔ for actin (red). Right panels: Bright-field images. In the 55-s frame, the border of the phagocytic cup is marked by a closed arrowhead. In the 59-s frame the end of the released bacterium is indicated by an open arrowhead. On the upper right a section of another cell is seen. Time is indicated in seconds. Bar, 10 μm

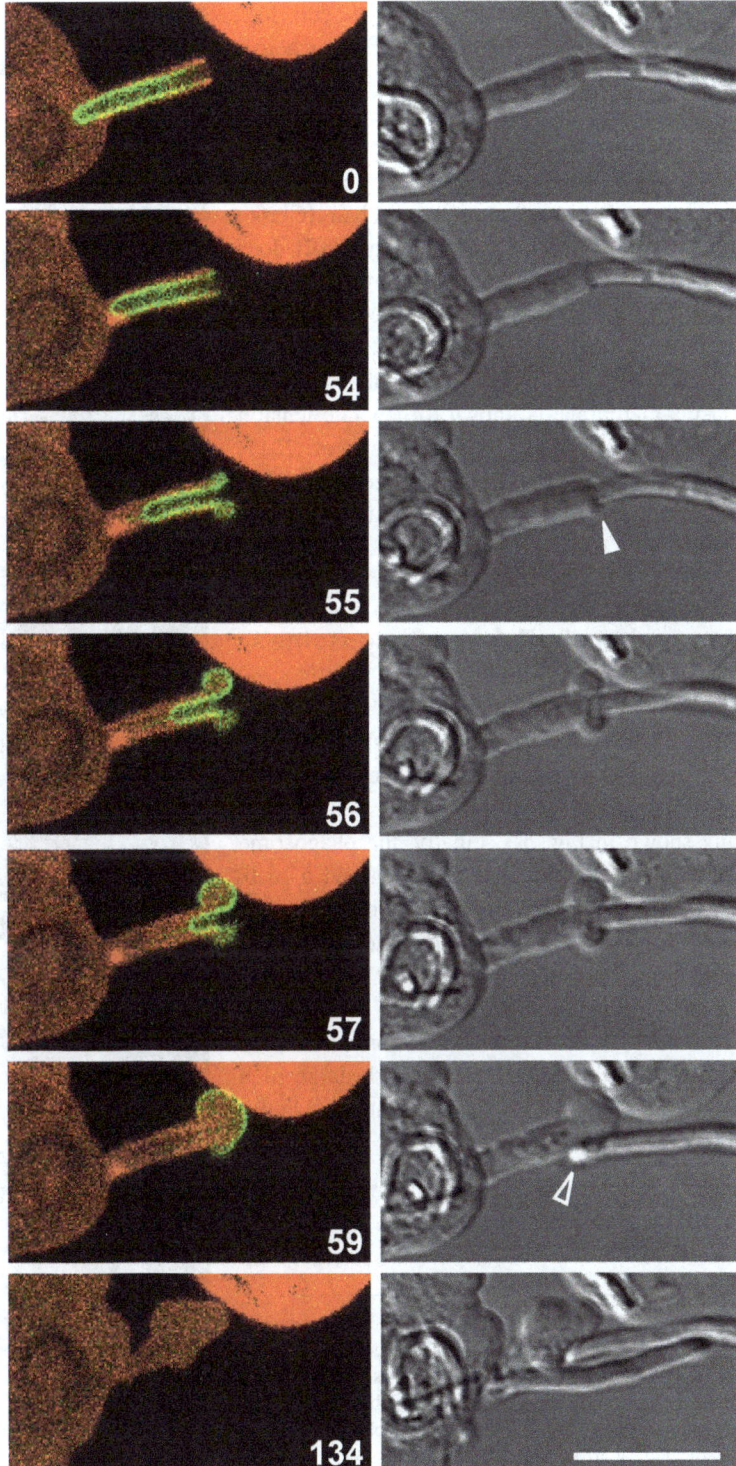

II to be associated with the phagocytic cup (Fig. 5.4 and on-line as Movie 5.3). Only when a bleb was formed after ejection of the particle, myosin-II accumulated at the cortex of the bleb when it was retracted (148–157 s frames). These results are in accord with previous ones obtained using bipartite yeast particles as bait (Clarke et al. 2010a). They suggest that myosin-II does not directly act on the phagosome in either stabilizing the shape of its extended membrane or in applying force for retraction to release the particle. However, myosin-II localized on the plasma membrane is known to generate tension (Charras and Paluch 2008; Collier et al. 2017; Pasternak et al. 1989), which may be responsible for the rapid release of the particle and the formation of a bleb.

Conversion of a Tubular Phagosome into a Bleb

The 130–133 s frames of Fig. 5.4 show an actin ring that stayed in place when the phagosome membrane was pushed through it, similar to the ring shown in Fig. 5.3. Finally, the phagosome membrane detached from actin formed a bleb. This is the common way of bleb formation upon release of a particle: the tubular extension of the phagosome first shortens, and after its retraction the bleb is formed at the border of the cell. Exceptionally, the entire tubular extension may be directly converted into a particularly large bleb by turning the phagosome membrane inside-out (Fig. 5.5).

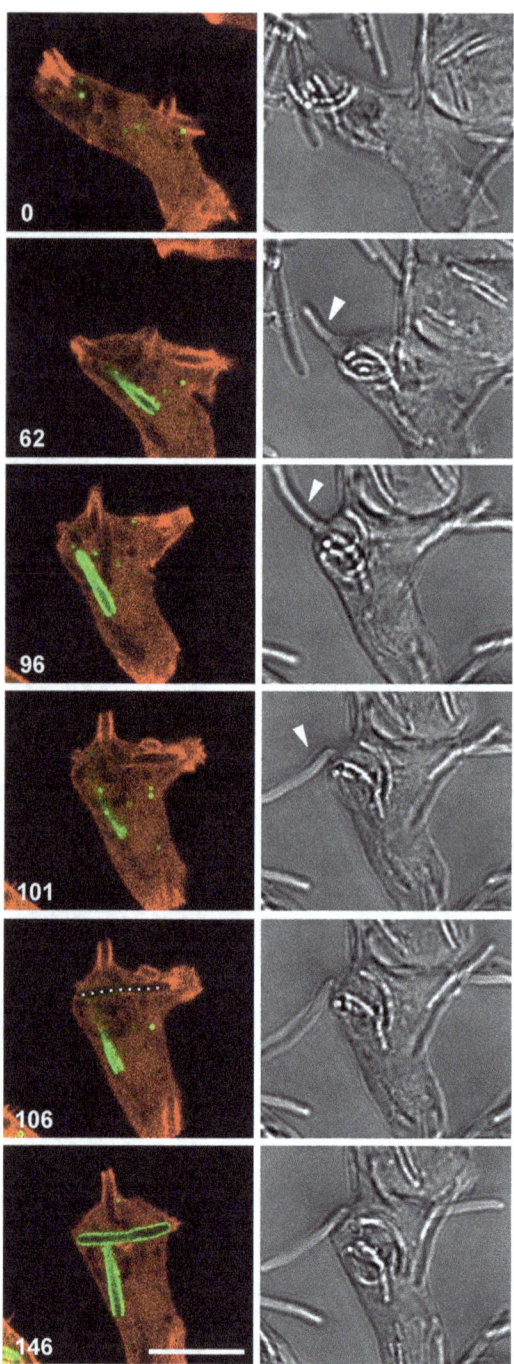

Fig. 5.6 (continued) also engaged with other bacteria, one of them internalized into a PI(3)P-decorated phagosome. Left panels: Fluorescence images. In the 106-s frame the bacterium taken up is indicated by a dotted line. Right panels: Bright-field images. The bacterium released is indicated by arrowheads. Time is indicated in seconds. The images are shifted upwards between the 62-s frame and the 96-s frame. Bar, 10 μm

Fig. 5.6 Simultaneous uptake and release. The cell expresses GFP-2FYVE as a label for PI(3)P (green) and mRFP-LimEΔ for actin (red). One bacterium is released on the left, while another one is engulfed on the right, the phagosome becoming decorated with PI(3)P. The cell is

Simultaneous Uptake and Release of Particles

As an evidence that the decision to release a particle is made on the basis of the individual phagosome rather than in response to a global signal, we observed cells that were engaged with multiple bacteria, as the one in Fig. 5.6. While releasing a bacterium the cell incorporates another one, which is normally processed as indicated by full 2FYVE decoration and subsequent depletion of the label. The coincidence of release and uptake suggests that the decision is made on the level of the individual phagosome.

Struggle for a Bacterium

If each of the two ends of a long bacterium is engaged by another cell, one cell has to give up. In the example shown in Fig. 5.7, the bacterium partially taken up by the cell on the lower right is attacked twice by the cell on the upper left. Both attempts turned out to be unsuccessful, the latter cell releasing the particle, which finally is completely taken up by its competitor.

Discussion

The uptake of bacteria that are longer than the diameter of the phagocyte requires an efficient uptake mechanism since the bacteria become bent within the phagosome or stick out membrane projections opposite to the site of uptake (Fig. 5.1). The data reported here on uptake and release of long cylindrical particles are comparable to previous results obtained with bipartite yeast particles, where the phagocyte has also the choice of engulfing or releasing the particle (Clarke et al. 2010a). However, in dealing with a bipartite particle, there is a third choice: to stop at the furrow region and accumulate actin there to bite the particle into two pieces (Fig. 5.8).

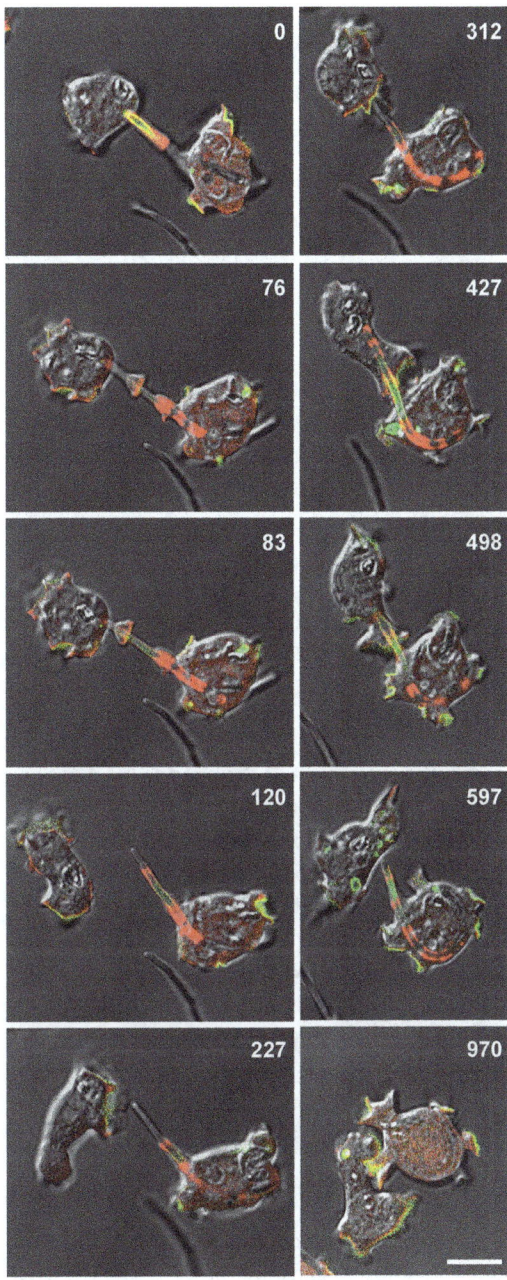

Fig. 5.7 Competition for a long bacterium. The cells express GFP-PHcrac as a label for PIP3 (green) and mRFP-LimEΔ for filamentous actin (red). The cell on the upper left tries twice to engulf a bacterium, which is finally incorporated by the cell on the lower right. Time is indicated in seconds. Bar, 10 μm

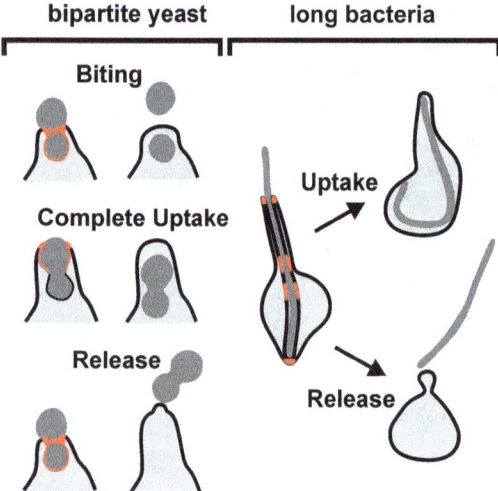

Fig. 5.8 Diagrammatic comparison of decision making during the uptake of bipartite yeast particles, where the phagocytes have three choices (Clarke et al. 2010a), and during the uptake of long bacteria as reported here, where the phagocytes have the two choices of engulfment or release

Characteristic for the uptake of long bacteria are tubular extensions of the phagocytic cup that, with an actin-rich border, propagate along the bacterium. In this respect, the response of *Dictyostelium* cells to filamentous bacteria resembles that of macrophages (Möller et al. 2012; Naufer et al. 2018). The velocity of 2–6 μm per minute for cup extension in *Dictyostelium* is somewhat lower than that in macrophages, which reach 8 μm per minute. One principal difference might be that *Dictyostelium* cells accumulate actin when they attach to a particle along its axis (on the left of Fig. 5.4), whereas macrophages have to attach to one end of the particle to do so (Möller et al. 2012).

The ingestion of cylindrical particles is well suited to study phagosome processing. We used the long phagosomes produced to visualize the delivery of PI(3)P to the incipient phagosome, as recognized by GFP-2FYVE. The labeling began at the base of the cup already before closure of its orifice (Fig. 5.2), and persisted for about 15 min.

In macrophages, PI(3)P appears to be supplied to the phagosome in two steps: first by delivery through pre-existing vesicles and second, after closure of the cup, by synthesis on the cup membrane (Ellson et al. 2001). PI(3)P is required in macrophages for endosome-phagosome fusion (Jeschke et al. 2015) and persists in the phagosomal membrane until acidification of the lumen, which coincides with dissociation of the PI3-kinase Vps 34 from the membrane (Naufer et al. 2018).

Simultaneous uptake and release of particles argues against a global signal or state of the cell that determines how a particle is handled. Our data rather suggest a balance between uptake and release on the level of the individual phagosome, in accord with a zipper mechanism of phagocytosis (Griffin et al. 1975, 1976). A bleb, often formed at the end of release, indicates that expulsion of the particle is driven by membrane tension that at the opening of the cup is not counteracted by resistance of an actin layer.

Acknowledgements We thank Martin Spitaler and his team at the Imaging Facility of the Max Planck Institute of Biochemistry for cooperation, Petra Fey and dictyBase for providing information, and Marina Freudenberg, MPI for Immunobiology and Epigenetics, Freiburg/Br., for the *E. coli* Re mutant, strain F 515.

The Max Planck Society funded this work.

References

Bozzaro S (2013) The model organism Dictyostelium discoideum. Methods Mol Biol 983:17–37. https://doi.org/10.1007/978-1-62703-302-2_2

Bozzaro S, Peracino B, Eichinger L (2013) Dictyostelium host response to Legionella infection: strategies and assays. Methods Mol Biol 954:417–438. https://doi.org/10.1007/978-1-62703-161-5_26

Cardenal-Muñoz E, Barisch C, Lefrançois LH, López-Jiménez AT, Soldati T (2018) When Dicty met Myco, a (not so) romantic story about one amoeba and its intracellular pathogen. Front Cell Infect Microbiol 7:529–529. https://doi.org/10.3389/fcimb.2017.00529

Charras G, Paluch E (2008) Blebs lead the way: how to migrate without lamellipodia. Nat Rev Mol Cell Biol 9(9):730–736. https://doi.org/10.1038/nrm2453

Clarke M, Engel U, Giorgione J, Müller-Taubenberger A, Prassler J, Veltman D, Gerisch G (2010a) Curvature recognition and force generation in phagocytosis. BMC Biol 8:154. https://doi.org/10.1186/1741-7007-8-154

Clarke M, Maddera L, Engel U, Gerisch G (2010b) Retrieval of the vacuolar H-ATPase from phagosomes revealed by live cell imaging. PLoS One 5(1):e8585–e8585. https://doi.org/10.1371/journal.pone.0008585

Collier S, Paschke P, Kay RR, Bretschneider T (2017) Image based modeling of bleb site selection. Sci Rep 7(1):6692–6692. https://doi.org/10.1038/s41598-017-06875-9

Dunn JD, Bosmani C, Barisch C, Raykov L, Lefrançois LH, Cardenal-Muñoz E, López-Jiménez AT, Soldati T (2018) Eat prey, live: Dictyostelium discoideum as a model for cell-autonomous defenses. Front Cell Infect Microbiol 8:1906. https://doi.org/10.3389/fimmu.2017.01906

Ellson CD, Anderson KE, Morgan G, Chilvers ER, Lipp P, Stephens LR, Hawkins PT (2001) Phosphatidylinositol 3-phosphate is generated in phagosomal membranes. Curr Biol 11(20):1631–1635. https://doi.org/10.1016/S0960-9822(01)00447-X

Fischer M, Haase I, Simmeth E, Gerisch G, Müller-Taubenberger A (2004) A brilliant monomeric red fluorescent protein to visualize cytoskeleton dynamics in Dictyostelium. FEBS Lett 577(1–2):227–232. https://doi.org/10.1016/j.febslet.2004.09.084

Gerisch G (1960) Zellfunktionen und Zellfunktionswechsel in der Entwicklung von Dictyostelium discoideum. Wilhelm Roux Arch Entwickl Mech Org 152(5):632–654. https://doi.org/10.1007/BF00582043

Griffin FM Jr, Griffin JA, Silverstein SC (1976) Studies on the mechanism of phagocytosis. II. The interaction of macrophages with anti-immunoglobulin IgG-coated bone marrow-derived lymphocytes. J Exp Med 144(3):788–809.https://doi.org/10.1084/jem.144.3.788

Griffin FMJ, Griffin JA, Leider JE, Silverstein SC (1975) Studies on the mechanism of phagocytosis. I. Requirements for circumferential attachment of particle-bound ligands to specific receptors on the macrophage plasma membrane. J Exp Med 142(5):1263–1282. https://doi.org/10.1084/jem.142.5.1263

Jeschke A, Zehethofer N, Lindner B, Krupp J, Schwudke D, Haneburger I, Jovic M, Backer JM, Balla T, Hilbi H, Haas A (2015) Phosphatidylinositol 4-phosphate and phosphatidylinositol 3-phosphate regulate phagolysosome biogenesis. Proc Natl Acad Sci U S A 112(15):4636. https://doi.org/10.1073/pnas.1423456112

Möller J, Luehmann T, Hall H, Vogel V (2012) The race to the pole: how high-aspect ratio shape and heterogeneous environments limit phagocytosis of filamentous Escherichia coli bacteria by macrophages. Nano Lett 12(6):2901–2905. https://doi.org/10.1021/nl3004896

Moores SL, Sabry JH, Spudich JA (1996) Myosin dynamics in live Dictyostelium cells. Proc Natl Acad Sci U S A 93(1):443–446. https://doi.org/10.1073/pnas.93.1.443

Müller-Taubenberger A, Ishikawa-Ankerhold HC (2013) Fluorescent reporters and methods to analyze fluorescent signals. Methods Mol Biol 983:93–112. https://doi.org/10.1007/978-1-62703-302-2_5

Naufer A, Hipolito VEB, Ganesan S, Prashar A, Zaremberg V, Botelho RJ, Terebiznik MR (2018) pH of endophagosomes controls association of their membranes with Vps34 and PtdIns(3)P levels. J Cell Biol 217(1):329. https://doi.org/10.1083/jcb.201702179

Pasternak C, Spudich JA, Elson EL (1989) Capping of surface receptors and concomitant cortical tension are generated by conventional myosin. Nature 341(6242):549–551. https://doi.org/10.1038/341549a0

Swart AL, Harrison CF, Eichinger L, Steinert M, Hilbi H (2018) Acanthamoeba and Dictyostelium as cellular models for Legionella infection. Front Cell Infect Microbiol 8:61–61. https://doi.org/10.3389/fcimb.2018.00061

Thanedar S, Margolin W (2004) FtsZ exhibits rapid movement and oscillation waves in helix-like patterns in Escherichia coli. Curr Biol 14(13):1167–1173. https://doi.org/10.1016/j.cub.2004.06.048

Zähringer U, Lindner B, Seydel U, Rietschel ET, Naoki H, Unger FM, Imoto M, Kusumoto S, Shiba T (1985) Structure of de-O-acylated lipopolysaccharide from the Escherichia coli Re mutant strain F 515. Tetrahedron Lett 26(51):6321–6324. https://doi.org/10.1016/S0040-4039(01)84588-3

Membrane Tension and the Role of Ezrin During Phagocytosis

Rhiannon E. Roberts, Sharon Dewitt, and Maurice B. Hallett

Abstract

During phagocytosis, there is an apparent expansion of the plasma membrane to accommodate the target within a phagosome. This is accompanied (or driven by) a change in membrane tension. It is proposed that the wrinkled topography of the phagocyte surface, by un-wrinkling, provides the additional available membrane and that this explains the changes in membrane tension. There is no agreement as to the mechanism by which unfolding of cell surface wrinkles occurs during phagocytosis, but there is a good case building for the involvement of the actin-plasma membrane crosslinking protein ezrin. Not only have direct measurements of membrane tension strongly implicated ezrin as the key component in establishing membrane tension, but the cortical location of ezrin changes at the phagocytic cup, suggesting that it is locally signalled. This chapter therefore attempts to synthesise our current state of knowledge about ezrin and membrane tension with phagocytosis to provide a coherent hypothesis.

Keywords

Ezrin · Calpain · Cell surface area · Membrane tension · Cell cortex

Introduction

The plasma membrane tension and the control of the apparent expansion of the plasma membrane surface area are two key components of phagocytosis. It is only recently that our understanding of these events has improved, but there is already increasing evidence that these two events are linked and that they are controlled during phagocytosis. The molecular mechanism and the molecular identity of the components and how they are signalled during phagocytosis are still unresolved. However, there is strong evidence that ezrin, a protein which links the cortical actin network to the plasma membrane is involved.

In this chapter, the possible mechanisms for membrane expansion and controlling membrane tension during phagocytosis will be discussed. Also the molecular characteristics of ezrin and the evidence for its involvement in controlling membrane tension and the apparent membrane expansion that occurs during phagocytosis will be presented.

R. E. Roberts · M. B. Hallett (✉)
School of Medicine, Cardiff University, Cardiff, UK
e-mail: hallettmb@cf.ac.uk

S. Dewitt
School of Dentistry, Cardiff University, Cardiff, UK

© Springer Nature Switzerland AG 2020
M. B. Hallett (ed.), *Molecular and Cellular Biology of Phagocytosis*, Advances in Experimental Medicine and Biology 1246, https://doi.org/10.1007/978-3-030-40406-2_6

Apparent Expansion of Cell Surface Area During Phagocytosis

It is obvious that during the uptake of particles, there must be an increase in the apparent surface area of the cell (see Chap. 1: Fig. 1.2). However, the extent of the increase in cell surface area may not be so obvious. Neutrophils have a capacity to internalise a large number of particles (Hallett and Dewitt 2007) or internalise a single massive object with diameter almost equal to the cell diameter (Herant et al. 2005) both of which "fill" the internal space. This requires an increase the apparent surface area of neutrophils by approximately an additional 100–200 % (Herant et al. 2005: Hallett and Dewitt 2007; Lee et al. 2015). Macrophages have been estimated to increase their surface area by up to 600% (Lam et al. 2009).

A related phenomenon, often called "frustrated phagocytosis" where the radius of curvature is infinite (ie the cell spreads on a planar surface) may result in an even larger increase in surface area (Dewitt and Hallett 2007). As there is no point that closure of a phagosome can occur, presumably "frustrated phagocytosis" induces the maximum possible membrane expansion. Neutrophils and macrophages under these circumstances, "spread out" on the surface doubling or trebling their surface area. Both types of phagocytosis (limited phagosome and infinite phagosome size) requires about the same additional surface area. For example, the surface area of the largest ingestible spheres in phagocytosis is about the same as the largest surface area increase of a spread macrophage (Cannon and Swanson 1992). The massive apparent membrane expansion in both types of phagocytosis also occurs within 100–200 s. Watching a cell spread out on a surface or undergo phagocytosis, it appears as if the cell membrane simply stretches. However, this is not possible as the plasma membrane has little ability to stretch. It is estimated that the membrane may only stretch about 4% of its length (Evans and Skalak 1979; Waugh 1983: Hamill and Martinac 2001). The plasma membrane is a phospholipid bilayer and thus essentially a sandwich across

its thickness of water/lipid/water which are hydrophilic/hydrophobic/hydrophilic (see Fig. 6.1). This structure is strong in the water/lipid/water direction due to the hydrophobic filling in the hydrophilic "sandwich". However, laterally there is little adhesive force to keep adjacent phospholipids in place. Once the lateral stretching exceeds a point where water can enter the bilayer, the structure and lateral strength of the bilayer is lost and the bilayer breaks down (Fig. 6.1). This limits stretch to a maximum of about 4% of its length (Evans and Skalak 1979: Waugh 1983: Hamill and Martinac 2001), a figure which is far too low to provide the 100% of the cell surface area required for phagocytosis. For this reason, there must be a reservoir of membrane from which additional membrane can be added to increase the surface area of the cell.

Exocytotic Addition to the Plasma Membrane

One mechanism for expanding the cell surface area is the addition of vesicular membrane to the plasma membrane by fusion of intracellular vesicles with the plasma membrane (Fig. 6.2b). This is an exocytotic event. Simple geometry can calculate that the fusion of 100 vesicles of 1 μm in diameter to the plasma membrane of a spherical cell of 10 μm in diameter will double to cell surface area (or 200 vesicles to treble the cell surface area). The volume occupied by these vesicles would only be 10% of the cell volume and thus could easily be accommodated. It has been shown that during spreading of fibroblasts, the increase in cell surface area was accompanied by exocytosis (Gauthier et al. 2009). Although fibroblasts are not specialist phagocytic cells, this may be considered to be a form of "frustrated phagocytosis". It has also been shown that plasma membrane tension controls this exocytosis (Gauthier et al. 2011). This is in keeping with the "membrane tension hypothesis" proposed by Sheetz that high tension in the plasma membrane favours the recruitment of membrane to the surface (ie exocytosis) whereas low tension in the plasma membrane favours its retrieval (ie endocytosis).

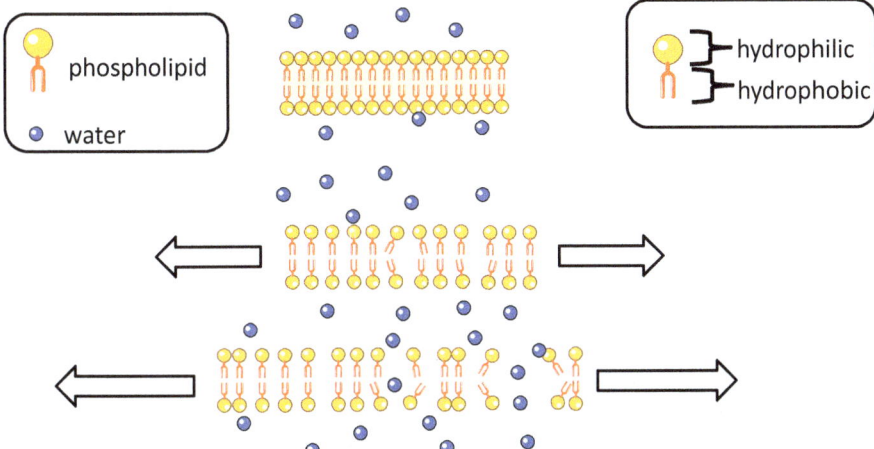

Fig. 6.1 The effect of lateral force (stretching) in phospholipid bilayer. The upper image shows the organisation of phospholipid and water molecules (on the left) to form a phospholipid bilayer (central), with the hydrophilic and hydrophobic regions of the phospholipid indicated on the right. The middle image shows the effect of moderate stretching. The phospholipid organisation of the bilayer structure begins to be lost, as individual phospholipid molecules part. The lower image shows the effect of stretching beyond its breaking point. The bilayer breaks down as the hydrophilic head groups of the phospholipids are attracted to water encroaching within the bilayer and bilayer rupture occurs (allowing water to pass through the membrane). (The graphics were compiled using elements provided by Servier Medical Art (https://smart.servier.com/category/cellular-biology/intracellular-components/))

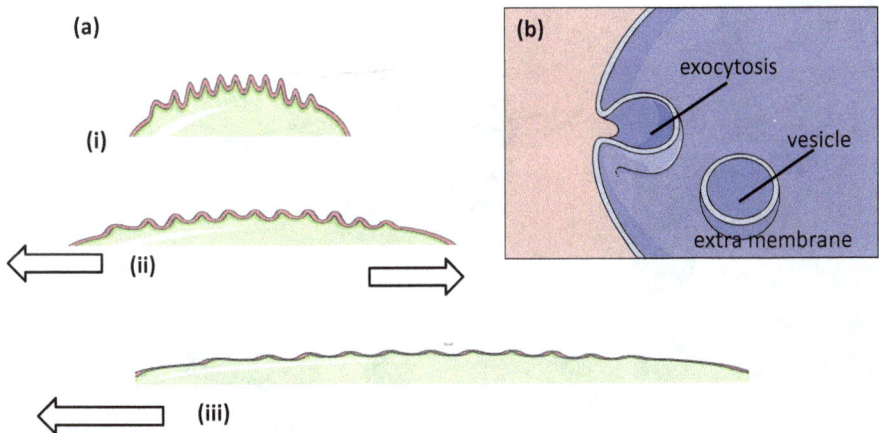

Fig. 6.2 Mechanisms of cell surface expansion. (**a**) A cartoon showing the progressive unfolding of cell surface wrinkles and the apparent increase in the surface area., with (*i*) the resting condition with wrinkles: (*ii*) the wrinkle height is reduced and the surface area increases: (*iii*) nearing full unfolding of the wrinkles and maximum increase in cell surface area. (**b**) A cartoon showing exocytosis of an intracellular vesicle. The vesicular membrane becomes incorporated in the cell surface area following fusion of the vesicle with the plasma membrane. (The graphics were compiled using elements provided by Servier Medical Art (https://smart.servier.com/category/cellular-biology/intracellular-components/))

In this way, a homeostatic mechanism ensures that the membrane tension is maintained within physiologically appropriate the limits.

The nature of the "membrane reservoir" was also investigated by Sheetz using a method whereby a membrane-attached bead was pulled vertically by laser tweezers to form a tether (ie a thin out-pulling of the plasma membrane) and the force determined as an indicator of membrane tension (Raucher and Sheetz 1999). Using chick embryo fibroblasts, they found that the force on the bead remained constant regardless of the length of the pulled tether and thus was buffered by the "membrane reservoir". Pulling past a threshold resulted in a sudden exponential rise in force suggesting that the membrane reservoir was depleted. This finding suggests that in addition to exocytosis there was another "membrane reservoir", namely the folded and wrinkled cell surface.

Cell Surface "Wrinkles"

The wrinkled cell surface may thus represent another "membrane reservoir" for additional cell surface area. The mechanism for increasing the apparent cell surface area would be by the release of cell surface wrinkles and microridges). Scanning electron microscopy (SEM) images of a number of cells show that the cell surface is not smooth (Fig. 6.3a). Phagocytic cells such as neutrophils have a wrinkled cell surface on which there are many longitudinal ridges (microridges) (Fig. 6.3a). An SEM study on the effect of phagocytosis on the surface morphology of macrophages, showed that the surface folds (wrinkled density) was reduced significantly after phagocytosis (Petty et al. 1981). The number and extent of these surface structures was also decreased on human neutrophils during phagocytosis as observed by SEM and by sdFRAP

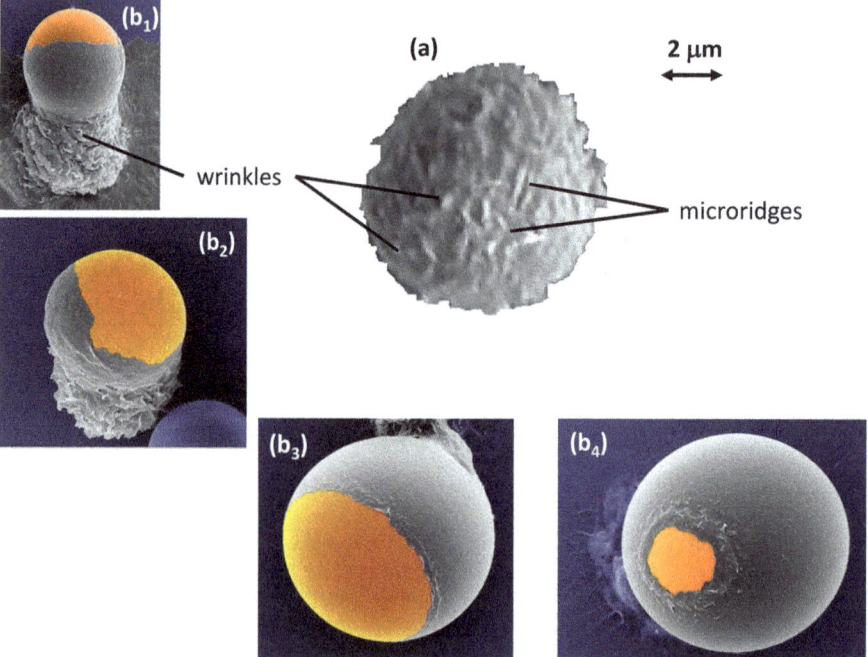

Fig. 6.3 Cell surface topography in "spherical" and phagocytotic cells. Scanning electron micrographs (SEM) of human neutrophils are shown. (**a**) An unstimulated spherical human neutrophil with surface structure (wrinkles and microridges) indicated. (**b$_{1, 2, 3, 4}$**) A series of SEMs of human neutrophils attempting phagocytosis of spherical bead of different sizes, showing the wrinkled cell body and the smooth extending phagocytic cup around the target. In b$_1$ the wrinkles are indicated. The colour version of this figure is available in the e-book version of this volume on-line. (The images in (**b**) are reproduced from: Francis and Heinrich (2018) with kind permission of Yale J Mol Biol)

(subdomain Fluorescence recovery after photobleaching), a technique which interrogates the cell surface topography of living cells (Al Jumaa et al. 2017).

During phagocytosis, the details of the kinetics of the cell surface membrane expansion are important in understanding the cell surface unwrinkling process. Directed phagocytosis (ie using a micropipette to deliver a phagocytic target to the cell) permits the entire phagocytic process from particle contact to phagosome closure can be monitored (Dewitt and Hallett 2002; Herant et al. 2005). The progression of the pseudopodia from human neutrophils from the phagocytic cup extending around the target, do not increase at a single rate. Measured as the length of the phagocytic pseudopodia, there are three phases (Fig. 6.4a): (i) an initial rapid phase, followed by (ii) a slow phase and finally (iii) a second rapid phase which continues until completion of the closed phagosome (Dewitt and Hallett 2002; Herant et al. 2005). The cytosolic Ca^{2+} signal involved in the initiation of phagocytosis (see Chap. 8) occurred after the first two phases, including the initial rapid (but limited) extending of pseudopodia to form the phagocytic cup (Fig. 6.4b). Thus there was no cytosolic free Ca^{2+} signal during the first

membrane expansion phase (Dewitt and Hallett 2002). In contrast, there was a large cytosolic Ca^{2+} signal at the time of the second acceleration phase of pseudopodial extension (Dewitt and Hallett 2002; and Fig. 6.4b). The first fast phase, required no Ca^{2+} signalling and was possibly a physical phenomenon involving binding of the particle to "slack membrane". Whereas the final phase, which was signalled by Ca^{2+}, involved the release of additional membrane.

Cox et al. (1999) used the fluorescent probe FM 1-43, which only fluoresces significantly once in the lipid bilayer, to monitor the actual cell surface area of mouse macrophages. During frustrated phagocytosis, the FM 1-43 signal only increased by 15% during the initial 10 min (rising to 25% at 30 min). This increase was small compared to the expected surface area increase of 100% or more, and suggested that exocytosis was a small component of the "additional" membrane. The remainder of the apparent surface area increase was without exocytosis and may thus have been by changes in the cell surface topography (eg unfolding of cell surface wrinkles).

Masters et al. (2013) showed that RAW 264.7 macrophages undergoing frustrated phagocytosis

Fig. 6.4 Kinetics of phagocytic pseudopodia engulfment. (**a**) Data which shows the three phases of phagocytic pseudopodia engulfment with the two rapid phases indicated together with the slower phase. (**b**) The same data as shown in (**a**) with the timing of the cytosolic Ca^{2+} signal superimposed. (cytosolic Ca^{2+} was simultaneously monitored). (© 2002 Dewitt SD and Hallett MB. Originally published in J. Cell Biol. https://doi.org/10.1083/jcb.200206089)

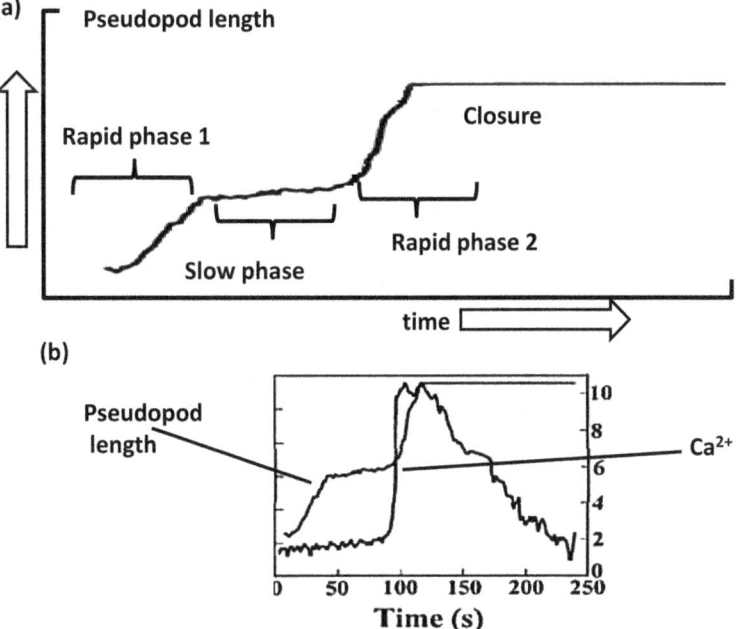

resulted in a two stage cell spreading, a fast initial phase (which they called P1) and a slower late phase of spreading (which they called P2). Using the fluorescent probe FM 1-43 to monitor cell surface area, after Cox et al. (1999), they compared FM 1-43 measurable surface area to the apparent increased cell surface area during frustrated phagocytosis. In phase 2, there was a large increase in FM 1-43 fluorescence, as expected from exocytosis driven membrane expansion (Masters et al. 2013). However, during the first phase (P1) in which rapid cell spreading occurred, there was no detectable increase in FM 1-43 intensity (Masters et al. 2013). The majority of the cell expansion occurred in P1, yet there was a lack of increased FM 1-43 fluorescence, showing that no new membrane had been added to the cell surface. This was again consistent with the unfolding of the cell surface wrinkles. The lack of FM 1-43 fluorescence in P1, showed that the rapid increase in contact area was not the result of exocytosis and was probably therefore due the release of cell surface structures (wrinkles, microridges etc) to provide more available surface area without the addition of new membrane.

A similar conclusion was reached in the rapid cell spreading (frustrated phagocytosis) of human neutrophils neutrophils (Dewitt et al. 2013). In this study, two membrane markers were used together, DiI loaded into the cell membrane at the start of the experiment to act as a control and FM 1-43 in the solution to detect the addition of new membrane during cell spreading. This ratio approach enabled detection of changes in the FM 1-43/DiI intensity ratio, indicating that new (nonstained DiI but FM 1-43 stained) membrane had been added to the cell surface Also the approach enabled visualisation of the location of the event by ratio imaging without artefacts associated with changes in cell shape (Dewitt et al. 2009). However, no new membrane was added to the cells during any phase of neutrophil spreading, despite a large and rapid increase in cell area. It was again concluded that the apparent increase in cell surface area was the not the result of exocytosis, but unwrinkling of the cell surface structures (Dewitt et al. 2013).

Another direct measurement of cell surface area, is the measurement of the total cell membrane capacitance (Cm). As lipid membranes and cell membranes fortuitously have a capacity of very close to 1 $\mu F/cm^2$ (or 10 $fF/\mu m^2$), additional cell membrane added to the surface area, will cause the capacitance of the cell to increase. When a phagosome is formed, so that the membrane forming the phagosome is no longer part of the cell surface, the capacitance of the cell will step down (by an increment equivalent to the surface are of the phagosome). This measure thus provides a quantitation of the roles of unwrinkling (ie no increase in the actual surface area) during phagocytosis and exocytosis (adding additional membrane). The theoretical outcomes of the two extremes where either all or none of the membrane for phagocytosis is provided by exocytosis, is shown in Fig. 6.5a, b. The result of phagocytosis of 0.8 μm particles by J774 macrophages on membrane capacitance (Fig. 6.5c) as recorded by Di et al. (2002). The average step decrease was 23fF/event. This would result from the formation of a phagosome with surface area of 2.3 μm^2 corresponding to an enclosed particle of diameter 0.86 μm. As the particles were 0.8 μm in diameter, this suggesting that each step was the internalisation of a single particle. There was also a small rise in Cm before phagosome closure (Fig. 6.5c), indicating some additional new membrane, but this was insufficient to provide a significant extra membrane for phagosome formation (hence the downward staircase), showing that the major effect was a decrease in cell surface area at each phagocytic event. These measurements therefore point to unfolding of the cell surface wrinkles, at each phagocytotic event, as the main source of additional available membrane for phagocytosis.

Membrane Tension and Cell Shape Changes

It is clear that membrane tension is an important factor in determining the ability of cell to dynamically change shape (Sheetz and Dai 1996). For

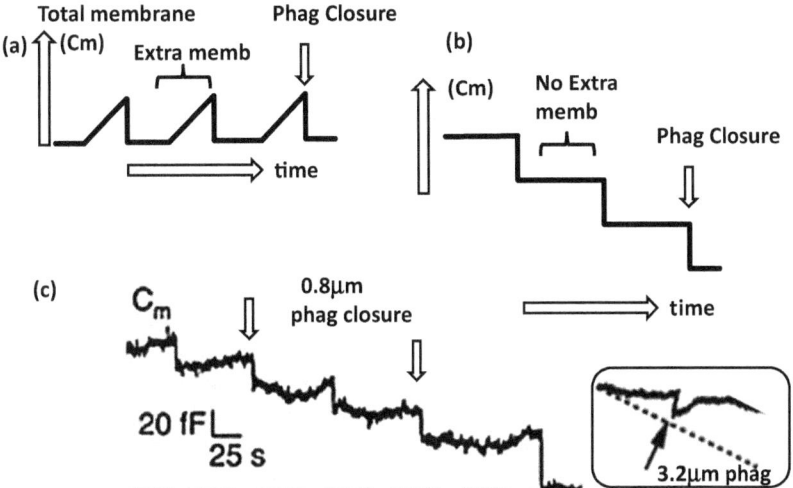

Fig. 6.5 Membrane capacitance changes during phagocytosis. Membrane capacitance (Cm) changes of J774 marcophages during phagocytosis reflect the cell surface area. In (**a**) and (**b**), the theoretical outcomes of measuring membrane capacitance are shown when (**a**) the membrane required for phagocytosis is provided by adding new membrane to the plasma membrane (ie by exocytosis) during the pre-phagocytic phase indicated by "extra memb"; and (**b**) without the additional membrane (ie by releasing membrane on the cell surface held in wrinkles to become available for phagocytosi (indicated "No extra memb" in the pre-phagocytic phase). In (**c**) the actual result of phagocytosis of particles (0.8 μm in diameter) is shown. The sudden fall in Cm as the phagosome closes off from the plasma membrane is indicated in each section as "phag closure". (Reprinted by permission from Springer/Nature: Nature Cell Biol. 4(4):279–85: Di A, Krupa B, Bindokas VP, Chen Y, Brown ME, Palfrey HC, Naren AP, Kirk KL, Nelson DJ.© 2002)

example, cell spreading and lamellipodia extension in NIH 3T3 cells (a mouse fibroblast line) are regulated by membrane tension (Raucher and Sheetz 2000). As changing the cell surface area would have an effect its surface tension, the two parameters, surface area and tension are linked. In a phagocytic cell line, HL60 neutrophils, a remarkable experiment highlights the roles of membrane tension and plasma membrane availability (Houk et al. 2012). HL60 neutrophils can be forced to adopt very extended morphologies, with the front (nucleus-free) and motile part of the cell forming pseudopodia separated by only by a thin cytoplasmic thread from rest of the cell body, which is immotile and inert (Houk et al. 2012). The extent of the excess of plasma membrane in these cells is apparent from length of the connecting thread which can be 25 μm (dimeter 1 μ). This stretching of the cell (with increased surface area) will increase the tension in the membrane. Presumably as the surface wrinkles are pulled apart (as shown by Herant et al. (2005). Houk et al. proposed to sever the interconnecting cytoplasmic thread arguing that this was an experimentum crucis. They write that "*both diffusion-based inhibition mechanisms (for giving this cell polarity) predict that the cell body will remain inactive after severing unless the resynthesis/turnover rate is high. In contrast, a tension-based inhibition mechanism would allow the cell body to reanimate because the cell could relax into a reduced-tension morphology after severing.*" They found that within 70 seconds of severing the thin interconnecting cytoplasmic thread, pseudopodia formed in the cell body. Presumably this was because membrane which, before severing, was within the thin thread was made available to the cell body by severing (Houk et al. 2012). This gives a graphic demonstration of how membrane tension prevents pseudopodia formation, and how releasing that tension permits pseudopodia to form.

Wrinkles and Membrane Tension

A striking feature of scanning electron microscopy images of cells undergoing phagocytosis is that the plasma membrane which forms the phagocytic cup and the extending pseudopodia are devoid of wrinkles and microridges (Bessis 1973; Francis and Heinrich 2018), despite the highly wrinkled cell body just microns away (Fig. 6.3b). This situation is also apparent in living neutrophils interrogated by subdomain fluorescence recovery after photobleaching, or sdFRAP (Al Jumaa et al. 2017). This optical technique measures the time taken by a membrane-associated fluorophore to diffuse from the boundary of a zone of photo-bleaching to a subdomain a define 2D distance distant from the bleach front. The 3D distance for diffusion over this 2D distance may include the membrane undulations, the peaks and valleys of microridges and wrinkles. Thus the timing for diffusion gives information about the actual (3D) diffusion distance, and thus the smoothness of the surface over which diffusion has occurred. In neutrophils, the diffusion distance of the cell body is significantly greater than that expected for a flat surface (but not unexpected for a highly wrinkled surface), whereas at the phagocytic cup, it is consistent with diffusion over a smooth membrane surface (Al Jumaa et al. 2017). Thus, phagocytic cells both fixed for SEM and living, have cell surface topographies which are demonstrably altered locally during phagocytosis. It appears as if either phagocytosis has pulled the wrinkles apart or that the wrinkles are pulled apart to allow phagocytosis. Clearly, membrane tension is involved in these changes.

The elegant experiments of Herant et al. (2005) gave direct measurements of membrane tension during phagocytosis by single human neutrophils. They used a methodology based on micropipette aspiration (Hochmuth 2000) which involves aspirating a small part of the cell into a wide micropipette and accurately measuring the length of the part of cell within the pipette to give a readout of the cell cortex tension (Fig. 6.6a). Essentially, the cortical tension of a partially aspirated neutrophil can be calculated from the Laplace law (Evans and Yeung 1989). While holding a neutrophil in the aspiration pipette, the phagocytic target held in a second micropipette was presented and phagocytosis induced (Fig. 6.6b). In this way, both the progression of phagocytosis and the accompanying changes in surface tension were monitored (Fig. 6.6c). The technique also showed the same three phases of phagocytosis as previously reported (Dewitt and Hallett 2002), the initial fast phase of pseudopod extension, followed by the slow phase and finally the second fast phase leading to phagosome closure (Fig. 6.6c). Herant et al. (2005) found that there was no change in cortical tension during the first rapid fast, despite pseudopodia reaching around approx. 40% the length of the target. However, in the second phase of pseudopodia extension, tension rises steeply at the onset continuing until phagosome closure (Fig. 6.6c). The interpretation of these findings was that as there was no increase in membrane tension during the first phase, the "slack" in the wrinkled membrane was simply taken up easily without increased tension. This may be because in every wrinkle, the molecular mechanism holding the wrinkles in place left a certain amount of slack; or because some wrinkles were held more tightly than others and some were held more loosely (ie slack). Once all the slack is used, a signalling event (probably cytosolic Ca^{2+} see Chap. 9) is required to loosen further wrinkles allowing them to be pulled apart as the membrane tension increases. Hence the second phase of pseudopod extension is accompanied by an increase in tension. These data clearly point to membrane unwrinkling as a mechanism for providing the extra membrane required for phagocytosis and further explain the three phase of phagocytosis. A further discovery made was that although the cortical tension was low during aspiration to mechanically expand the cell surface area (provided the expansion of surface area was less than about 30%), during phagocytosis, tension was low even after a surface area expansion was up to 80%. This may point to an effect of signalling during phagocytosis, which reduces wrinkle tightness. Interestingly, with lymphocytes, which are non-phagocytic cells,

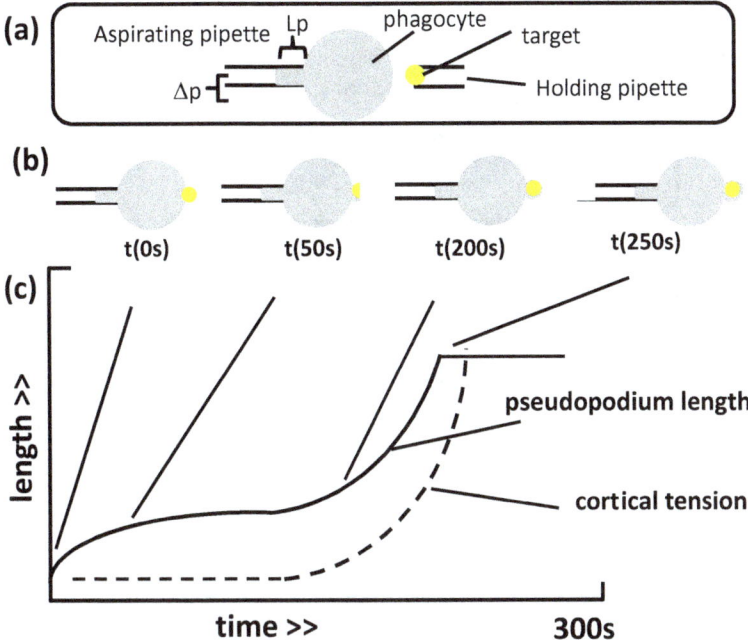

Fig. 6.6 Cortical tension changes during the phases of phagocytosis. (**a**) shows the experiment technique used to obtain the cortical tension measurements, when Δp is the pressure required to aspirate part of the neutrophil into the pipette, and Lp is the length of the aspirated cytoplasm. (**b**) Cartoons showing contact of the cell with the target; phagocytic cup formation; pseudopod exten-sion; and closure of the phagosome shown. The timing of each stage is indicated on the data shown in (**c**) by a connecting line. (**c**) A graphic showing the three phases of phagocytic pseudopodia extension (pseudopodia length) and the simultaneous measurement of cortical tension, as indicated. (The complete data (which the graphic loosely depicts) is from Herant et al. 2005)

aspiration could also only expand the surface area to a similar amount as neutrophils, before rupture ie an additional 25% for lymphocytes, 30% for neutrophils (Guillou et al. 2016. This was also only a small fraction of the surface area available during cell spreading (an additional 150%), suggesting the release of further membrane was achieved through an active signalling process, rather than simply an additional puling force. By monitoring the cortical tension during phagocytois in neutrophil, Herant et al. (2005) showed that mechanism underlying phagocytosis was consistent with a phagocytosis-signalling loosening of the wrinkles which made additional membrane in wrinkles more easily available.

It is interesting to note that Raucher and Sheetz (1999) using the laser trap to pull a cell surface tether also has three phases (Fig. 6.7). After an initial phase to form a tether of about 1 μm, further pulling on the tether elongates the tether to about 7 μm (elongation phase) without an increase in force (Fig. 6.7). This was identified as the "reservoir" and may correspond to the unwrinking of slack wrinkles, in a similar way to the increase of phagocytic pseudopodia length without an increase in cortical tension (Fig. 6.6c: Herant et al. 2005). Pulling the tether beyond 7 μm (the exponential phase) was associated with an increase in force (Fig. 6.7), and may correspond to the pulling apart of the "molecular Velcro" holding the firm wrinkles.

The tension measurements made by Herant et al. (2005) lead them to conclude that "(1) wrinkles are stabilized by membrane-cytoskeleton-membrane bonds. (2) wrinkles see the same surface tension. (3) The baseline surface tension is set by the strength of the weakest wrinkle. (4) When the surface tension increases, the weakest wrinkle is unravelled preferentially thus releasing membrane from the wrinkle compartment to the

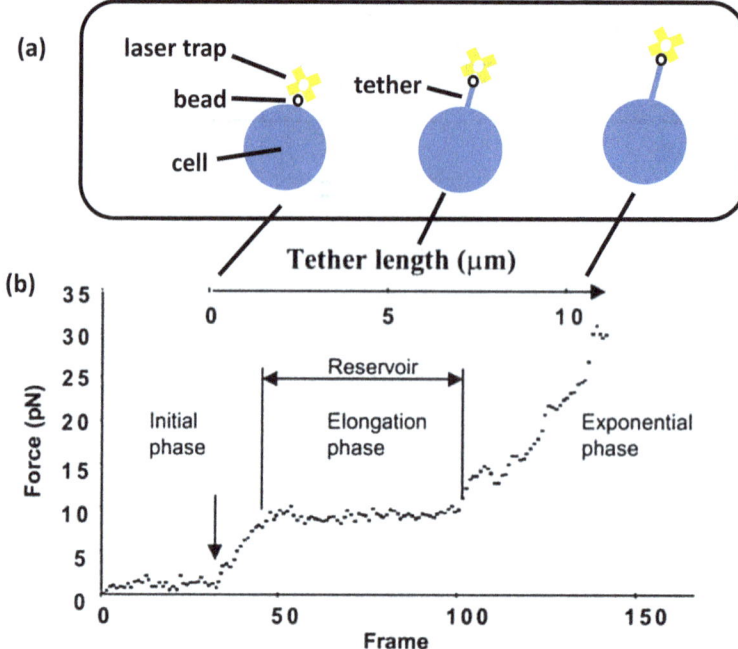

Fig. 6.7 Membrane tension and the membrane reservoir. (**a**) A cartoon showing the principle of the method of Raucher and Sheetz (1999) for monitoring membrane tension. A bead attached to the cell membrane is held in a lase trap. The laser trap is moved to pull out a membrane tether. The force required to do so is a read out of the membrane tension. (**b**) Data from Raucher and Sheetz showing the three phases of tension change as the tether is elongated. The location of the graph of the cartoon version of the three stages in (**a**) are indicated. (**b**) is reprinted from Biophysical Journal, vol 77, Raucher and Sheetz, Characteristics of a membrane reservoir buffering membrane tension. (1992–2002 Copyright (1999), with permission from Elsevier)

covering compartment". They could have added that their work points to a mechanism for allowing cells to change shape rapidly, (eg form pseudopodia, undergo phagocytosis or cell spreading). Weakening or destroying the "membrane-cytoskeleton-membrane bonds" which hold the wrinkles in place would act as a permissive switch for pseudopodia formation in the same way that severing the interconnecting cytoplasmic thread in the experiment of Houk et al. (2012) cause pseudopodia formation in the previous inert cell body.

Molecular Maintenance of Membrane Tension

There is thus increasing evidence, as outlined in the previous section, that the membrane reservoir and control of membrane tension (at least for smaller objects) for phagocytosis is the wrinkled cell surface. The question therefore reduces to establishing the mechanisms for maintaining cell surface wrinkles during inaction and releasing them for phagocytosis. It is clear that the wrinkled morphology must be maintained by an interaction between the plasma membrane and the underlying cortical actin network Sheetz (2001), such that the wrinkled cell surface holds an excess of cell membrane (ie an excess of the minimum required to enclose a smooth cell of the same geometry).

The cortical actin network is formed by actin polymerisation with branch points provided, in humans and other mammals, by a small protein of 502 amino acids called WASps (Wiskott–Aldrich Syndrome protein) and by its analogues in amoeba and other cell types. This cortical actin network is also cross-linked to the plasma membrane by a number of proteins, including talin and members of the ERM (Ezrin-Radixin-Moesin) family. Of these cross-linking proteins, ezrin has been strongly implicated in maintaining cell surface structures such as microvilli (and wrinkles), and will be discussed in detail.

Ezrin

Ezrin is an important cytoskeletal protein which has been implicated in the control of membrane tension (and in releasing the membrane reservoir within the wrinkled cell surface for the apparent membrane expansion required for phagocytosis (Robert and Hallett 2019). Ezrin (previously known as p81 and cytovillin) is a membrane-cytoskeletal linker protein and is member of the ERM family, so named by taking the initials of its family members ie **E**zrin **R**adixin and **M**oesin. Interestingly, Ezrin takes it name from Ezra Cornell, founder of Cornell University where the protein was first purified in 1983 (Bretscher 1983). Ezrin exists in two configurations, a "closed" or folded and inactive form: and an "open" or active form (Gautreau et al. 2000; Viswanatha et al. 2013). In some cells, there is a pool of "inactive" ezrin within the cytosol (Gautreau et al. 2000; Viswanatha et al. 2013), which is folded such that the N and C termini are not available for binding until phosphorylated on threonine 567 (Gautreau et al. 2000; Viswanatha et al. 2013). Although this is a potential route for regulating ezrin binding, its role in regulating the wrinkled cell surface may not be critical in phagocytosis. In neutrophils, phosphorylation of ezrin was prevented using a pharmacological inhibitor of ezrin phosphorylation, NSC668394 (a cell-permeable quinoline that binds to ezrin inhibiting phosphorylation at Thr567 with an $IC_{50} = 8.1\ \mu M$). However, there was little effect on phagocytosis (Roberts et al. 2018) suggesting that phosphorylation is not the regulatory step in neutrophils. The open conformation of ezrin binds phosphatidylinositol 4,5-bisphosphate (PIP2) (Yonemura et al. 1999: Fievet et al. 2004) at the plasma membrane at its N-terminal ERM-domain (Fig. 6.8a). In the open configuration, the actin binding domain at the C-terminal

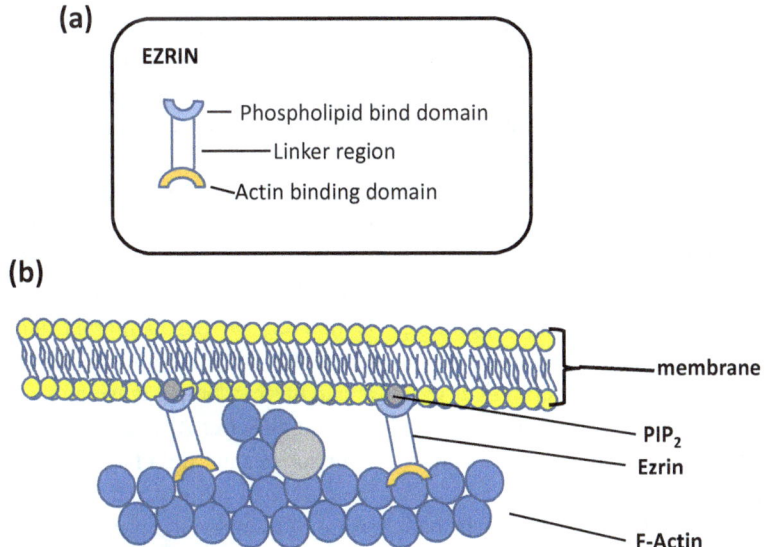

Fig. 6.8 Ezrin cross-linking the membrane to the cortical actin network. (**a**) shows a depiction of the three main elements of an active ezrin molecule, with the PIP2 and actin binding domains indicated, separated by the linker region. (**b**) shows how ezrin cross-linksPIP2 head groups in the lipid bilayer of the plasma membrane with poly-merised actin lying beneath the membrane. (The graphics of Fig. 6.7b were compiled using elements provided by Servier Medical Art (https://smart.servier.com/category/cellular-biology/intracellular-components/ and were originally published in Int. J. Mol. Sci. 20: Article Number: 1383 (2019))

is available to bid the cortical actin networks (Algrain et al. 1993; Viswanatha et 2013). With the C-terminal actin-binding domain of ezrin bound to the F-actin of the cortical cytoskeleton and the N-terminus to the PIP_2 at the plasma membrane, ezrin thus acts as crosslinking molecule linking the cortical actin network to the plasma membrane (Fig. 6.8b).

The linker region between the two bindings regions on ezrin at the N and C termini is a substrate for the Ca^{2+} activated protease calpain, and may thus provide a more important regulatory route for phagocytosis (see Chap. 9). Ezrin is thus an important molecule for consideration when formulating theories about cell membrane tension and considering the cell cortex (Fehon et al. 2010).

As its other name, cytovilln, suggests, ezrin is particularly associated with microvilli or other cell surface structures. Ezrin was first purified from the microvilli of epithelial cells (Bretscher 1983). It is associated with a number of cell surface structures on a variety of cell types (Berryman et al. 1993, 1995, Bonilha et al. 1999). For example, knocking out ezrin expression in mice has profound effects on the microvilli morphology, with malformed, and shortened villi (Saotome et al. 2004; Casaletto et al. 2011). This points to potential role of ezrin in maintaining cell surface structures such as wrinkles and microridges and the maintenance of membrane tension.

Ezrin and Membrane Tension

In additional to morphological data suggesting that ezrin is key to maintaining the cell surface topography, there is also direct biophysical evidence that ezrin cross-linking the cortical actin network to the plasma membrane has an impact on membrane tension.

Diz-Munoz et al. 2010 found that ezrin had an large effect on membrane tension in zebra fish mesodermal cells. These cells provided a useful model as they were derived from embryos injected with a dominant negative nonphosphorylatable version of *ezrin* (DN*Ezrin* T564A ezrin) or a combination of morpholinos

against *ezrin* and *moesin-a* to inactivate ERM protein function. Using atomic force microscopy, they measured static tether force (ie membrane tension); together with adhesion energy and dynamic tether force. All these parameters were significantly reduced in DNezrin cells and morpholino ezrin/moesin combination cells (Diz-Munoz et al. (2010). With all parameters, there was little extra reduction with the ezrin/mosein combination, so it can be inferred that the effect on membrane tension was mainly due to the reduction in active ezrin. They also reported the DNezrin and the ezrin/moesin morpholino combination increased bleb size and frequency, as may be expected from the weaker interaction between the plasma membrane and the underlying actin cytoskeleton. Again these effects were mainly (or totally) due to ezrin reduction. Diz-Munoz et al. interestingly estimated the number of cytoskeletal membrane cross-links in control cells at about 600 cross-linking molecules/μm^2, or a mean lateral separation between ezrin molecules of 41 nm (Diz-Munoz et al. (2010).

Another approach adopted by Liu et al. (2012) was the production of transgenic mice expressing a modified super-active ezrin. By substituting threonine 567 with glutamic acid, a constitutively active form of ezrin (phosphomimetic protein T567E) is produced. This phosphomimetic form of ezrin was expressed the lymphocytes of transgenic mice (Liu et al. 2012). They directly measured the membrane tension of these cells by the optical trap approach of Raucher and Sheetz (1999) ie measuring the force required to maintain a thin membrane tether ($\sim$100-nm diameter) pulled from the cell surface. The apparent membrane tension was markedly increased by about 70% (Liu et al. 2012). Although lymphocytes are not phagocytic, they found that the ability of the isolated lympohcyes to change shape was also significantly reduced in vitro and defective homing in vivo. Interestingly, the "homing response" of lymphocytes triggered by chemokines, has been reported to involve a reduction in cell surface microvilli of lymphocytes as a result of dephosphorylation of ezrin (Brown et al. 2003). The ezrin phosphomimetic protein would obviously not be effected by this mechanism and the microvilli would remain (and inhibit the response to chemokines). Liu et al. looked at this idea but failed to detect any differences in lymphocyte microvilli in the transgenic cells or after chemokine stimulation (Liu et al. 2012, Fig 5S).

Rouven Brückner et al. (2015) used Madin-Darby canine kidney cells (MDCK cells) as a model epithelial cell. They adopted two approaches to modify ezrin plasma membrane interactions: firstly to increase the number of ezrin links to the plasma membrane by injection phosphatidylinositol 4,5-bisphosphate (PIP2) into the inner leaflet of the bilayer, It has previously been shown that microinjection of PIP_2 recruits more ezrin to the interface between plasma membrane and cortex (Braunger et al. 2014). Secondly, they reduced the effectiveness of ezrin by injecting neomycin to mask PIP_2, by pharmacological inhibition of phosphorylatin of ezrin using NSC668394 to reduce the amount of open/active ezrin (see above) and by short interference RNA (siRNA) against ezrin to reduce the amount of erzin. They measured membrane tension by indentation of the cell surface subsequent pulling of tethers using an atomic force microscope (AFM). They reported that 3 h after microinjection of PIP_2 micelles into the cell, increasing the number of PIP_2 binding sites for ezrin on the inner leaflet of the bilayer, the membrane stiffness increased about threefold. Neomycin, which masked PIP_2 and so made it unavailable for ezrin binding, weakened the connection between plasma membrane and underlying cytoskeleton. A similar result was obtained with NSC668394 and ezrin siRNA. Each method of reducing the effectiveness of ezrin crosslinking to the plasma membrane reduced the tether force (and hence membrane tension. As with the other studies, this points to a central role of ezrin in maintaining membrane tension.

Ezrin and Phagocytosis

As discussed earlier, the changes in cortical tension in neutrophils undergoing phagocytosis (see Fig. 6.6) strongly suggest tension in the neu-trophil membrane is maintained by its wrinkled surface but that applied force can pull the wrinkles apart. It is as if there were a "molecular Velcro" holding the wrinkles in place (Herant et al. 2005). During phagocytosis, the holding strength of this "molecular Velcro" is weakened. It is proposed that ezrin may constitute the molecular identity of this "molecular Velcro". Ezrin connects the membrane to the underlying cortical actin (Fig. 6.8) and so may be able to maintain various cell surface configurations, including wrinkles and microridges (Fig. 6.9).

Regulation Membrane Tension by Ezrin During Phagocytosis

Immunostaining shows that ezrin is abundant at the neutrophil periphery (Fig. 6.10a; Elumalai et al. 2011; Elumalai 2012). However, in neutrophils which have undergone phagocytosis, immunostaining of ezrin is absent from to the phagocytic cup and extending pseudopodia (Fig. 6.10b; Elumalai 2012). In RAW 24.7 macrophages, fluor-tagged ezrin locates at the cell periphery and in membrane surface structures (Fig. 6.10c, d; Elumalai 2012). There is also a similar localised loss which can be observed dynamically (Fig. 6.11). Ezrin is lost from the site of phagocytosis at the early stage cup formation and is completely absent before the phagosome is formed (Fig. 6.11; Elumalai 2012; Roberts et al. 2017, 2020). It is expected from the effect of ezrin (Braunger et al. 2014), this localised loss of ezrin would result in a localised decrease in membrane tension as a result of unwrinkling the membrane and providing a localised supply of additional membrane. When cytosolic Ca^{2+} is measured simultaneously with ezrin, it was found that the onset of the loss of peripheral ezrin coincides with the Ca^{2+} signal (Elumalai 2012; Roberts et al. 2017, 2020). Together these findings suggest that ezrin release from the crosslinking function facilitates localised pseudopodia formation.

There are (at least) two possible mechanism for the release of ezrin. The first is related to its requirement for membrane PIP_2. It is known

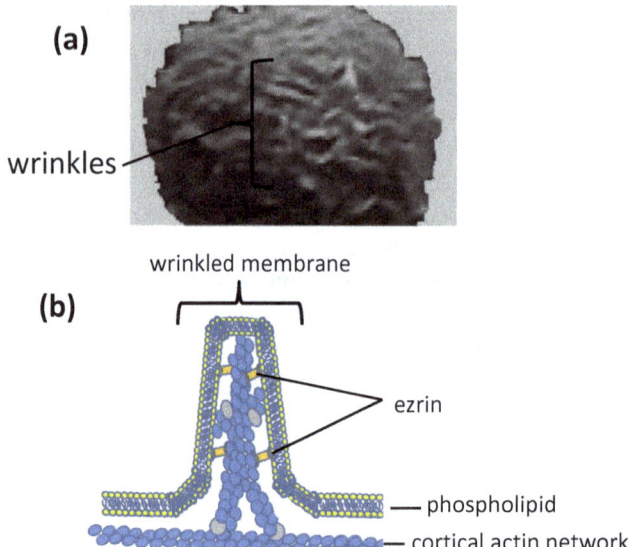

Fig. 6.9 Ezrin cross-linking forming membrane wrinkles. (**a**) A typical scanning electron micrograph (SEM) of a human neutrophil showing its wrinkled cell surface, as indicated. (**b**) The molecular arrangement which would hold the wrinkled membrane in place is shown with ezrin cross-linking polymerised f-actin and the plasma membrane phospholipid bilayer, as shown in Fig. 6.7b. The graphics of Fig. 6.5b were compiled using elements provided by Servier Medical Art. (https://smart.servier.com/category/cellular-biology/intracellular-components/ and were originally published in Int. J. Mol. Sci. 20: Article Number: 1383 (2019))

that in some cells, phospholipase C is activated by an elevation of cytosolic Ca^{2+} (Rhee 2001; Thore et al. 2005). Also PIP_3 is produced locally at the phagocytosis cup in macrophages (Swanson 2008; Zhang et al. 2010), HL60 netrophils (Dewitt et al. 2006) and amoeba (Dormann et al. 2004). Thus it may be anticipated that PIP_2 levels would locally fall. Without firm binding of ezrin via PIP_2, it may dissociate (Coscoy et al. 2002: Raucher et al. 2000) and so no longer function as an effective cross-linker, thus making wrinkled surface membrane available. However, it has been shown that in macrophages, during phagocytosis there is an early increase in PIP_2 at the phagosomal cup (Botelho et al. 2000), at a time when ezrin is dissociating. They report that PIP_2 only reduces later at the end stage, when the phagosome seals. These data suggests that ezrin loss from the cell periphery is unrelated to the amount of PIP_2 in the phagocytic cup and and that the release of ezrin from the membrane is not the result of a reduction in PIP_2 binding.

The second possibility is that a Ca^{2+} activated enzyme liberates ezrin. The Ca^{2+} activated

cytosolic protease calpain may play a key role, as ezrin (unlike moesin) is a substrate for the this protease. Cleavage between the actin binding and PIP_2 binding domains C of ezrin at the cell cortex would be a simple mechanism for releasing the restraint holding the wrinkled membrane in place. The evidence for this mechanism has recently been provided using constructions of ezrin in which the C and N termini were tagged with different fluors (Roberts et al. 2020). During phagocytosis, there was a separation of the two fluors specifically in the phagocytic cup and phagocytic pseudopodia (Roberts et al. 2020). The consequences of ezrin cleavage at these sites are discussed in Chap. 9 (The role of the role of calpain in phagocytosis).

Whatever the exact mechanism for the release of cortical ezrin, its timing and Ca^{2+} dependence suggests that this is an important step. Once the wrinkles of the cell membrane are released and becomes available for the formation of pseudopodia, actin polymerization can supply the protrusion force by the Brownian ratchet mechanism (see next section) and pseudopodia would extend.

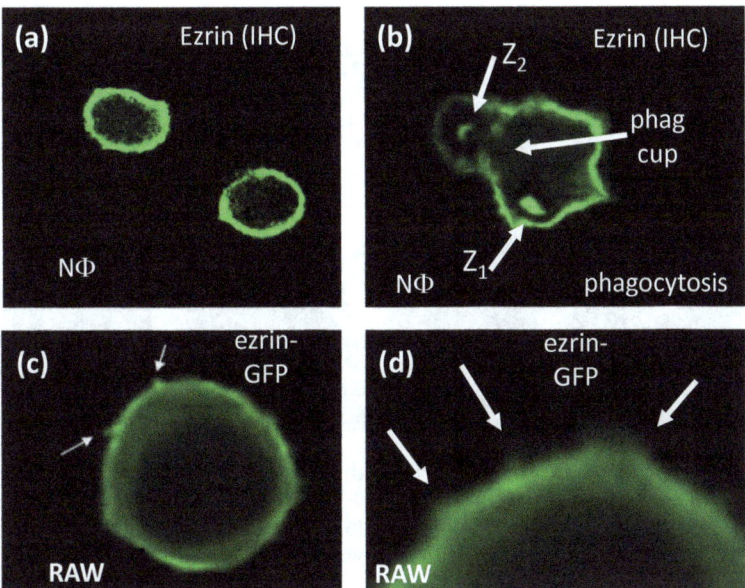

Fig. 6.10 Ezrin location in phagocytes. (**a** and **b**) show the immunohistochemical (IHC) location of ezrin in human neutrophils (NΦ). (**a**) show the obvious peripheral location in spherical (resting) neutrophils. (**b**) shows the location in a neutrophil undergoing phagocytosis of C3bi-opsonised zymosan particles (the central portion of which has an auto-fluorescent signal). Z_1 is an fully internal particle which has ezrin left on the phagosomal membrane, despite abundant ezrin which has laterally diffused back into the outer adjacent membrane. Z_2 is a newly enclosed zymosan particle having reduced ezrin in the phagocytic pseudopodia and not ezrin at the phagocytic cup (base) indicated (phag cup). Note that elsewhere on the cell surface ezrin is uniformly present. (**c** and **d**) show confocal microscopy images of RAW246.7 macrophages expressing ezrin-GFP within. (**c**) shows the peripheral location of ezrin in these cell which is similar to human neutrophils (see Fig. 6.7a). The arrows indicate that surface features (larger but equivalent to wrinkles on neutrophils) also have ezrin. (**d**) shows a higher magnification of the RAW 246.7 cell edge with irregular features of similar size to neutrophil wrinkles containing ezrin

Actin Push and the Brownian Ratchet

In non-muscle cells, such as phagocytes, actin exerts a pushing force on the plasma membrane, rather than the contractile force of actin-myosin in muscle cells. This force is achieved by polymerisation of monomeric actin (g-actin)pushing against the plasma membrane. There is a significant amount of polymerised actin beneath the cell cortex (Pollard 1990; Stossel 1982). WASp provides branch points on the linear f-actin chain which provides anchorage for filamentous actin (f-actin) to grow towards the plasma membrane. The growing tips of these actin filaments continue until they reach the plasma membrane (Fig. 6.12). Polymerisation stops because additional actin monomers can only be added between the tip of the f-actin polymer and the plasma membrane if Brownian motion fluctuation open up a sufficient gap (Peskin et al. 1993). This may occur if the plasma membrane is not held under tension and is free enough that its distance relative to the f-actin tip fluctuates sufficiently for an additional actin monomer to enter (Fig. 6.12). Monomeric actin will then be added to the existing f-actin chain. This additional actin monomer prevents the plasma membrane from returning to its original position, so the membrane has been pushed out a little (Fig. 6.12). This mechanism is a Brownian ratchet (Peskin et al. 1993) with each small step being a one-way ratchet. Its theoretical maximum rate of unimpeded pushing force against the plasma membrane is 0.75 µm/s, with an actin concentration of 25 µM (Peskin et al. 1993). This exceeds that required for the fast rate of neutrophil spreading, at about 10 µm/100 s (i.e., 0.1 µm/s). The process continues until the membrane tension exceeds that which allows Brown-

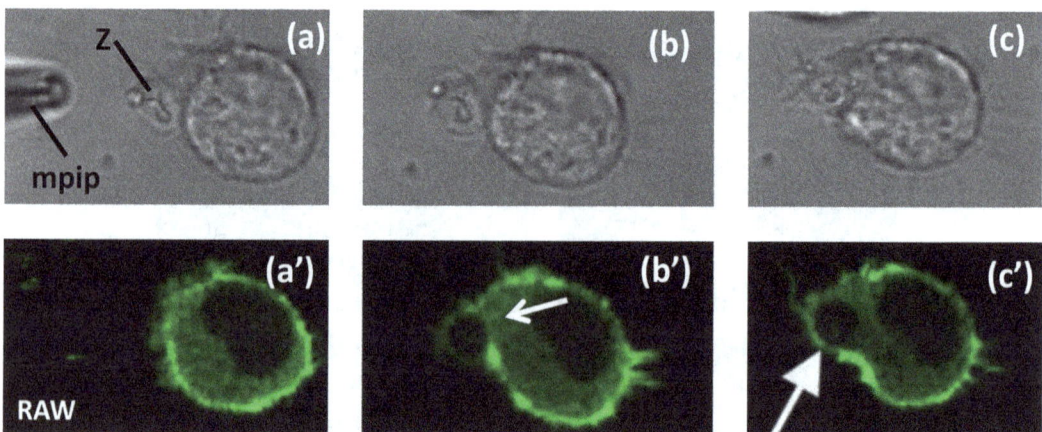

Fig. 6.11 Dynamic loss of ezrin at the phagocytic cup. RAW 246.7 macrophages expressing ezrin-GFP wer presented with mouse Cbi opsoinised zymosan particles. (**a**–**c**) shows the phase contrast images of the process of phagocytosis. (**a′**–**c′**) shows the confocal fluorescence image. (**a** and **a′**) show the distribution of ezrin immediately after presentation of the zymosan target (indicated by "Z") using a micropipette (indicated by "mpip") image (**a**).

(**b** and **b′**) shows the stage at which phagocytic pseudopodia are engulfing the particle (equivalent to that shown in by the neutrophil in Fig. 6.9b). The arrow indicates the loss of ezrin at the phagocytic cup. (**c** and **c′**) shows completion of phagocytosis where the phagosome is devoid of ezrin and the arrow indicates the return of ezrin to the outer membrane by lateral diffusion

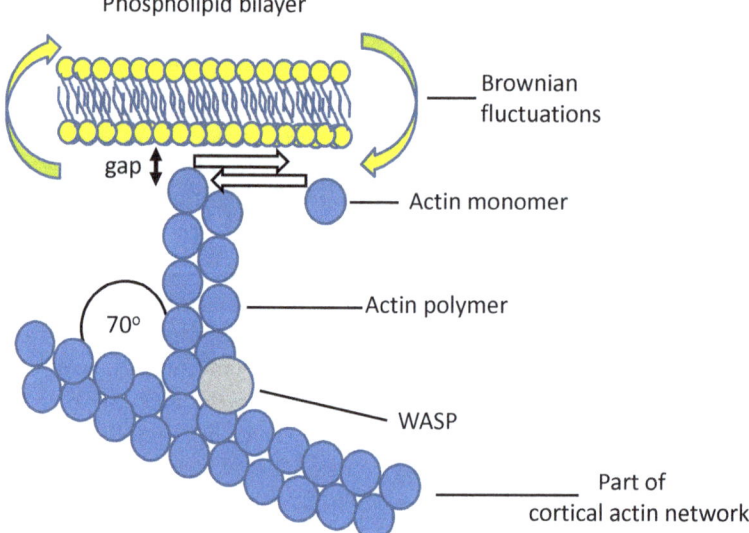

Fig. 6.12 Brownian ratchet and actin mediated plasma membrane protrusion. The graphic show Brownian ratchet–driven actin polymerisation at the plasma membrane. The gap between the tip of polymerising actin and the plasma membrane is only large enough to permit the addition of another actin monomer, if the fluctuations in the position of the plasma membrane are large

enough. It is this Brownian fluctuation which is limited by membrane tension. The graphics of Fig. 6.5b were compiled using elements provided by Servier Medical Art. (https://smart.servier.com/category/cellular-biology/intracellular-components and were originally published in Int. J. Mol. Sci. 20: Article Number: 1383 (2019))

ian fluctuations in the position of the membrane. However, once the membrane tension is reduced (when the linkage holding the wrinkles in place is severed), the Brownian ratchet mechanism will continue to push against the plasma membrane forming protrusions and pseudopodia. This effect probably underlies an observation made on stretched HL60 neutrophils (Houk et al. 2012). The highly stretched cell morphology exerted a high tension at the cell body, sufficient to prevent the Brownian ratchet and actin polymerisation. When the tether was broken by laser cutting, the cell body immediately began to ruffle. Presumably, the Brownian ratchet was freed to polymerise actin against the plasma membrane and produce protrusions. Thus, two important functions of the microridges and wrinkles exist; (i) to maintain a membrane reservoir and (ii) to limit cortical actin polymerisation. If ezrin is lost and the wrinkles are allowed to unfold, there will be both the additional membrane required for phagocytosis and pseudopodia formation would occur as a result of actin polymerisation and the driving force of the Brownian ratchet.

Conclusion and Synthesis

Although there is clearly a lot of work to be done, a preliminary hypothesis that may be drawn from the accumulating wealth of data about membrane tension, ezrin and phagocytosis. The hypothesis of "membrane supply and demand (MSD) "is that phagocytosis is limited by the amount of available cell surface membrane. For phagocytosis of large objects, or infinitely large in the case of frustrated phagocytosis (cell spreading), additional membrane can come from two sources, (i) unfolding of cell surface wrinkles until depleted and then by (ii) exocytosis of intracellular vesicles. Both events require signalling to release wrinkles and to trigger exocytosis. The two events must also be co-ordinated. With medium sized targets (or multiple small objects). Unfolding of cell surface wrinkles is sufficient to provide the additional membrane required for phagosome formation (Fig. 6.13). This must be signalled by intracellular mechanism that release firmly held wrinkles, as there is insufficient membrane within the "slack" wrinkle pool. With very small targets, there may be enough slack in the wrinkled cell surface to provide the small amount of additional membrane without the need for signalling the release of extra wrinkles or exocytosis.

This hypothesis puts understanding the signalling of the release of wrinkles as an important priority because there is a potential therapeutic benefit in inhibiting it. In inflammatory disease, phagocytes (neutrophils) leave the blood and inappropriately invade an area of the body, such as a joint space. They do this by spreading out on the endothelial lining of the blood vessels at a particular location (signalled by locally generated TNF), thereby (i) becoming stationary at the high TNF site and are no longer subject to the force of flowing blood which pushes spherical cells

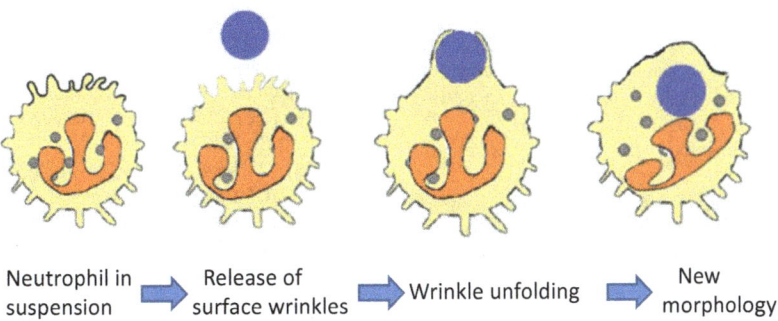

Neutrophil in suspension → Release of surface wrinkles → Wrinkle unfolding → New morphology

Fig. 6.13 The role of wrinkles in phagocytosis. The cartoon shows a simplified scheme by which unfolding of cell surface wrinkles provides the additional extra membrane (and relaxation of membrane tension) necessary for phagocytic pseudopodia progression and completion of phagocytosis. (It was adapted from a figure originally published in Intl. J. Mol. Sci 20: Article Number: 1383 (2019))

rapidly past; and (ii) moving between endothelial cell and so enter the extravascular space. It is the massive accumulation of neutrophils (and other inflammatory cells) at an inappropriate site which causes the problem, So inhibiting the ability of neutrophils to spread out may thus be beneficial in inflammatory disease. However, phagocytosis of infecting microbes by neutrophils is crucial, and blanket inhibition of neutrophil activity would be unwelcome. The MSD hypothesis suggests that by inhibiting the signalling of unfolding of cell surface wrinkles, neutrophil spreading would be inhibited most, and phagocytosis of small target such a bacteria may be unaffected. This would give an ideal anti-inflammatory effect.

It is thus important to establish the signalling mechanism. In this chapter, we have presented the evidence for a role for ezrin in the wrinkling and unfold mechanism. In Chap. 9, evidence is presented that it is an interaction with calpain, a Ca^{2+} activated protease, that is the key.

References

Al Jumaa MA, Dewitt S, Hallett MB (2017) Topographical interrogation of the living cell surface reveals its role in rapid cell shape changes during phagocytosis and spreading. Sci Rep 7:9790

Algrain M, Turunen O, Vaheri A, Louvard D, Arpin M (1993) Ezrin contains cytoskeleton and membrane binding domains accounting for its proposed role as a membrane-cytoskeletal linker. J Cell Biol 120:129–139

Berryman M, Franck Z, Bretscher A (1993) Ezrin is concentrated in the apical microvilli of a wide variety of epithelial cells whereas moesin is primarily found in endothelial cell. J Cell Sci 105:1025–1043

Berryman M, Gary R, Bretscher A (1995) Ezrin oligomers are major cytoskeletal components of placental microvilli: a proposal for their involvement in cortical morphogenesis. J Cell Biol 131:1231–1242

Bessis M (1973) Living blood cells and their ultrastructure. Springer, Berlin

Bonilha VL, Finnemann SC, Rodriguez-Boulan E (1999) Ezrin promotes morphogenesis of apical microvilli and basal infoldings in retinal pigment epithelium. J Cell Biol 47:1533–1544

Botelho RJ, Teruel M, Dierckman R, Anderson R, Wells A, York JD, Meyer T, Grinstein S (2000) Localized biphasic changes in phosphatidylinositol-4,5-bisphosphate at sites of phagocytosis. J Cell Biol 151:1353–1367

Braunger JA, Brückner BR, Nehls S, Pietuch A, Gerke V, Mey I, Janshoff A, Steinem C (2014) Phosphatidylinositol 4,5-bisphosphate alters the number of attachment sites between ezrin and actin filaments. J Biol Chem 289:9833–9843

Bretscher A (1983) Purification of an 80,000-dalton protein that is a component of the isolated microvillus cytoskeleton, and its localization in nonmuscle cells. J Cell Biol 97:425–432

Brown MJ, Nijhara R, Hallam JA, Gignac M, Yamada KM, Erlandsen SL, Delon J, Kruhlak MJ, Shaw S (2003) Chemokine stimulation of human peripheral blood T lymphocytes induces rapid dephosphorylation of ERMs which facilitates loss of microvilli and polarization. Blood 102:3890–3899

Cannon GJ, Swanson JA (1992) The macrophage capacity for phagocytosis. J Cell Sci 101:907–913

Casaletto JB, Saotome I, Curto M, McClatchey A (2011) Ezrin-mediated apical integrity is required for intestinal homeostasis. Proc Natl Acad Sci U S A 108:11924–11929

Coscoy S, Waharte F, Gautreau A, Martin M, Louvard D, Mangeat P, Arpin M, Amblard F (2002) Molecular analysis of microscopic ezrin dynamics by two-photon FRAP. Proc Natl Acad Sci U S A 99:12813–12818

Cox D, Tseng CC, Bjekic G, Greenberg S (1999) A requirement for phosphatidylinositol 3-kinase in pseudopod extension. J Biol Chem 274:1240–1247

Dewitt S, Hallett MB (2002) Cytosolic free Ca^{2+} changes and calpain activation are required for ß2 integrin-accelerated phagocytosis by human neutrophils. J Cell Biol 159:181–189

Dewitt S, Hallett MB (2007) Leukocyte membrane "expansion": a central mechanism for leukocyte extravasation. J Leukoc Biol 81:1160–1164

Dewitt S, Tian W, Hallett MB (2006) Localised PtdIns(3,4,5)P$_3$ or PtdIns(3,4)P$_2$ at the phagocytic cup is required for both phagosome closure and Ca^{2+} signalling in HL60 neutrophils. J Cell Sci 119:443–451

Dewitt S, Darley RL, Hallett MB (2009) Translocation or just location? Pseudopodia affect fluorescent signals. J Cell Biol 184:197–203

Dewitt S, Francis RJ, Hallett MB (2013) Ca^{2+} and calpain control membrane expansion during the rapid cell spreading of neutrophils. J Cell Sci 126:4627–4635

Di A, Krupa B, Bindokas VP, Chen Y, Brown ME, Palfrey HC, Naren AP, Kirk KL, Nelson DJ (2002) Quantal release of free radicals during exocytosis of phagosomes. Nat Cell Biol 4:279–285

Diz-Munoz A, Krieg M, Bergert M, Ibarlucea-Benitez I, Muller DJ, Paluch E, Heisenberg C-P (2010) Control of directed cell migration in vivo by membrane-to-cortex attachment. PLoS Biol 8:e1000544

Dormann D, Weijer G, Dowler S, Weijer CJ (2004) In vivo analysis of 3-phosphoinositide dynamics during Dictyostelium in phagocytosis and chemotaxis. J Cell Sci 117:6497–6509

Elumalai GL (2012) Cytosolic signalling and behaviour of oral neutrophils "Search for biochemical memory". PhD thesis, Cardiff University. http://orca.cf.ac.uk/43089/

Elumalai GL, Dewitt S, Hallett MB (2011) Ezrin and talin relocates from the plasma membrane to cytosol during neutrophil extravasation. Eur J Clin Investig 41(Suppl. 1):47–47

Evans EA, Skalak R (1979) Mechanics and thermodynamics of biomembranes. CRC Press, Boca Raton

Evans E, Yeung A (1989) Apparent viscosity and cortical tension of blood granulocytes determined by micropipet aspiration. Biophys J 56:151–160

Fehon RG, McClatchey AI, Bretscher A (2010) Organizing the cell cortex: the role of ERM proteins. Nat Rev Mol Cell Biol 11:276–287

Fievet BT, Gautreau A, Roy C, Del Maestro L, Mangeat P, Louvard D, Arpin M (2004) Phosphoinositide binding and phosphorylation act sequentially in the activation mechanism of ezrin. J Cell Biol 164:653–659

Francis EA, Heinrich V (2018) Mechanistic understanding of single-cell behavior is essential for transformative. Advances in biomedicine. Yale J Biol Med 91:279–289

Gauthier NC, Rossier OM, Mathur A, Hone JC, Sheetz M (2009) Plasma membrane area increases with spread area by exocytosis of a GPI-anchored protein compartment. Mol Biol Cell 20:3261–3272

Gauthier NC, Fardin MA, Roca-Cusachs P, Sheetz MP (2011) Temporary increase in plasma membrane tension coordinates the activation of exocytosis and contraction during cell spreading. Proc Natl Acad Sci U S A 108:14467–14472

Gautreau A, Louvard D, Arpin M (2000) Morphogenic effects of ezrin require a phosphorylation-induced transition from oligomers to monomers at the plasma membrane. J Cell Biol 150:193–203

Guillou L, Babataheri A, Saitakis M, Bohineust A, Dogniaux S, Hivroz C, Barakat A, Husson J (2016) T-lymphocyte passive deformation is controlled by unfolding of membrane surface reservoirs. Mol Biol Cell 27:3574–3582

Hallett MB, Dewitt S (2007) Ironing out the wrinkles of neutrophil phagocytosis. Trends Cell Biol 17:209–214

Hamill OP, Martinac B (2001) Molecular basis of mechanotransduction in living cells. Physiol Rev 81:685–740

Herant M, Heinrich V, Dembo M (2005) Mechanics of neutrophil phagocytosis: behavior of the cortical tension. J Cell Sci 118:1789–1797

Hochmuth RM (2000) Micropipette aspiration of living cells. J Biomechanics 33:15–22

Houk AR, Jilkine A, Mejean CO, Boltyanskiy R, Dufresne ER, Angenent SB, Altschuler SJ, Wu LF, Weiner OD (2012) Membrane tension maintains cell polarity by confining signals to the leading edge during neutrophil migration. Cell 148:175–188

Lam J, Herant M, Dembo M, Heinrich V (2009) Baseline mechanical characterization of J774 macrophages. Biophys J 96:248–254

Lee C-Y, Thompson GR, Hastey CJ, Hodge GC, Lunetta JM, Pappagianis D, Heinrich V (2015) Coccidioides endospores and spherules draw strong chemotactic, adhesive, and phagocytic responses by individual human neutrophils. PLoS One 10:e0129522

Liu Y, Belkina NV, Park C, Nambiar R, Loughhead SM, Patino-Lopez G, Ben-Aissa K, Hao J-J, Kruhlak MJ, Qi H, von Andrian UH, Kehrl JH, Tyska MJ, Shaw S (2012) Constitutively active ezrin increases membrane tension, slows migration, and impedes endothelial transmigration of lymphocytes in vivo in mice. Blood 119:445–453

Masters TA, Pontes B, Viasnoff V, Li Y, Gauthier NC (2013) Plasma membrane tension orchestrates membrane trafficking, cytoskeletal remodeling, and biochemical signaling during phagocytosis. Proc Natl Acad Sci U S A 110:11875–11880

Peskin CS, Odell GM, Oster GF (1993) Cellular motions and thermal fluctuations: the Brownian ratchet. Biophys J 65:316–324

Petty HR, Haffeman DD, McConnell HM (1981) Disappearance of macrophage surface folds after antibody-dependent phagocytosis. J Cell Biol 89:223–229

Pollard T (1990) Actin. Curr Opin Cell Biol 2:33–40

Raucher D, Sheetz MP (1999) Characteristics of a membrane reservoir buffering membrane tension. Biophys J 77:1992–2002

Raucher D, Sheetz MP (2000) Cell spreading and lamellipodial extension rate is regulated by membrane tension. J Cell Biol 148:127–136

Raucher D, Stauffer T, Chen W, Shen K, Guo SL, York JD, Sheetz MP, Meyer T (2000) Phosphatidylinositol 4,5-bisphoshate functions as a second messenger that regulates cytoskeleton-plasma membrane adhesion. Cell 100:221–228

Rhee SG (2001) Regulation of phosphoinositide-specific phospholipase. Annu Rev Biochem 70:281–312

Roberts RE, Hallett MB (2019) Neutrophil cell shape change: mechanism and signalling during cell spreading and phagocytosis. Int J Mol Sci 19:1383–1398

Roberts RE, Vervliet T, Bultynck G, Parys JB, Hallett MB (2017) Dynamics of ezrin location at the plasma membrane: relevance to neutrophil spreading. Eur J Clin Invest 47(Suppl 1):148–148

Roberts RE, Elumalai GL, Hallett MB (2018) Phagocytosis and motility in human neutrophils is competent but compromised by pharmacological inhibition of ezrin phosphorylation. Curr Mol Pharmacol 11:305–315

Roberts RE, Martin M, Marion S, Elumalai GL, Lewis K, Hallett MB (2020) Ca^{2+} activated cleavage of ezrin visualised dynamically in living cells during phagocytosis and "cell surface area expansion". J Cell Sci 133: jcs236968. https://doi.org/10.1242/jcs.236968

Rouven Brückner B, Pietuch A, Nehls S, Rother J, Janshoff A (2015) Ezrin is a major regulator of membrane tension in epithelial cells. Sci Rep 5:14700. https://doi.org/10.1038/srep14700

Saotome I, Curto M, McClatchey AI (2004) Ezrin is essential for epithelial organization and villus morphogenesis in the developing intestine. Dev Cell 6:855–864

Sheetz MP (2001) Cell control by membrane-cytoskeleton adhesion. Nat Rev Mol Cell Biol 2(5):392–396

Sheetz MP, Dai JW (1996) Modulation of membrane dynamics and cell motility by membrane tension. Trends Cell Biol 6:85–89

Stossel T (1982) The structure of cortical cytoplasm. Philos Trans R Soc Lond Ser B 299:275–289

Swanson JA (2008) Shaping cups into phagosomes and macropinosomes. Nat Rev Mol Cell Biol 9:639–649

Thore S, Dyachok O, Gylfe E, Tengholm A (2005) Feedback activation of phospholipase C via intracellular mobilization and store-operated influx of Ca^{2+} in insulin-secreting β-cells. J Cell Sci 118:4463–4471

Viswanatha R, Wayt J, Ohouo PY, Smolka MB, Bretscher A (2013) Interactome analysis reveals ezrin can adopt multiple conformational states. J Biol Chem 288:35437–35451

Waugh RE (1983) Effects of abnormal cytoskeletal structure on erythrocyte membrane mechanical properties. Cell Motil 3:609–622

Yonemura S, Tsukita S, Tsukita S (1999) Direct involvement of ezrin/radixin/moesin (ERM)-binding membrane proteins in the organization of microvilli in collaboration with activated ERM proteins. J Cell Biol 145:1497–1509

Zhang Y, Hoppe AD, Swanson JA (2010) Coordination of Fc receptor signaling regulates cellular commitment to phagocytosis. Proc Natl Acad Sci U S A 107:19332–19337

Molecular Mechanisms of Calcium Signaling During Phagocytosis

7

Paula Nunes-Hasler ⓘ, Mayis Kaba, and Nicolas Demaurex ⓘ

Abstract

Calcium (Ca^{2+}) is a ubiquitous second messenger involved in the regulation of numerous cellular functions including vesicular trafficking, cytoskeletal rearrangements and gene transcription. Both global as well as localized Ca^{2+} signals occur during phagocytosis, although their functional impact on the phagocytic process has been debated. After nearly 40 years of research, a consensus may now be reached that although not strictly required, Ca^{2+} signals render phagocytic ingestion and phagosome maturation more efficient, and their manipulation make an attractive avenue for therapeutic interventions. In the last decade many efforts have been made to identify the channels and regulators involved in generating and shaping phagocytic Ca^{2+} signals. While molecules involved in store-operated calcium entry (SOCE) of the STIM and ORAI family have taken center stage, members of the canonical, melastatin, mucolipin and vanilloid transient receptor potential (TRP), as well as purinergic P2X receptor families are now recognized to play significant roles. In this chapter, we review the recent literature on research that has linked specific Ca^{2+}-permeable channels and regulators to phagocytic function. We highlight the fact that lipid mediators are emerging as important regulators of channel gating and that phagosomal ionic homeostasis and Ca^{2+} release also play essential parts. We predict that improved methodologies for measuring these factors will be critical for future advances in dissecting the intricate biology of this fascinating immune process.

Keywords

TRPC · TRPM · TRPML · TRPV · Phagocyte · Ion channels

Abbreviations

2-APB	2-Aminoethoxydiphenyl borate
5-BDBD	5-(3-Bromophenyl)-1,3-dihydro-2H-Benzofuro[3,2-e]-1,4-diazepin-2-one
ADP	adenosine diphosphate
ADPR	adenosine diphosphate ribose
AM	acetoxymethyl ester
AMTB	(N-(3-Aminopropyl)-2-[(3-methylphenyl) methoxy]-N-

P. Nunes-Hasler (✉)
Department of Pathology and Immunology, University of Geneva, Geneva, Switzerland
e-mail: paula.nunes@unige.ch

M. Kaba · N. Demaurex
Department of Cellular Physiology and Metabolism, University of Geneva, Geneva, Switzerland

© Springer Nature Switzerland AG 2020

M. B. Hallett (ed.), *Molecular and Cellular Biology of Phagocytosis*, Advances in Experimental Medicine and Biology 1246, https://doi.org/10.1007/978-3-030-40406-2_7

	(2-thienylmethyl)benzamide hydrochloride)	IgG	immunoglobulin G
ANO6	Anoctamin 6 (also called TMEM16F)	IP$_3$	inositol trisphosphate
		IP$_3$R	inositol trisphosphate receptor (also called ITPR)
ARC	arachidonate-regulated channel	ITAM	immunoreceptor tyrosine-based activation motifs
ASPH	aspartyl/asparaginyl beta- hydroxylase		
ATP	adenosine triphosphate	Kd	dissociation constant
BAPTA-AM	1,2-Bis(2-aminophenoxy)ethane-N,N,N′, N′-tetraacetic acid tetrakis(acetoxymethyl ester)	LPS	lipopolysaccharide
		LRC	leukotriene-regulated channel
		MCS	membrane contact site
		MCU	mitochondrial calcium uniporter
BST1	bone marrow stromal cell antigen 1	MEF	mouse embryonic fibroblast
		NAADP	nicotinic acid adenine dinucleotide phosphate
CAD	calcium release-activated channel activation domain		
		NAD	nicotinamide adenine dinucleotide
cADPR	cyclic adenosine diphosphate ribose	NADPH	nicotinamide adenine dinucleotide phosphate
Cav-1	caveolin-1		
CFTR	cystic fibrosis transmembrane conductance regulator	NCX	sodium calcium exchanger
		NFAT	nuclear factor of activated T cells
cPLA2	cytosolic phospholipase A2	NHERF1	Na$^+$/H$^+$ exchange regulatory cofactor 1 (also called SLC9A3R1)
CR	complement receptor		
CRACR2A	calcium release-activated channel regulator A (also called EFC4B)	NHERF2	Na$^+$/H$^+$ exchange regulatory cofactor 1 (also called SLC9A3R2)
CRACR2B	calcium release-activated channel regulator (also called EFC4A)	Nir	Pyk2 N-terminal domain-interacting receptor 1 (also called PITPNM)
Cx	connexins	NK1	neuropeptide tachykinin receptor (also called substance P receptor, or TACR1)
DAG	diacylglycerol		
DAMP	damage-associated molecular pattern		
		NO	nitrous oxide
DC	dendritic cell	NOX2	nicotinamide adenine dinucleotide phosphate oxidase 2
DNA	deoxyribonucleic acid		
ER	endoplasmic reticulum	OAG	1-oleoyl-2-acetyl-sn-glycerol
Esyt	extended synaptotagmin (also called FAM62)	ORP	oxysterol-binding protein-related protein (also called OSBPL)
Fc	fragment crystallizable	PA	phosphatidic acid
FCCP	carbonyl cyanide-4-(trifluoro methoxy) phenylhydrazone	Panx	pannexins
		PARP	poly (ADP-ribose) polymerase
FcγR	Fc gamma receptor	PC	phosphatidyl choline
FcγRIIA	Fc gamma receptor IIA (also called FCGR2A, or CD32)	PI(3,4,5)P2	phosphatidylinositol 3,4,5-tris phosphate
fMLP	N-formylmethionyl-leucyl-phenylalanine (also called fMLF)	PI(3,5)P2	phosphatidylinositol 3,5-bisphosphate
		PI(4,5)P2	phosphatidylinositol 4,5-bisphosphate
GPCR	G-protein coupled receptor		
HEK	human embryonic kidney cells	PI3K	phosphatidylinositol 3 kinase
HMOX-1	heme oxygenase 1 (also called HO-1)	PIRT	phosphoinositide interacting regulator of TRP

PKC	protein kinase C
PLCγ	phospholipase C gamma
PLD	phospholipase D
PMCA	plasma membrane calcium ATPase
POST	partner of stromal interaction molecule 1 (also called TMEM20 or SLC35G1)
RNF24	RING finger protein 24
ROS	reactive oxygen species
Rosco	(R)-roscovitine (also called Seliciclib)
RyR	ryanodine receptors
S1P	sphingosine-1-phosphate
SARAF	store-operated calcium entry associated factor (also called TMEM66)
SERCA	sarco/endoplasmic reticulum calcium ATPase
SESD1	SEC14 domain and spectrin repeat-containing protein 1
SK	sphingosine kinase
SNARE	soluble N-ethylmaleimide-sensitive-factor attachment protein receptors
SOAR	STIM1 Orai1-activating region
SOCE	store-operated calcium entry
Src	sarcoma kinase
STIM	stromal interaction molecule
Syk	spleen tyrosine kinase
SytVII	synaptotagmin VII
TMBIM1	transmembrane Bax inhibitor motif (also called LFG3, or RECS1)
TPC	two-pore channel
TRPC	transient receptor potential canonical
TRPM	transient receptor potential melastatin
TRPML	transient receptor potential mucolipin (also called MCOLN)
TRPP1	transient receptor potential polycystic 1 (also called PKD2)
TRPV	transient receptor potential vanilloid
VAMP7	vesicle-associated membrane protein 7 (also called synaptobrevin-like protein 1 SYBL1)
v-ATPase	vacuolar-type H^+-ATPase
VGCC	voltage-gated calcium channel

Introduction

Phagocytosis is a receptor-mediated cellular process essential for tissue homeostasis and critical for innate and acquired immunity. Following receptor engagement, apoptotic cells, microbes or other large foreign particles (>1 μm) are engulfed into a membrane-enclosed intracellular compartment called phagosome, generated through dynamic membrane and cytoskeletal remodeling. Phagosomes containing the internalized targets then undergo a series of maturation steps which collectively lead to changes in protein and lipids composition, fusion with components of the endocytic system and lysosomes, acidification, and production of reactive oxygen species (ROS), turning the vacuole into a degradative compartment (Vieira et al. 2002). The internalized material is either destroyed and its nutrients reutilized or expelled or, in the case of antigen presenting cells, processed material can be displayed at the cell surface to evoke further immune responses. In the worst case, certain intracellular pathogens can highjack phagosomes and turn them into a protected compartment that fosters pathogen proliferation. Thus, understanding the molecular signals and mechanism that govern phagocytosis does not only establish the basis of a fundamental cell biological process but is paramount for understanding the pathogenesis of certain infectious agents and for the development of more effective immunotherapies.

Calcium (Ca^{2+}) is a key second messenger involved in regulation of variety of cellular processes in immune cells such as migration, adhesion, cytokine production, secretion and gene transcription (Clapham 2007). That Ca^{2+} signals accompany phagocytosis has been observed for nearly 4 decades, indeed as soon as Ca^{2+} sensitive dyes became available, and it is now well documented that both global as well as localized, periphagosomal Ca^{2+} signals occur during phagocytosis (Nunes and Demaurex 2010;

Westman et al. 2019). However, the exact role that Ca^{2+} plays, and the requirement of Ca^{2+} for both ingestion and different maturation steps has been debated, and the literature is fraught with contradictory reports (Nunes and Demaurex 2010; Westman et al. 2019). As argued by Westman and colleagues (2019), perhaps one of the leading reasons behind the controversies is that phagocytosis is not a singular phenomenon, but rather a course of action that is recognizable morphologically but can be achieved by a variety of underlying molecular mechanisms. These may differ depending on cell type and the physiological context or status of the cell in question, that can influence which pathways are engaged and what signals are therefore required. Another reason lies in the caveats associated with the methodologies used to measure and manipulate Ca^{2+} signals, developed long before the identification of the channels and regulatory molecules involved in generating phagocytic Ca^{2+} signals. Only in the past decade were finer genetic tools added to existing chemical and pharmacological methods of measuring and buffering Ca^{2+} elevations that allow the identification of the molecular requirements of these Ca^{2+} signals. Perhaps a generalized consensus can now be made that while not strictly required under all circumstances, Ca^{2+} signals can make ingestion and maturation more efficient, and their importance is highlighted by pathogens that specifically silence these signals in order to promote their own survival (Nunes and Demaurex 2010). In the present chapter, we briefly discuss the Ca^{2+}-dependent processes mediating phagocytic ingestion and maturation, which have been extensively reviewed (Nunes and Demaurex 2010; Westman et al. 2019). We review in more detail the current knowledge on the molecular mechanisms that generate Ca^{2+} signals during phagocytosis, focusing on channels and regulators that have been identified over the past decade.

Overview of Ca^{2+} Signaling

While 98% of body Ca^{2+} is contained within the skeleton, extracellular Ca^{2+} concentrations are tightly maintained around 1.1–1.4 mM by the balance of parathyroid hormone, vitamin D and calcitonin and their actions on bone and kidneys (Hurwitz 1996). Intracellular Ca^{2+} concentrations are also tightly regulated, and cellular energy is spent on maintaining steep Ca^{2+} gradients in different intracellular compartments (Clapham 2007). Cytosolic Ca^{2+} is maintained around 10–100 nM by ATP or electrochemical gradient-driven pumps and exchangers such as PMCA and NCX that expel Ca^{2+} outside the cell or sequester it into intracellular Ca^{2+} stores. The endoplasmic reticulum (ER) is the largest and perhaps most important Ca^{2+} store, whose free Ca^{2+} levels are maintained around 0.4–1 mM. This is mainly achieved by ATP-driven SERCA pumps that counteract a constitutive Ca^{2+} leak that may occur through the translocon complex (Lang et al. 2011; Hammadi et al. 2013), presenilins (Bezprozvanny 2013), RyRs (Santulli et al. 2017) or potentially through TMBIM proteins (Liu 2017). Regulated ER Ca^{2+} release occurs mainly through IP_3R and RyRs (Santulli et al. 2017), but also through TRPP1 channels (Koulen et al. 2002), which are all ER Ca^{2+} channels that are also activated by cytosolic Ca^{2+} itself. In addition to IP_3-mediated IP_3R activation, cyclic adenosine diphosphate ribose (cADPR) as well as the lipid sphingosine-1-phosphate (S1P) can lead to ER Ca^{2+} release, through RyR activation or an unknown mechanism, respectively (Lee and Zhao 2019; Bootman et al. 2002; Guse and Wolf 2016). Acidic organelles such as late endosomes and lysosomes maintain Ca^{2+} levels between 10–500 μM (Yang et al. 2019) and are arguably the second most important Ca^{2+} store. The mechanisms by which lysosomes uptake Ca^{2+} are poorly understood and may involve Ca^{2+}/H^+ exchange (Christensen et al. 2002; Morgan et al. 2011), as well as transfer of Ca^{2+} directly from the ER (Penny et al. 2015; Kilpatrick et al. 2016). However, several regulated lysosomal Ca^{2+} release channels have been identified including TPC, TRPM and TRPML family channels, as well as TMBIM1 (Kilpatrick et al. 2017; Brailoiu et al. 2009; Lisak et al. 2016; Patel and Cai 2015). Mitochondria can take up Ca^{2+} from the cytosol or directly from the ER via the MCU channel and are thought to serve as cytosolic Ca^{2+} buffers rather than stores (Csordas

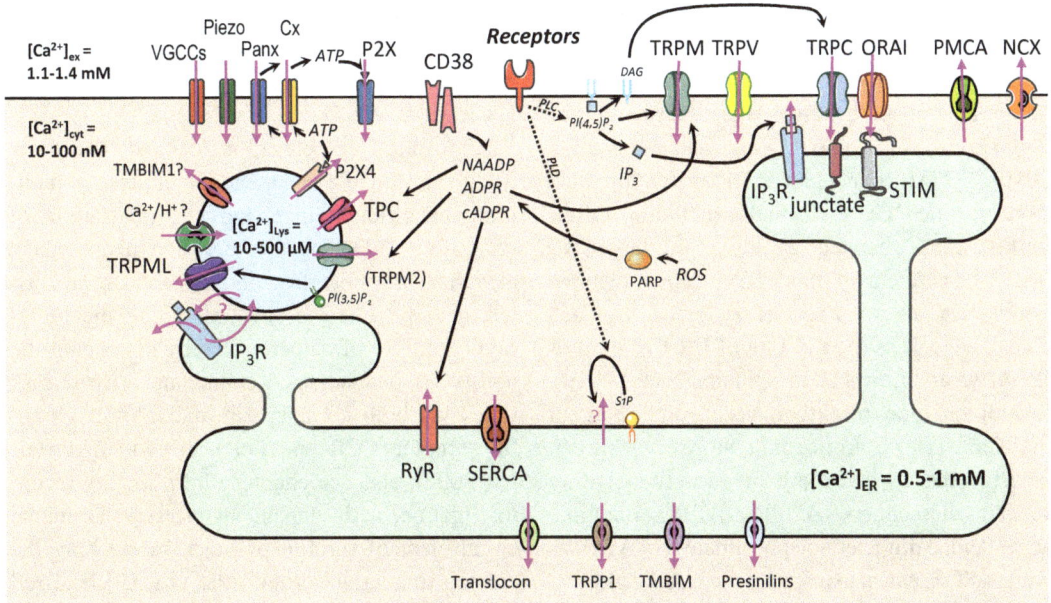

Fig. 7.1 Molecular mechanisms of Ca^{2+} signaling in phagocytes. Ca^{2+} is present at millimolar concentrations in the extracellular environment, is maintained at nanomolar concentrations in the cytosol, and is stored at micromolar concentrations in intracellular stores such as the ER and lysosomes. The pink arrows denote the direction of Ca^{2+} flux. Ca^{2+} is extruded via PMCA and NCX. Ca^{2+} is pumped into the ER via SERCA pumps and passively leaks via the translocon, presinilins, TRPP1 and possibly TMBIM proteins. Regulated ER Ca^{2+} release occurs via RyR, IP$_3$R, and unknown S1P-sensititive receptors. Ca^{2+} is imported into lysosomal stores possibly via Ca^{2+}/H^+ exchange and is released by TPC and TRPML family channels as well as by TRPM2, P2X4 and possibly TM-BIM1 proteins. Plasma membrane Ca^{2+} influx channels include members of the VGCC, Piezo, Panx, Cx, P2X, TRPM, TRPV, TRPC and ORAI families. TRPC and ORAI channels function in store-operated Ca^{2+} entry, driven by regulators in the ER including STIM, junctate and IP$_3$R channels. Metabolites that activate Ca^{2+} channels include ATP extruded via Cx and Panx that activates P2X receptors at the plasma membrane and lysosomes as well as PI(3,5) P2 that activates TRPML channels. CD38 receptors produce NAADP, ADPR and cADPR which activate TPC, TRPM2 and RyR receptors respectively. PARP enzymes also produce ADPR in response to ROS. A variety of PLC-dependent receptors lead to the cleavage of PI(4,5)P2 to produce IP$_3$ and DAG. DAG activates TRPC channels, and PI(4,5)P2 additionally regulates TRPM channels. PLD-dependent receptors can activate IP$_3$R-independent ER Ca^{2+} release by stimulating S1P

et al. 2012; De Stefani et al. 2016; Demaurex and Guido 2017). Golgi and nuclei also have elevated levels of Ca^{2+} (Pizzo et al. 2011) although their function as bona-fide stores or in regulated Ca^{2+} release has not been defined. (See Fig. 7.1).

Because of the large Ca^{2+} concentration gradient between the extracellular space or intracellular stores and the cytosol, Ca^{2+} signals occur as rapid, sharp increases in intracellular Ca^{2+} concentrations upon the opening of Ca^{2+}-permeable channels. In excitable cells such as neurons and muscles a variety of voltage-gated Ca^{2+} channels (VGCCs) mediate entry upon membrane depolarization. In non-excitable phagocytic cells however, despite expression of VGCCs, depolarization alone generates little Ca^{2+} currents and instead Ca^{2+} entry is thought to be dominated by a process termed store-operated Ca^{2+} entry (SOCE, also called capacitive Ca^{2+} entry) (Prakriya and Lewis 2015). In this process receptor activation first leads to the release of Ca^{2+} from the ER, often from IP$_3$R, that generates an initial Ca^{2+} spike. Since stores are finite reservoirs, extensive activation can lead to depletion of ER Ca^{2+} which in turn activates luminal ER Ca^{2+} sensors of the STIM family of proteins and junctate/ASPH. These transmembrane ER proteins remodel the ER, shaping and translocating to structures called membrane contact sites (MCS), which are sites of close (< 20 nm) apposition between membranes held together by protein tethers that allow protein-protein interactions and non-vesicular transport across the intermembrane gap (Prinz et al. 2019; Ong and Ambudkar 2019). STIM and junctate activate so-called store-operated Ca^{2+} channels, including ORAI family Ca^{2+} channels as the main target, as well as members of the TRPC cation channel family, leading to a secondary cytosolic Ca^{2+} signal, as well as the refilling of ER Ca^{2+} stores (Prakriya and Lewis 2015; Qiu and Lewis 2019). In addition, more recently it has become apparent that Ca^{2+} release from acidic stores, which can be triggered by a variety of stimuli including second messengers ADPR/cADPR, nicotinic acid adenine dinucleotide phosphate (NAADP) as well as ROS, can also instigate the release of Ca^{2+} from the ER via a Ca^{2+}-induced Ca^{2+} release mechanism dependent on IP_3R and ER-lysosome MCS (Kilpatrick et al. 2016), linking acidic store Ca^{2+} signaling to SOCE. Finally, phagocytes express a variety of ligand-gated and mechano- or temperature-sensitive Ca^{2+}-permeable channels including several other members of the TRP cation channel family, such as TRPV, TRPM and TRPML, nucleotide-activated P2X family cation channel receptors, pannexins (Panx), connexins (Cx), and Piezo proteins (Heng et al. 2008), which can in principle link a panoply of stimuli to the generation of initiating Ca^{2+} signals as well as SOCE.

Phagocytic Ca^{2+} Signals

Phagocytosis is initiated by binding of plasma membrane phagocytic receptors to corresponding ligands present on the particle surface, and an initial global Ca^{2+} signal is often observed upon particle binding. Indeed, imaging of neutrophils in zebrafish has now confirmed that a brief global Ca^{2+} spike occurs upon bacterial ingestion *in vivo* in this cell type (Beerman Rebecca et al. 2015). Thereafter, oscillatory global Ca^{2+} spikes as well as localized periphagosomal rings or hotspots that follow maturing phagosomes have been observed as early as cup-stage and as late as 90 min af-

ter ingestion (Nunes-Hasler and Demaurex 2017; Sawyer et al. 1985; Dewitt and Hallett 2002; Nunes et al. 2012; Guido et al. 2015; Nunes-Hasler et al. 2017) (See Fig. 7.2).

Cytosolic Ca^{2+} elevations enhance the activity of several Ca^{2+}-dependent targets to facilitate the efficient ingestion and maturation of phagosomes. During the ingestion phase, Ca^{2+} elevations promote the activation and lateral clustering of β2 integrins by unleashing the proteolytic activity of calpain, thereby enhancing the ability of phagocytes to bind and engulf particles via both complement and immunoglobulin receptors (CR and FcR) [(Dewitt and Hallett 2002); also see chapter 9]. Following receptor ligation, actin-dependent protrusions enable engulfment of foreign particles by pushing the plasma membrane around the target. Localized Ca^{2+} elevations then promote actin depolymerization at the base of the phagocytic cup, likely by enhancing the activity of the actin-severing protein gelsolin (Serrander et al. 2000; Larson et al. 2005). Actin dissolution at cups favors the docking of endosomes to the base of nascent phagosomes and allows particle internalization to proceed. Ca^{2+} elevations at entry sites then mediate the subsequent fusion of endosomes to the nascent phagosomes by activating a family of soluble N-ethylmaleimide-sensitive-factor attachment protein receptors (SNARE) complexes, the Ca^{2+}-dependent synaptotagmins (Syt) II, V, VII and XI (Lindmark et al. 2002; Vinet et al. 2008). Annexins, Ca^{2+} dependent lipid binding proteins that promote membrane aggregation and fusion (Gerke et al. 2005) have also been implicated in mediating particle recognition, binding and internalization (Yona et al. 2006; Yu et al. 2019). It should be noted however, that these Ca^{2+}-dependent steps are not necessarily limiting and can be bypassed, depending on the cell or phagocytic pathway engaged. For instance, FcR-mediated phagocytosis and actin dynamics were shown to be Ca^{2+} independent in murine macrophages (Greenberg et al. 1991; Di Virgilio et al. 1988; reviewed in Westman et al. 2019).

During the phagosome maturation phase, Ca^{2+} elevations also promote the tethering of adaptor proteins to endomembranes and their

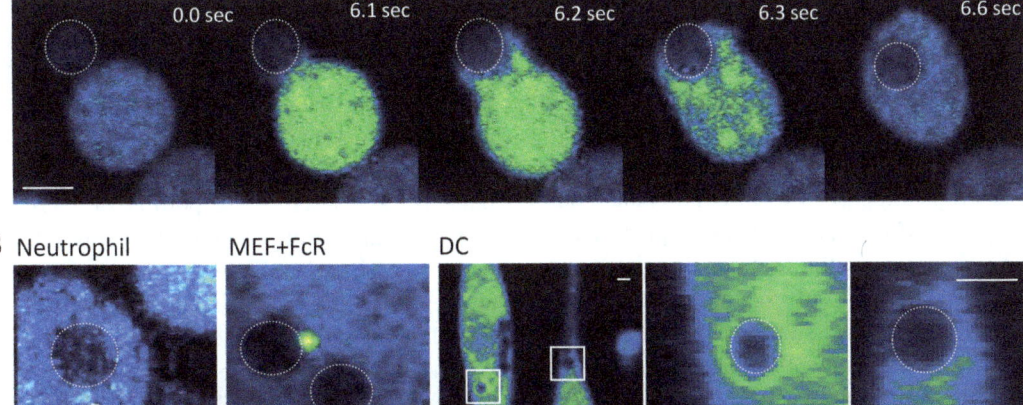

Fig. 7.2 Examples of global and local phagocytic Ca^{2+} signals. (**a**) Timelapse video of mouse neutrophils loaded with Fluo8 Ca^{2+}-sensitive dye show a rapid global signal upon ingestion of a phagocytic target. (**b**) The left panel shows a periphagosomal Ca^{2+} hotspot near a phagosome in a mouse neutrophil loaded with Fluo8. The middle panel shows a robust periphagosomal Ca^{2+} hotspot in phagocytic MEF cells expressing FcγRIIA-cmyc recep-tors, loaded with Fluo8 and 5 μM BAPTA-AM to help visualize hotspots. The right panels show ring and rod-shaped localized periphagosomal Ca^{2+} signals in DCs loaded with Fluo8 and 2.5 μM BAPTA-AM to help visualize hotspots. Phagocytic targets are outlined in a white dotted line. Fluo8 fluorescence is pseudocolored to highlight zones of high fluorescence in green/yellow, where low fluorescence is blue. White bars are 3 μm

subsequent fusion with phagosomes. These processes are mediated by the sequential activity of the prototypical Ca^{2+} transducer calmodulin, which promotes the docking of endosomes, while the fusion steps are driven by the Ca^{2+}-dependent activity of Syts and possibly by annexins as discussed above (reviewed in Westman et al. 2019). A Ca^{2+}-dependent increase in actin depolymerization rates also promotes phago-lysosome fusion, since the peri-phagosomal cortical actin ring needs to be dissolved to enable endolysosomes to access the phagosomal membrane (Nunes et al. 2012). Another important contribution of pro-phagocytic Ca^{2+} elevations is to promote the assembly and activation of the nicotinamide adenine dinucleotide phosphate (NADPH) oxidase NOX2 on phagosomes, leading to the production of large quantities of superoxide anions within the vacuole. Several Ca^{2+}-dependent targets contribute to NOX2 ac-tivation. Ca^{2+} elevations activate Ca^{2+}-sensitive isoforms of protein kinase C (PKCα and PKCβII) that phosphorylate the cytosolic subunits of the oxidase p47[phox] and p40[phox]. Concomitantly, the S100 Ca^{2+}-binding protein A8 and A9 are recruited to phagosomes and activate the membrane-bound flavocytochrome b₅₅₈ compo-nent of the oxidase (Brechard et al. 2008; Steinck-wich et al. 2011). Ca^{2+} elevations also activate cytosolic phospholipase A2 (cPLA2) to promote the release of arachidonic acid from membrane phospholipids. Arachidonic acid binds with high affinity to S100A8/A9 complexes, increasing NADPH oxidase activity and stimulating super-oxide anion production (Kerkhoff et al. 2005). Together, these Ca^{2+}-dependent steps favor the efficient assembly and activation of the NADPH oxidase on phagocytic vacuoles, and failure to generate local and global Ca^{2+} elevations leads to impaired intraphagosomal superoxide production (Zhang et al. 2014). Finally, the Ca^{2+} elevations occurring throughout the phagocytic process have long-term effects as the Ca^{2+}-dependent phosphatase calcineurin can promote the nuclear translocation of the transcriptional factor EB, thereby boosting lysosome biogenesis and cell metabolism towards a more efficient phagocytic phenotype (Gray et al. 2016).

The coordinated activation of all these sequential phagocytic steps requires precise control of local and global Ca^{2+} elevations in both time and space. Using a micromanipulation technique to present complement-opsonized zymosan particles to neutrophils, Dewitt et al. could resolve phagocytic Ca^{2+} signals into two separate temporal components, with a first signal initiated by integrin engagement at the point of particle contact and a second global signal occurring at the time of phagosomal closure that temporally correlated with the activation of NADPH oxidase (Dewitt and Hallett 2002). Based on these observations, they proposed that Ca^{2+} elevations control the timing of oxidase activation while subsequent localized events determine the site of oxidase activation. Subsequently, long-lasting localized Ca^{2+} elevations were observed in the vicinity of phagosomes that persisted for several minutes (Nunes et al. 2012), reviewed in (Demaurex and Nunes 2016). Of note, the presence of a periphagosomal cortical actin skeleton might contribute to the persistence of Ca^{2+} microdomains around phagosomes. Actin-based confinement of Ca^{2+} elevations has been described during Shigella invasion and attributed to diffusional hindrance of IP_3R accumulating at bacterially induced actin foci by a diffusion-capture process (Tran Van Nhieu et al. 2013). Whether a similar mechanism contributes to the long-lasting periphagosomal Ca^{2+} elevations remains to be tested, but the causal relationship between periphagosomal Ca^{2+} microdomains and actin severing has been established (Nunes et al. 2012). It is tempting to speculate that the localized Ca^{2+} microdomains initiated at STIM1-generated ER-phagosomal MCS (discussed further below) determine the position of the sites of phagosomal NADPH oxidase activity. The complexity and versatility of the Ca^{2+} signals associated with phagocytosis is fascinating and exemplifies how the spatiotemporal encoding of Ca^{2+} signals can differentially regulate a range of Ca^{2+}-dependent effector functions during phagosome ingestion and maturation.

A likely suspect for triggering the initial Ca^{2+} spike is the engagement of phagocytic receptors. A variety of phagocytic receptors have been identified to date and were reviewed recently (Freeman and Grinstein 2014). The most studied and perhaps best understood phagocytic receptors are of the Fc-gamma receptor family (FcγR) that bind to the conserved Fc portion of IgGs, as well as complement receptor 3 (CR3) which are $\alpha_M\beta_2$ integrins that binds to the complement molecule C3bi. Three classes of FcγR; FcγRI, FcγRII and FcγRIII, have been described each of which has several isoforms that are differentially expressed in different cell types, however the initial signal transduction events are fairly similar between different isoforms (Nunes and Demaurex 2010; Freeman and Grinstein 2014). Subsequent to particle binding, FcRs cluster, which leads to phosphorylation of tyrosine residues present in the ITAM (immunoreceptor tyrosine-based activation motifs) by Src family kinases. Phoshorylated residues in the ITAM serve as docking sites for many proteins including spleen tyrosine kinase (Syk) family kinases and PI3K, which in turn have numerous effectors, that are again cell-type dependent. A notable Syk effector is PLCγ, a phospholipase that hydrolyzes PI(4,5)P2 into diacylglycerol (DAG) and inositol trisphosphate (IP_3) which can then diffuse and activate IP_3R on the ER thus resulting in Ca^{2+} release from stores. Sufficient IP_3R-mediated release can then trigger SOCE. Interestingly, FcR engagement in certain contexts does not lead to IP_3 production and has instead been linked to the activation of PLD, a phospholipase that cleaves phosphatidyl choline (PC) into choline and phosphatidic acid (PA) (Nunes and Demaurex 2010). PA in turn activates sphingosine kinase (SK), generating S1P, which induces ER-Ca^{2+} release through an unknown channel, as well as activation of TRPC5 (Nunes and Demaurex 2010; Xu et al. 2006; Taha et al. 2006). CR3 signal transduction also occurs via Src, Syk and PI3K, although Ca^{2+} transients resulting from CR3 ligation more likely arises as the result of the activation of the PLD pathway, rather than via IP_3-mediated ER Ca^{2+} release (Nunes and Demaurex 2010).

In addition to phagocytic receptors, phagocytic Ca^{2+} signals could potentially be triggered by the activation of CD38 or its close homolog CD157/BST1. CD38 and CD157 are receptors and multifunctional enzymes that are expressed in multiple cell types including phagocytic immune cells, and that are upregulated when cells are activated, for example by bacterial products such as lipopolysaccharide (LPS) and formylated peptides (Quarona et al. 2013). CD38 is expressed at the plasma membrane in both type II and type III topological configurations such that its catalytic domain is present both outside and inside the cell (Lee and Zhao 2019). CD38 and CD157 are enzymes of the β-NAD^+-metabolizing ADP-ribosyl cyclase family, and can generate cADPR, ADPR as well NAADP which are all Ca^{2+} second messengers as mentioned above. ADPR can also be generated by poly (ADP-ribose) polymerase (PARP) which is activated by ROS and DNA damage (Miller 2006). cADPR activates RyR mediating ER-Ca^{2+} release, ADPR activates TRPM2 channels and NAADP activates TPC channels. Both TPCs and TRPM2 can mediate lysosomal Ca^{2+} release and can therefore potentially trigger Ca^{2+}-induced Ca^{2+} release from the ER as well as influx from the PM either indirectly by activating SOCE or directly as in the case of TRPM2 (Galione and Chuang 2020). Indeed, CD38 has been shown to contribute to N-formylmethionyl-leucyl-phenylalanine (fMLP)-induced Ca^{2+} signaling in neutrophils and to promote phagocytosis and bacterial clearance in macrophages and neutrophils (Partida-Sanchez et al. 2003; Lischke et al. 2013). However, the contribution of CD38/CD157 to phagocytic Ca^{2+} signals have not yet been assessed directly.

Finally, one can also consider that ROS production, extracellular ATP release, mechanical stimulation and temperature changes during fever are all additional stimuli that may occur concomitantly with phagocytosis and could contribute to triggering phagocytic Ca^{2+} signals in a physiological context. Over the past 10 years a great deal of progress has been made in identifying the channels and regulators involved in generating Ca^{2+} signals during phagocytosis and we review

major findings on this topic in the next section (see summary in Fig. 7.3).

Ca^{2+}-Permeable Channels Involved in Phagocytic Ca^{2+} Signals

ORAI Channels, STIM Proteins and Junctate

Since SOCE is a primary Ca^{2+} influx pathway in non-excitable cells, and release of Ca^{2+} from the ER is a consequence of phagocytic receptor engagement and cellular activation, it is not surprising that after the identification of the core molecular machinery underlying SOCE in 2005–2006, numerous studies have focused on deciphering the role of ORAI channels and STIM proteins, the principle actors in this process, in phagocytic cells. In mammalians there are 3 ORAI isoforms, ORAI1, ORAI2 and ORAI3, that are highly Ca^{2+}-selective channels primarily resident on the plasma membrane (PM) gated by a direct interaction of both N- and C- cytosolic termini with STIM proteins at MCS (Trebak and Putney 2017). ORAI1 can also assemble into heteromers with ORAI3, which then form channels that are activated independently of store depletion by arachidonate and leukotrienes, and are therefore termed ARC or LRC for arachidonate-regulated or leukotriene-regulated channel respectively (Zhang et al. 2018). There are two STIM isoforms, STIM1 and STIM2, each exhibiting different alternatively spliced variants. They are single pass transmembrane proteins residing primarily in the ER, which have a luminal EF-hand Ca^{2+}-sensing domain and a cytosolic domain containing the CAD/SOAR (calcium release-activated channel activation domain/ STIM1 Orai1-activating region) domain that unfolds upon activation to interacts with ORAI channels across the MCS gap (Prakriya and Lewis 2015). ORAI1 is the dominant isoform mediating SOCE, forming homomeric hexamers that interact chiefly with STIM1. ER Ca^{2+} depletion is the primary trigger for STIM activation, oligomerization, translocation to MCS and ORAI gating, but both STIM and ORAI

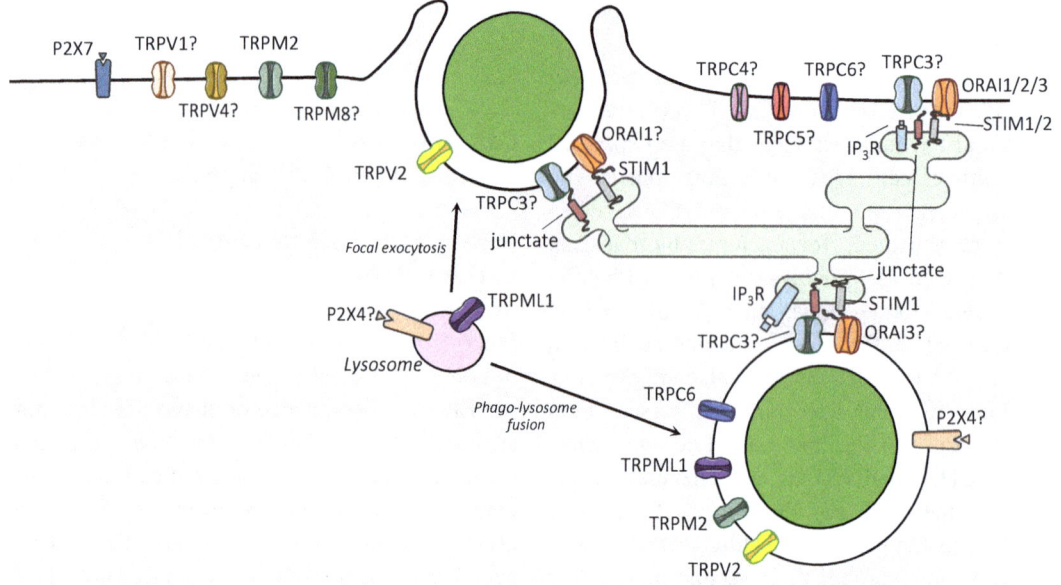

Fig. 7.3 Ca^{2+} channels potentially mediating phagocytic Ca^{2+} signals. The plasma membrane SOCE channels ORAI1/2/3 activated by STIM1/2 and junctate mediate global phagocytic Ca^{2+} signals. STIM1 and junctate also regulate localized phagocytic Ca^{2+} signals at membrane contact sites between the ER and nascent or mature phagosomes. ORAI1 may be more important at phagocytic cups while ORAI3 may be more important in mature phagosomes. TRPC channels that associate with STIM/ORAI complexes such as TRPC3/4/5/6 may additionally contribute either as SOCE channels or independently. ER-Ca^{2+} release additionally plays an important role in both global and local phagocytic Ca^{2+} signals. TRPV2 mediate Ca^{2+} signals during cup formation. Additionally, P2X7, TRPV1, TRPV4, TRPM2 and TRPM8 may mediate Ca^{2+} influx at the plasma membrane during phagocytosis. Lysosomal Ca^{2+} release via TRPML1 and possibly P2X4 promotes focal exocytosis of lysosomes and phago-lysosome fusion. TRPC6, TRPM2, TRPML1, TRPV2 and possibly P2X4 localize to phagosomes and may mediate phagosomal Ca^{2+} extrusion

activity are also regulated by phosphorylation, cysteine modification by ROS, as well as interaction with several accessory proteins including CRACR2A and CRACR2B in the case of ORAI, and junctate, junctophilin-4, POST, Surf4, SARAF, and UNC93B1 (Prakriya and Lewis 2015; Srikanth et al. 2013; Fujii et al. 2012; Maschalidi et al. 2017) in the case of STIM. STIM proteins can also activate members of the TRPC family of channels, including TRPC1, 3, 4, 5 and 6, although this activity is often indirect or dependent on ORAI1 (Curcic et al. 2019; Bavencoffe et al. 2017). In the absence of both STIM proteins junctate, which is also an ER transmembrane protein that has a luminal Ca^{2+}-sensing domain, and that has been reported to interact with TRPC2, TRPC3 and TRPC5, can also independently mediate SOCE (Guido et al. 2015; Treves et al. 2004, 2010; Stamboulian et al. 2005).

Essentially all studies investigating the impact of STIM1 or ORAI1 ablation or downregulation on global SOCE signals in phagocytes showed reduced signaling, although to varying extents from moderate to strong, while interference with STIM2, ORAI2 or ORAI3 expression had much milder effects as might be expected (reviewed recently Demaurex and Nunes 2016). Nevertheless, downregulation of STIM1 alone was sufficient to decrease phagocytosis or phagosome maturation in several reports (Nunes et al. 2012; Nunes-Hasler et al. 2017; Zhang et al. 2014; Clemens et al. 2017; Sogkas et al. 2015a, b; Lim et al. 2017; Michaelis et al. 2015; Heo et al. 2015), suggesting that the STIM/ORAI system may be a good molecular target to modulate phagocytosis. However, not all studies found phagocytic defects upon interference with STIM/ORAI (Elling et al. 2016; Vaeth et al.

2015), and in the only report that examined STIM1/STIM2 double-knock out macrophages and dendritic cells (DCs), phagocytic ingestion was surprisingly intact, although phagosomal maturation was not extensively explored (Vaeth et al. 2015). Interestingly, we found that STIM1 knock-out was sufficient to significantly reduce localized periphagosomal Ca^{2+} hotspots in neutrophils, phagocytic fibroblasts and DCs (Nunes et al. 2012; Nunes-Hasler et al. 2017), whereas STIM1/STIM2 double knock-out did not further decrease these signals in phagocytic fibroblasts (Guido et al. 2015). Notably, although global Ca^{2+} influx was reduced only by 20% in this neutrophil $stim1^{-/-}$ conditional knock-out model, localized signals were hampered by more than 50%. The fact that this was sufficient to impair phagocytic function highlights the importance of localized phagocytic Ca^{2+} signals. Altogether, these results indicate that the unique character of STIM/ORAI gating, which can only happen at particular sites where two membranes come together, provides a platform for restricting signaling to this particular periphagosomal location within the cell. Having said that, although overexpressed ORAI channels were observed to localize to phagosomes in an isoform-specific manner, with ORAI1 preferentially accumulating at cups and ORAI3 at mature phagosomes (Nunes et al. 2012); and the specific ORAI blocker GSK7975A reduced periphagosomal hotspots (Nunes-Hasler et al. 2017), ORAI function was not directly genetically assessed. Nevertheless, using specific (GSK7975A) and non-specific (La^{3+}, 2-APB) channel inhibitors our results suggested that both the opening of Ca^{2+} channels on the phagosomal membrane and ER Ca^{2+} release contributed more or less equally to these localized signals.

It should be noted that although SOCE is undoubtedly important for phagocytic function and STIM/ORAI downregulation can have a major impact on limiting Ca^{2+} entry after store depletion, Ca^{2+} signals arising from store release are intact, and cells can still generate global Ca^{2+} oscillations by release and uptake into stores. Indeed, our finding that overexpression of junctate could completely rescue phagocytic defects in STIM-depleted cells (Guido et al. 2015) exemplifies how increased ER Ca^{2+} release can easily compensate for the loss of SOCE. Moreover, the fact that double knock-out of STIM proteins was not sufficient to abrogate phagocytic Ca^{2+} signals suggests that other channels influence these signals.

TRPC Channels

Canonical members of the TRP ion channel family, consisting of TRPC1-7, where TRPC2 is a pseudogene in humans, are homo- and hetero-tetrameric non-selective cation channels and thus mediate entry of both Ca^{2+} and Na^+ upon activation with differing permeabilities depending on the channel composition (Curcic et al. 2019). TRPC channels are multimodal and all can also be activated by different stimuli independently of store depletion (Curcic et al. 2019; DeHaven et al. 2009). However, most have been implicated in participating and shaping SOCE, where TRPC1,2,4,5 and 6 can bind directly to STIM1 (Worley et al. 2007; Jardin et al. 2009). In addition, it has now been shown that all TRPC can be activated by DAG the other biproduct of PLC activation in addition to IP_3 (Curcic et al. 2019; Mederos et al. 2018). Thus, all TRPC family members could conceivably contribute to phagocytic Ca^{2+} signals originating from either PLC-dependent receptor activation or store depletion, and recently evidence for the participation of TRPC3,4,5 and 6 have emerged as detailed below.

TRPC3

In addition to direct activation by DAG, TRPC3 is gated by its interaction with junctate as well as with IP_3R receptors, and it exhibits store-dependent translocation to the plasma membrane that is also regulated by Cav-1, Homer1, VAMP2 and RNF24 (Treves et al. 2004, 2010; Kiselyov et al. 1998; Kim et al. 2006; de Souza and Ambudkar 2014). Indeed, TRPC3 has been reported to control Ca^{2+} waves in response to FcE receptor stimulation in mast cells (Cohen et al. 2009), and to contribute to sustained cytosolic

Ca^{2+} signals in phagocytic microglia (Mizoguchi et al. 2014). Moreover, it was recently shown that phagocytosis of apoptotic cells was reduced by 50% in $trpc3^{-/-}$ macrophages, highlighting the importance of this channel for phagocytic function in this context (Tano et al. 2011). It will be interesting to determine in the future exactly how TRPC3 participates in generating phagocytic Ca^{2+} signals: whether its activation is store-dependent or mediated by DAG; if store-dependent activation contributes, then whether this depends on STIM proteins or if alternative activation via junctate/IP$_3$R interactions occurs; and finally, whether it mediates STIM-independent localized phagocytic Ca^{2+} signals.

TRPC4/5

TRPC4 and 5 are more closely related than other TRPC isoforms, sharing 65% homology and they form heteotetramers with each other as well as with TRPC1 (Fu et al. 2015). TRCP4 is expressed as two isoforms α and β and the α isoform interacts with IP$_3$R (Fu et al. 2015; Mery et al. 2001). TRPC4 has many interacting partners and multiple factors including Ca^{2+} itself, calmodulin, IP$_3$R, PI(4,5)P2, Gαi as well as SESTD1 have been implicated in channel gating (Fu et al. 2015). TRPC5 is equally multimodal, with a large variety of activators including S1P, reduced thioredoxin, as well as Ca^{2+}, among others, in addition to store-operated gating via direct binding to STIM1 (Mederos et al. 2018; Zholos 2014). It is interesting to note that PI(4,5)P2 inhibits TRPC4α whereas the β isoform requires this phospholipid for activation, while both positive and negative roles for PI(4,5)P2 were reported for TRPC5 (Otsuguro et al. 2008; Trebak et al. 2009). In addition, TRPC4 and 5 were initially thought to be either DAG-insensitive or inhibited by DAG, however it was recently discovered that DAG sensitivity is blocked under resting conditions by TRPC4/5 association with the PDZ scaffolding proteins NHERF1/2 (Mederos et al. 2018). Whether TRPC4/5 regulates phagosomal Ca^{2+} signals remains to be determined, but the complex regulation of TRPC4/5 activation by PLC activity may explain why TRPC4/5 activation by thioredoxin correlated with reduced

phagocytic activity in macrophages in a recent report (Pereira et al. 2018).

TRPC6

TRPC6 has 70–80% identity with TRPC3 and TRPC7 and can form heterotetramers with these closer homologs as well as with TRPC1 (Bouron et al. 2016). As mentioned above, TRPC6 can be directly activated by DAG as well as DAG analogs, and has been classically known as operating downstream of PLC-dependent G-protein coupled receptors (GPCRs). TRPC6 is also regulated by direct binding to phosphoinositides, with PI(3,4,5)P2 promoting activation, whereas similar to TRPC5 both positive and negative regulation by PI(4,5)P2 have been reported (Bouron et al. 2016). TRPC6 can bind directly to ORAI channels and ORAI/STIM complexes, and operate as both a store-dependent and store-independent channel depending on its interaction partners and cellular context (Brechard et al. 2008; Jardin et al. 2009; Bouron et al. 2016). In addition, mechanical stretch, ROS, and Ca^{2+} itself among other stimuli, can activate the channel (Bouron et al. 2016). TRPC6 has a wide variety of interaction partners, a major class of which are molecules involved in trafficking, and plasma membrane insertion from intracellular vesicles or sequestration into the ER can influence TRPC6 activity (Bouron et al. 2016).

Interestingly Riazanski and colleagues recently showed that TRPC6 is directly inserted into the phagosomal membrane and promotes phagosomal acidification and bacterial killing in alveolar macrophages (Riazanski et al. 2015). In this study, the authors utilize a pharmacological GPCR activator called (R)-roscovitine (Rosco) as well as DAG analog OAG to activate currents and Ca^{2+} oscillations that were absent in $trpc6^{-/-}$ cells, although the impact of TRPC6 deletion on cytosolic Ca^{2+} signals during phagocytic ingestion was not assessed. The same TRPC6 activators as well as classic SOCE inducer fMLP and Ca^{2+} ionophore A23187 increased TRPC6 trafficking to the plasma membrane, and Rosco and OAG increased phagosomal TRPC6 expression. Intriguingly, the authors propose that TRPC6 mediates Ca^{2+} efflux from phagosomes and that this facilitates v-ATPase-mediated H^+ accumulation by providing charge compensation. In support of this model, the authors found that TRPC6 activation rescues acidification deficiency in CFTR-depleted cells, and that conversely CFTR inhibition in TRPC6-null cells aggravates acidification defects, demonstrating that, consistent with a charge compensation function, TRPC6 and CFTR have overlapping roles in promoting phagosomal acidification. To measure Ca^{2+} efflux the authors allowed phagocytosis to occur in the presence of the low-affinity Ca^{2+} sensitive dye calcein as well as the proton ionophore FCCP and observed a decrease in calcein fluorescence in response to Rosco and OAG that was absent in TRPC6-null cells. However, it is unclear why depleting the proton gradient would be necessary to observe Ca^{2+} efflux, and since no calibration was provided it is uncertain whether changes in Ca^{2+} or the loss of dye were being observed. Whether phagosomal Ca^{2+} efflux can be confirmed by more precise methods, whether Na^+ efflux plays a role and whether TRPC6 influences global and localized phagocytic Ca^{2+} signals are all interesting questions raised by this study that merit investigation.

TRPM Channels

TRPM2

The functions of members of the melastatin branch of TRP ion channels are perhaps a little less well understood than those of TRPC channels. There are 8 members, TRPM1-8 where the function of TRPM2 is the most well-studied (Harteneck 2005; Syed Mortadza et al. 2015; Venkatachalam and Montell 2007). TRPM2, (formerly named LTRPC2 and TRPC7) is a Na^+ and Ca^{2+} permeable cation channel that additionally has an ADP-ribose pyrophosphatase at its C-terminus. While TRPM2 cleaves ADPR, it is also activated by the nucleotide, and notably it is potently activated by ROS, arachidonic acid as well as intracellular Ca^{2+} itself (Syed Mortadza et al. 2015; Venkatachalam and Montell 2007). TRPM2 is highly expressed

in phagocytic immune cells, mainly at the plasma membrane but also in lysosomes in certain cell types, and has been linked to a variety of immune functions including adhesion, migration, inflammasome activation and cytokine production (Syed Mortadza et al. 2015). In a pioneering study by Di and colleagues, Na^+ flux through TRPM2 was shown to participate in a negative feedback loop where NADPH oxidase-induced ROS activates the channel, which in turn dampens ROS production by opposing charge compensation (Di et al. 2011). Notably, it was then demonstrated by the same group that, similar to TRPC6, TRPM2 activated bactericidal activity *in vitro* and *in vivo* as well as phagosomal acidification in macrophages (Di et al. 2017), which was proposed to act, again, via charge compensation. Here, the authors confirmed that TRPM2 localizes to the phagosomal membrane, and elegantly showed TRPM2-dependent currents activated by H_2O_2 and ADPR on isolated phagosomes. TRPM2 was not present in lysosomes and had no effect on phago-lysosome fusion, but the authors detected a higher fluorescence of the Ca^{2+} dye Fluo-3 within phagosomes, concluding that higher intraphagosomal Ca^{2+} levels were detected in TRPM2-depleted cells. However, it is unclear how the membrane permeant Fluo3-AM accumulated within phagosomes, as this is not observed with other AM dyes. Moreover, since Fluo3 is not ratiometric it is difficult to distinguish whether the higher dye fluorescence is due to changes in Ca^{2+} concentration or to dye accumulation. In addition, Fluo3, like all Ca^{2+}-sensitive dyes, is also pH sensitive, and pH-related changes to the dye's Kd were not considered. The defect in bacterial killing of $trpm2^{-/-}$ macrophages was subsequently confirmed by two other groups. In one case TRPM2 deletion induced susceptibility to *Listeria monocytogenes* infection *in vivo* (Knowles et al. 2011). A second study confirmed that TRPM2 deletion exacerbated polymicrobial sepsis as in (Di et al. 2017), and showed that while phagocytic ingestion was equivalent, LPS-induced global Ca^{2+} signals and bacterial killing were specifically impaired in TRPM2-deficient macrophages (Qian et al. 2014). The authors of this latter report found that treatment with hemin, a drug that increased expression of the Ca^{2+}-induced cytoprotective/antioxidant enzyme heme oxygenase 1 (HMOX-1), was sufficient to restore bacterial killing to TRPM2-null macrophages, although effects of hemin on phagosomal pH were not assessed. Finally, an intriguing study showed that H_2O_2 mediated oxidation of the Met214 residue renders TRPM2 sensitive to heat, and that TRPM2 is involved in increasing the phagocytic activity of macrophages during fever, but whether this activity is Ca^{2+}-dependent was not assessed (Kashio et al. 2012).

Whereas macrophage phagosomes acidify rapidly, neutrophil phagosomes remain neutral or alkaline for the first 30 min after phagocytic ingestion because of the much higher levels of NADPH oxidase activity, which produces superoxide molecules that not only neutralize phagosomal protons, but also inhibit V-ATPase (Nunes et al. 2013; Foote et al. 2019). Thus, it might be expected that TRPM2-mediated phagosomal charge compensation might be less relevant during high levels of NADPH activity. Indeed, TRPM2-deletion in neutrophils did not affect phagosomal pH or ROS production in response to a phagocytic stimulus (Foote et al. 2017), at least not during the first 20 min after ingestion. Whether later time points, when neutrophil phagosomes finally acidify, may be affected, and more generally whether TRPM2 contributes to either global or local cytosolic phagocytic Ca^{2+} signals remain to be defined.

TRPM8

TRPM8 channels are non-selective plasma membrane cation channels with a preference for Ca^{2+} permeation that are cold-sensitive, pressure-sensitive, and activated by ingredients of the mint family such as menthol and eucalyptol (1,8-cineol) (Gonzalez-Muniz et al. 2019). In addition, TRPM8 channels are positively modulated by PI(4,5)P2, a regulatory protein called PIRT, and are directly activated by testosterone (Gonzalez-Muniz et al. 2019). Although most-studied in the nervous system, the effectiveness of cold therapy on inflammation led Khalil and colleagues

to investigate the role of TRPM8 function in macrophages (Khalil et al. 2016). Agonist activation of TRPM8 with menthol enhanced whereas genetic deletion or pharmacological inhibition with AMTB reduced phagocytosis of beads and bacteria. Phagocytic Ca^{2+} signals were not studied, and the physiological stimuli of this intriguing channel during phagocytosis *in vivo* remain to be elucidated.

TRPML Channels

The mucolipin subfamily of TRP channels is comprised of 3 members TRPML1-3 that are non-selective cation channels permeable to Ca^{2+}, Na^+, Fe^{2+} and Zn^{2+} ions that are expressed mainly in the endolysosome system (Venkatachalam and Montell 2007; Venkatachalam et al. 2015). TRPML1 (also called MCOLN1) is the most well-studied isoform as mutations in this gene cause mucolipidosis IV, a lysosomal storage disease characterized by neurodegeneration. TRPML1 localizes to late endosomes and lysosomes, is activated by the endosomal phosphoinositide PI(3,5)P2, and has a biphasic response to luminal pH with maximal gating occurring around pH 4.5 (Venkatachalam et al. 2015). TRPML1 can traffic to the plasma membrane in response to cytosolic Ca^{2+} elevations, but it may not be functional there as it is inhibited by PI(4,5)P2 (Venkatachalam et al. 2015). In a seminal study, Samie and colleagues observed that TRPML1 promotes phagocytic ingestion of large targets (>5 μm) in macrophages (Samie et al. 2013). Phagocytic ingestion of large but not small targets was impaired in $trpml1^{-/-}$ macrophages, and treatment with a TRPML1 agonist ML-SA1 improved large target phagocytosis while treatment with its inhibitor ML-SAI1 has the opposite effect. The authors observed early focal exocytosis of lysosomes during phagocytic cup formation of large targets that was dependent on Ca^{2+}, SytVII, and VAMP7. In addition, TRPML1 currents were measured by patch clamp on isolated phagosomes and Ca^{2+} efflux from lysosomal TRPML1 was detected using the genetically encoded probe

GCaMP3 fused directly to TRPML1's cytosolic N-terminus. Overall, the data suggest that localized TRPML1-dependent lysosomal Ca^{2+} efflux promotes lysosomal fusion to nascent phagosomes. Indeed, this model was supported by a subsequent study which showed that for smaller targets TRPML1 expression promotes instead phagosome maturation and bacterial killing in macrophages, since phago-lysosome fusion occurs at later time points when smaller targets are employed (Dayam et al. 2015). The same group showed that TRPML1 contributes to phagosome maturation in neutrophils as well (Dayam et al. 2017). Thus, phago-lysosome fusion occurs at early and late time-points for large and small targets respectively, and TRPML1 is important in both cases. In the macrophage study, TRPML1-deficient lysosomes docked but did not fuse with phagosomes, and a short treatment with the Ca^{2+} ionophore ionomycin rescued TRPML1-dependent phagocytic defects. To image Ca^{2+} dynamics the authors employed Fluo4-AM, a single-wavelength Ca^{2+}-sensitive dye that showed cytosolic loading as well as compartmentalization in structures that overlapped lysosomal staining. Interestingly, whereas the authors failed to observe localized Ca^{2+} hotspots during phagocytosis, they detected an overall increase in cytosolic Ca^{2+} that was concomitant with an overall decrease in the compartmentalized fluorescence, which is consistent with TRPLM1-mediated lysosomal Ca^{2+} release. However, since the probe is single-wavelength and compartmentalized, it is difficult to distinguish between Ca^{2+}-dependent and loading-dependent changes in fluorescence. Moreover, as visualization of cytosolic Ca^{2+} hotspots requires loading of small amounts of Ca^{2+} chelators depending on cell type (Nunes et al. 2012; Nunes-Hasler et al. 2017; Luik et al. 2006), these results do not rule out the contribution of TRPML1 to localized periphagosomal Ca^{2+} dynamics. Since Ca^{2+} efflux through TRPML1 can trigger ER Ca^{2+} release and SOCE (Kilpatrick et al. 2016), it will be interesting to elucidate the contribution of TRPML1 to both global and localized phagocytic Ca^{2+} signals. Finally, it is notable that loss of TRPML1 expression leads to an increase in

lysosomal acidification (Soyombo et al. 2006), and future investigations determining whether TRPML1 also affects phagosomal pH, and how this ties in with charge compensation mediated by TRPC6 or TRPM2 will be important to form a comprehensive understanding of phagosomal ionic homeostasis and signaling.

TRPV Channels

TRPV1 and TRPV2

There are 6 members in the vanilloid family of TRP channels, TRPV1-6. TRPV1 is by far the most well-studied, as it is a non-selective cation channel activated by heat, protons, NO, and the "heat" ingredient in hot chilies capsaicin, that is involved in pain sensation, thermoregulation, osmoregulation, as well as inflammation, although its role in the latter is not completely understood (Samanta et al. 2018; Bujak et al. 2019; Sakaguchi and Mori 2019). TRPV1 deletion or pharmacological inhibition using SB366791 reduced phagocytosis of LPS-stimulated but not unstimulated macrophages (Fernandes et al. 2012). This effect was proposed to occur via TRPV1-dependent release of neuropeptide substance P and activation of its receptor NK1, a GPCR itself capable of stimulating PLC-dependent Ca^{2+} influx (Garcia-Recio and Gascon 2015). While Ca^{2+} signaling was not analyzed in this study, a recent report that TRPV1 as well as TRPV2 mediates Ca^{2+}-dependent activation of phagocytosis in microglia stimulated with cannabidiol (Hassan et al. 2014) suggests that Ca^{2+} influx via TRPV1 may contribute to the modulation of phagocytic ingestion. Indeed, the observation that TRPV1-mediated Ca^{2+} influx induced calpain activation in response to NO in macrophages (Zhao et al. 2016) suggests that further investigation on the role of TRPV1 in phagocytic Ca^{2+} signals may be fruitful.

On the other hand, a number of reports have now more convincingly linked TRPV2 activity to phagocytic function (Santoni et al. 2018). Despite its similarity to TRPV1 and a similar cation conductivity and function in pain, TRPV2 is not activated by vanilloids such as capsacin, and is only activated by extreme levels of heat that may not be physiologically relevant (Shibasaki 2016). TRPV2 channels are instead mechano-sensitive, osmo-sensitive, and activated by lysophosphatidylcholine, PI3K, cannabinoids, as well as translocation to the plasma membrane (Samanta et al. 2018; Shibasaki 2016; Penna et al. 2006; Muller et al. 2018). A pioneering study by Link and colleagues showed that TRPV2-deficient mouse macrophages showed a profound impairment in binding and ingestion of diverse phagocytic targets (Link et al. 2010). TRPV2 was observed to accumulate at phagocytic cups in a PI3K-, Src-, Akt-and PKC-dependent manner, but independently of Syk and PLC. Ca^{2+} currents in response to tetrahydrocannabinol were diminished, but not in response to the prototypical SOCE inducers ATP and fMLF. Congruent with these observations, it was TRPV2's role in membrane depolarization rather than Ca^{2+} influx that was linked to promoting phagocytic receptor clustering and particle binding, and depolarization was sufficient to overcome these defects in $trpv2^{-/-}$ cells. A more recent study confirmed reduction in phagocytosis in alveolar macrophages where TRPV2 was either deleted or its expression was reduced by exposure to cigarette smoke (Masubuchi et al. 2019). A subsequent report proposed that Ca^{2+} influx was indeed important for the effect of TRPV2 on phagocytosis in human macrophages, as they correlated phagocytic ingestion of *Pseudomonas aeruginosa* with a phagocytic Ca^{2+} signal that were both reduced by the selective TRPV2 inhibitor tranilast (Leveque et al. 2018). While the authors did not distinguish between the effects of Ca^{2+} and Na^+ entry, they did report an interesting increase in TRPV2 both at plasma membrane as well as at the phagosomal membrane during *P. aeruginosa* infection. Thus, whether TRPV2 contributes to phagosomal ion homeostasis or Ca^{2+} signaling remains an open question.

TRPV4

TRPV4 is a mechano-, osmo- and heat-sensitive-non-selective cation channel that is activated by PLA2 metabolites arachidonic acid and epoxyeicosatrienoic acids, it is sensitized by IP_3R and additionally regulated by calmodulin and Ca^{2+} itself (Fernandes et al. 2008; Verma et al. 2010; Garcia-Elias et al. 2014). TRPV4 function has been linked with numerous inflammatory conditions (Dutta et al. 2019) and was reported to contribute to ROS production in response to stretch in alveolar macrophages (Hamanaka et al. 2010). Recently, TRPV4 expression was shown to promote LPS-stimulated (but not basal phagocytosis) of several targets, an effect that required extracellular Ca^{2+} and was exacerbated by matrix stiffness, implying that mechanically activated TRPV4-mediated signals may be particu fibrosis (Scheraga et al. 2016). In DCs, TRPV4 was downregulated in response to LPS and dispensable for LPS-induced NFAT translocation, whereas $trpv4^{-/-}$ cells showed defects in FcR-dependent but not independent phagocytosis (Naert et al. 2019). Exactly how TRPV4 is stimulated, whether it is mechanically or otherwise, and whether its effects on phagocytosis are Ca^{2+} dependent will require further study.

P2X Channels

P2X7

P2X receptors are non-selective cation channels that are gated by extracellular ATP (North 2016). There are 7 subtypes P2X1-7 that are widely expressed in the immune system (North 2016; Linden et al. 2019). ATP is released from dying cells and as such is considered a damage-associated molecular pattern (DAMP) (Patel 2018). In addition, it may be released into the extracellular environment from immune cells themselves via pannexins or connexins in response to Ca^{2+} signals or membrane depolarization, among other stimuli (Valdebenito et al. 2018). In this manner, ATP can act as an autocrine or paracrine signal, responding to and activating Ca^{2+} signaling in turn either through direct P2X receptor gating or via P2Y, GPCRs that can generate PLC-dependent Ca^{2+} signals (Valdebenito et al. 2018). Thus, it is not surprising to find that purinergic signaling can modulate phagocytic activity.

P2X7 (formerly called P2Z) channels are the most well studied and are highly expressed in immune cells. In addition to non-selective cation permeation, they can also form larger pores permeable to hydrophilic molecules ≤ 900 Da upon extended activation, which can trigger cell death (Linden et al. 2019). They have been associated to inflammasome activation and to numerous inflammatory and autoimmune diseases (Linden et al. 2019). Several early studies using inhibitors as well as knock-out cells describe ATP stimulation to promote phago-lysosome fusion, phagosome acidification and killing of *Mycobacterium* species in macrophages in a manner at least partly dependent on Ca^{2+} and P2X7 receptor (Kusner and Adams 2000; Lammas et al. 1997; Fairbairn et al. 2001; Kusner and Barton 2001; Stober et al. 2001). P2Y receptor activation and ER Ca^{2+} release also contributed (Stober et al. 2001). Interestingly, by analyzing Ca^{2+} dependent mycobacterial killing using intra and extracellular Ca^{2+} chelators, Ca^{2+} ionophores, and ER Ca^{2+} release agents the authors highlighted the fact that ER Ca^{2+} release can compensate loss of extracellular Ca^{2+} influx (Stober et al. 2001). More recently, it was shown that stimulation of the Ca^{2+}-activated chloride channel and phospholipid scramblase Anoctamin 6 (ANO6) downstream of P2X7 activation contributes to phagocytosis and bacterial killing, although the exact mechanism for this effect requires further investigation (Ousingsawat et al. 2015). Also, in addition to cell autonomous effects on macrophage activation and phagosome maturation, an elegant study by Zumerle and colleagues employed photoactivatable caged IP$_3$ as well as extracellular addition of ATP catabolizing apyrase to eliminate paracrine signaling, and demonstrated the contribution of paracrine purinergic signaling involving P2X7 and P2X4 to macrophage phagocytosis of a varied number of targets (Zumerle et al. 2019). Finally, an intriguing study demonstrated that P2X7 receptors can also function as phagocytic scavenger receptors and confer phagocytic capability to HEK-293 cells when ectopically expressed (Gu et al. 2011). However, this effect is inhibited by its natural ligand ATP, and is Ca^{2+} dependent for some but not all targets (Gu et al. 2011; Perez-Flores et al. 2016).

P2X4

P2X4 receptors are non-selective cation channels with high Ca^{2+} permeability that are about a thousand times more sensitive to ATP than P2X7 (North 2016; Suurvali et al. 2017). It is expressed at the plasma membrane, but mainly localizes to the endocytic system, with prominent localization within lysosomes. The ATP sensitivity of P2X4 receptors is inhibited at low pH such that high levels of ATP within lysosomes do not constitutively activate the channel, and it has been postulated that they activate when lysosomes fuse with organelles of higher pH such as early endosomes (Suurvali et al. 2017; Murrell-Lagnado and Frick 2019). Indeed, studies using the genetically encoded Ca^{2+}-sensitive probe GECO fused on the cytosolic side to P2X4 suggest that P2X4 channels regulate lysosomal Ca^{2+} content as well as lysosomal Ca^{2+} release that promotes fusion with endosomes, although the sequence of events in a physiological setting remains unclear (Cao et al. 2015; Huang et al. 2014). Interestingly, in macrophages exposure to phagocytic targets lead to P2X4 translocation to both the phagosomal as well as the plasma membrane, and sensitization of ATP-induced currents (Qureshi et al. 2007; Stokes and Surprenant 2009). In addition, point-mutations in P2X4 and P2X7 associated with macular degeneration inhibited P2X7-mediated phagocytosis in HEK-293 and monocytes, but the effect of these mutations on Ca^{2+} influx was not examined (Gu et al. 2013). In another report, inhibiting P2X4 receptors with the selective antagonist 5-BDBD reversed the depressive effects of ethanol on microglial phagocytosis, although phagocytic Ca^{2+} signals were not measured (Gofman et al. 2014). Thus, while it is clear that P2X receptor signaling can influence phagocytic activity, more research will be required to determine whether P2X4 can regulate phagosomal Ca^{2+} levels, whether purinergic signaling contributes to phagocytic Ca^{2+} signaling, and if so what precise roles P2X receptors play.

Conclusions and Future Directions

Several themes emerge from the studies summarized in this chapter. The first centers around the importance of Ca^{2+} release from intracellular stores. While extensive studies on STIM proteins have now placed store-operated Ca^{2+} channels at the forefront of generating phagocytic Ca^{2+} signals, it is evident that inhibition of ORAI channels alone will likely not suffice to abrogate phagocytic function (if a reduction in this activity is the desired outcome), because store release can compensate, as evidenced by studies focusing on junctate (Guido et al. 2015) and P2X7 receptors (Stober et al. 2001). While IP$_3$Rs remain major players for mediating store release, other molecules emerging as novel regulators of ER release such as TRPP1 (Van Goethem et al. 2012), or TRPML1 active via lysosomal store release, may offer appealing alternatives to stimulate store release. On the other hand, direct stimulation of the activity of ORAI in order to promote phagocytic function may also represent an interesting avenue to pursue in the future as more selective activators of ORAI become available. As few studies have analyzed endogenous ORAI channels or their interaction with TRPC proteins in this context, future work in this area will be required to assess the utility of this approach.

Another trend that can be discerned is that several non-selective cation channels such as TRPC6, TRPM2, TRPML1 and potentially P2X4, localize to phagosomes and mediate phagosomal cation efflux, but what remains unclear is whether this activity is only playing a supportive role for charge compensation or ionic homeostasis within the phagosome during intense phagosome acidification or ROS production, or whether it contributes to Ca^{2+} signaling. As exemplified by the study on TRPM2 (Di et al. 2017) it will be of utmost importance to distinguish between role of Ca^{2+} and that of other major cations such as Na^+ and K^+, but also minor cations such as Mg^{2+}, Fe^{2+} and Zn^{2+}, which are often neglected. For this, the development of methodology to more easily or accurately track these ions will be critical for facilitating advances in this area. Indeed, inconclusive monitoring of intraphagosomal Ca^{2+} has been a drawback in several studies outlined above, namely because of the difficulty in accounting for changes in loading for single-wavelength probes or for considering the effects on pH on Ca^{2+}-sensitive probes. As the study of intralysosomal Ca^{2+} has gained increased attention in recent times, new tools in this area may be borrowed to decipher how Ca^{2+} fluxes across the phagosomal membrane shapes the intricate biology of this dynamic organelle (Lloyd-Evans and Waller-Evans 2019; Zhong et al. 2017; Morgan et al. 2015).

Finally, it is increasingly evident that Ca^{2+} channels have a complex relationship with membrane lipids, particularly with phosphoinositides, arachidonic acid and sphingolipids. Nearly all TRPs interact with phosphoinositides or are even directly activated by these lipids. While an enormous amount of research has focused on phospholipid kinases and phosphatases, more recently lipid transfer proteins acting at MCS such a Nir, ORP, or Esyt families are now being recognized for their role in the maintenance of the balance of plasma membrane and organellar lipids (Wong et al. 2017; Balla et al. 2019). Similarly, sphingolipids have been long recognized to play important roles (Weigert et al. 2019; Olson et al. 2016) yet still many basic questions such as how S1P mediates ER Ca^{2+} release remain unknown. It is likely that better tools to investigate and manipulate lipids will facilitate future research in this area.

In conclusion, considering the breadth of the literature accumulated over the past four decades, it is undeniable that Ca^{2+} signals are important to promote efficient phagocytosis in a large number of contexts, and mediate various downstream responses that streamline target elimination and immune defense. However, their study is challenging as Ca^{2+} signals are cell type dependent and further influenced by a variety of physical and chemical stimuli that may change depending on the physiological context of the phagocytic event. Identifying channels involved in generating phagocytic Ca^{2+} signals has been an extremely attractive endeavor for biomedical researchers in recent years, undoubtedly because ion channels are excellent targets for pharmacological stimulation or inhibition of phagocytic function, which can in principle respectively help eliminate ingested pathogens when phagocytic function is impaired or to limit excessive inflammation when it is damaging. As can be inferred from the discussion above, phagocytic cells express a large variety of Ca^{2+}-permeable channels, and while research in the past 10 years has significantly advanced our understanding of which channels participate in which functions, dissecting the contribution of each channel as well as the coordinated activity between these diverse players will require a great deal more study.

Acknowledgements We would like to thank Servier Medical Art (https://smart.servier.com) for graphics used in the figures. PNH is supported by a Novartis Foundation grant (#17B078), the 2019 Dr. Prof. Max Cloëtta Foundation Medical Researcher Scholarship, and Swiss National Science Foundation grant 310030_18909. ND is supported by Swiss National Science Foundation 310030_189042.

References

Balla T, Kim YJ, Alvarez-Prats A, Pemberton J (2019) Lipid dynamics at contact sites between the endoplasmic reticulum and other organelles. Annu Rev Cell Dev Biol 35:85–109. https://doi.org/10.1146/annurev-cellbio-100818-125251

Bavencoffe A, Zhu MX, Tian JB (2017) New aspects of the contribution of ER to SOCE regulation: TRPC proteins as a link between plasma membrane ion transport and intracellular Ca(2+) stores. Adv Exp Med Biol 993:239–255. https://doi.org/10.1007/978-3-319-57732-6_13

Beerman Rebecca WW, Matty Molly AA, Au Gina GG, Looger Loren LL, Choudhury Kingshuk RR, Keller Philipp JJ, Tobin David MM (2015) Direct in vivo manipulation and imaging of calcium transients in neutrophils identify a critical role for leading-edge calcium flux. Cell Rep 13(10):2107–2117. https://doi.org/10.1016/j.celrep.2015.11.010

Bezprozvanny I (2013) Presenilins and calcium signaling-systems biology to the rescue. Sci Signal 6(283):pe24. https://doi.org/10.1126/scisignal.2004296

Bootman MD, Berridge MJ, Roderick HL (2002) Calcium signalling: more messengers, more channels, more complexity. Curr Biol 12(16):R563–R565. https://doi.org/10.1016/S0960-9822(02)01055-2

Bouron A, Chauvet S, Dryer S, Rosado JA (2016) Second messenger-operated calcium entry through

TRPC6. Adv Exp Med Biol 898:201–249. https://doi.org/10.1007/978-3-319-26974-0_10

Brailoiu E, Churamani D, Cai X, Schrlau MG, Brailoiu GC, Gao X, Hooper R, Boulware MJ, Dun NJ, Marchant JS, Patel S (2009) Essential requirement for two-pore channel 1 in NAADP-mediated calcium signaling. J Cell Biol 186(2):201–209. https://doi.org/10.1083/jcb.200904073

Brechard S, Melchior C, Plancon S, Schenten V, Tschirhart EJ (2008) Store-operated Ca²⁺ channels formed by TRPC1, TRPC6 and Orai1 and non-store-operated channels formed by TRPC3 are involved in the regulation of NADPH oxidase in HL-60 granulocytes. Cell Calcium 44(5):492–506. https://doi.org/10.1016/j.ceca.2008.03.002

Bujak JK, Kosmala D, Szopa IM, Majchrzak K, Bednarczyk P (2019) Inflammation, cancer and immunity-implication of TRPV1 channel. Front Oncol 9:1087. https://doi.org/10.3389/fonc.2019.01087

Cao Q, Zhong XZ, Zou Y, Murrell-Lagnado R, Zhu MX, Dong XP (2015) Calcium release through P2X4 activates calmodulin to promote endolysosomal membrane fusion. J Cell Biol 209(6):879–894. https://doi.org/10.1083/jcb.201409071

Christensen KA, Myers JT, Swanson JA (2002) pH-dependent regulation of lysosomal calcium in macrophages. J Cell Sci 115(Pt 3):599–607

Clapham DE (2007) Calcium signaling. Cell 131(6):1047–1058. https://doi.org/10.1016/j.cell.2007.11.028

Clemens RA, Chong J, Grimes D, Hu Y, Lowell CA (2017) STIM1 and STIM2 cooperatively regulate mouse neutrophil store-operated calcium entry and cytokine production. Blood 130(13):1565–1577. https://doi.org/10.1182/blood-2016-11-751230

Cohen R, Torres A, Ma HT, Holowka D, Baird B (2009) Ca²⁺ waves initiate antigen-stimulated Ca²⁺ responses in mast cells. J Immunol 183(10):6478–6488. https://doi.org/10.4049/jimmunol.0901615

Csordas G, Varnai P, Golenar T, Sheu SS, Hajnoczky G (2012) Calcium transport across the inner mitochondrial membrane: molecular mechanisms and pharmacology. Mol Cell Endocrinol 353(1–2):109–113. https://doi.org/10.1016/j.mce.2011.11.011

Curcic S, Schober R, Schindl R, Groschner K (2019) TRPC-mediated Ca(2+) signaling and control of cellular functions. Semin Cell Dev Biol 94:28–39. https://doi.org/10.1016/j.semcdb.2019.02.001

Dayam RM, Saric A, Shilliday RE, Botelho RJ (2015) The phosphoinositide-gated lysosomal Ca(2+) channel, TRPML1, is required for phagosome maturation. Traffic 16(9):1010–1026. https://doi.org/10.1111/tra.12303

Dayam RM, Sun CX, Choy CH, Mancuso G, Glogauer M, Botelho RJ (2017) The lipid kinase PIKfyve coordinates the neutrophil immune response through the activation of the Rac GTPase. J Immunol 199(6):2096–2105. https://doi.org/10.4049/jimmunol.1601466

de Souza LB, Ambudkar IS (2014) Trafficking mechanisms and regulation of TRPC channels.

Cell Calcium 56(2):43–50. https://doi.org/10.1016/j.ceca.2014.05.001

De Stefani D, Rizzuto R, Pozzan T (2016) Enjoy the trip: calcium in mitochondria back and forth. Annu Rev Biochem 85:161–192. https://doi.org/10.1146/annurev-biochem-060614-034216

DeHaven WI, Jones BF, Petranka JG, Smyth JT, Tomita T, Bird GS, Putney JW Jr (2009) TRPC channels function independently of STIM1 and Orai1. J Physiol 587(Pt 10):2275–2298. https://doi.org/10.1113/jphysiol.2009.170431

Demaurex N, Guido D (2017) The role of mitochondria in the activation/maintenance of SOCE: membrane contact sites as signaling hubs sustaining store-operated Ca^{2+} entry. Adv Exp Med Biol 993:277–296. (PMID: 28900920). https://doi.org/10.1007/978-3-319-57732-6_15

Demaurex N, Nunes P (2016) The role of STIM and ORAI proteins in phagocytic immune cells. Am J Physiol Cell Physiol 310(7):C496–C508. https://doi.org/10.1152/ajpcell.00360.2015

Dewitt S, Hallett MB (2002) Cytosolic free Ca(2+) changes and calpain activation are required for beta integrin-accelerated phagocytosis by human neutrophils. J Cell Biol 159(1):181–189. https://doi.org/10.1083/jcb.200206089

Di Virgilio F, Meyer BC, Greenberg S, Silverstein SC (1988) Fc receptor-mediated phagocytosis occurs in macrophages at exceedingly low cytosolic Ca2+ levels. J Cell Biol 106(3):657–666. https://doi.org/10.1083/jcb.106.3.657

Di A, Gao XP, Qian F, Kawamura T, Han J, Hecquet C, Ye RD, Vogel SM, Malik AB (2011) The redox-sensitive cation channel TRPM2 modulates phagocyte ROS production and inflammation. Nat Immunol 13(1):29–34. https://doi.org/10.1038/ni.2171

Di A, Kiya T, Gong H, Gao X, Malik AB (2017) Role of the phagosomal redox-sensitive TRP channel TRPM2 in regulating bactericidal activity of macrophages. J Cell Sci 130(4):735–744. https://doi.org/10.1242/jcs.196014

Dutta B, Arya RK, Goswami R, Alharbi MO, Sharma S, Rahaman SO (2019) Role of macrophage TRPV4 in inflammation. Laboratory investigation; a journal of technical methods and pathology. https://doi.org/10.1038/s41374-019-0334-6

Elling R, Keller B, Weidinger C, Haffner M, Deshmukh SD, Zee I, Speckmann C, Ehl S, Schwarz K, Feske S, Henneke P (2016) Preserved effector functions of human ORAI1- and STIM1-deficient neutrophils. J Allergy Clin Immunol 137(5):1587–1591. e1587. https://doi.org/10.1016/j.jaci.2015.09.047

Fairbairn IP, Stober CB, Kumararatne DS, Lammas DA (2001) ATP-mediated killing of intracellular mycobacteria by macrophages is a P2X(7)-dependent process inducing bacterial death by phagosome-lysosome fusion. J Immunol 167(6):3300–3307. https://doi.org/10.4049/jimmunol.167.6.3300

Fernandes J, Lorenzo IM, Andrade YN, Garcia-Elias A, Serra SA, Fernandez-Fernandez JM, Valverde MA (2008) IP3 sensitizes TRPV4 channel to the mechano- and osmotransducing messenger 5′-6′-epoxyeicosatrienoic acid. J Gen Physiol 131(5):i2. https://doi.org/10.1085/JGP1315OIA2

Fernandes ES, Liang L, Smillie SJ, Kaiser F, Purcell R, Rivett DW, Alam S, Howat S, Collins H, Thompson SJ, Keeble JE, Riffo-Vasquez Y, Bruce KD, Brain SD (2012) TRPV1 deletion enhances local inflammation and accelerates the onset of systemic inflammatory response syndrome. J Immunol 188(11):5741–5751. https://doi.org/10.4049/jimmunol.1102147

Foote JR, Behe P, Frampton M, Levine AP, Segal AW (2017) An exploration of charge compensating ion channels across the phagocytic vacuole of neutrophils. Front Pharmacol 8:94. https://doi.org/10.3389/fphar.2017.00094

Foote JR, Patel AA, Yona S, Segal AW (2019) Variations in the phagosomal environment of human neutrophils and mononuclear phagocyte subsets. Front Immunol 10:188. https://doi.org/10.3389/fimmu.2019.00188

Freeman SA, Grinstein S (2014) Phagocytosis: receptors, signal integration, and the cytoskeleton. Immunol Rev 262(1):193–215. https://doi.org/10.1111/imr.12212

Fu J, Gao Z, Shen B, Zhu MX (2015) Canonical transient receptor potential 4 and its small molecule modulators. Sci China Life Sci 58(1):39–47. https://doi.org/10.1007/s11427-014-4772-5

Fujii Y, Shiota M, Ohkawa Y, Baba A, Wanibuchi H, Kinashi T, Kurosaki T, Baba Y (2012) Surf4 modulates STIM1-dependent calcium entry. Biochem Biophys Res Commun 422(4):615–620. https://doi.org/10.1016/j.bbrc.2012.05.037

Galione A, Chuang KT (2020) Pyridine nucleotide metabolites and calcium release from intracellular stores. Adv Exp Med Biol 1131:371–394. https://doi.org/10.1007/978-3-030-12457-1_15

Garcia-Elias A, Mrkonjic S, Jung C, Pardo-Pastor C, Vicente R, Valverde MA (2014) The TRPV4 channel. Handb Exp Pharmacol 222:293–319. https://doi.org/10.1007/978-3-642-54215-2_12

Garcia-Recio S, Gascon P (2015) Biological and pharmacological aspects of the NK1-receptor. Biomed Res Int 2015:495704. https://doi.org/10.1155/2015/495704

Gerke V, Creutz CE, Moss SE (2005) Annexins: linking Ca2+ signalling to membrane dynamics. Nat Rev Mol Cell Biol 6(6):449–461. https://doi.org/10.1038/nrm1661

Gofman L, Cenna JM, Potula R (2014) P2X4 receptor regulates alcohol-induced responses in microglia. J Neuroimmune Pharmacol 9(5):668–678. https://doi.org/10.1007/s11481-014-9559-8

Gonzalez-Muniz R, Bonache MA, Martin-Escura C, Gomez-Monterrey I (2019) Recent progress in TRPM8 modulation: an update. Int J Mol Sci 20(11). https://doi.org/10.3390/ijms20112618

Gray MA, Choy CH, Dayam RM, Ospina-Escobar E, Somerville A, Xiao X, Ferguson SM, Botelho RJ (2016) Phagocytosis enhances lysosomal and bactericidal properties by activating the transcription factor

TFEB. Curr Biol 26(15):1955–1964. https://doi.org/10.1016/j.cub.2016.05.070

Greenberg S, el Khoury J, di Virgilio F, Kaplan EM, Silverstein SC (1991) Ca(2+)-independent F-actin assembly and disassembly during Fc receptor-mediated phagocytosis in mouse macrophages. J Cell Biol 113(4):757–767. https://doi.org/10.1083/jcb.113.4.757

Gu BJ, Saunders BM, Petrou S, Wiley JS (2011) P2X(7) is a scavenger receptor for apoptotic cells in the absence of its ligand, extracellular ATP. J Immunol 187(5):2365–2375. https://doi.org/10.4049/jimmunol.1101178

Gu BJ, Baird PN, Vessey KA, Skarratt KK, Fletcher EL, Fuller SJ, Richardson AJ, Guymer RH, Wiley JS (2013) A rare functional haplotype of the P2RX4 and P2RX7 genes leads to loss of innate phagocytosis and confers increased risk of age-related macular degeneration. FASEB J 27(4):1479–1487. https://doi.org/10.1096/fj.12-215368

Guido D, Demaurex N, Nunes P (2015) Junctate boosts phagocytosis by recruiting endoplasmic reticulum Ca2+ stores near phagosomes. J Cell Sci 128(22):4074–4082. https://doi.org/10.1242/jcs.172510

Guse AH, Wolf IM (2016) Ca(2+) microdomains, NAADP and type 1 ryanodine receptor in cell activation. Biochim Biophys Acta 1863(6 Pt B):1379–1384. https://doi.org/10.1016/j.bbamcr.2016.01.014

Hamanaka K, Jian MY, Townsley MI, King JA, Liedtke W, Weber DS, Eyal FG, Clapp MM, Parker JC (2010) TRPV4 channels augment macrophage activation and ventilator-induced lung injury. Am J Physiol Lung Cell Mol Physiol 299(3):L353–L362. https://doi.org/10.1152/ajplung.00315.2009

Hammadi M, Oulidi A, Gackiere F, Katsogiannou M, Slomianny C, Roudbaraki M, Dewailly E, Delcourt P, Lepage G, Lotteau S, Ducreux S, Prevarskaya N, Van Coppenolle F (2013) Modulation of ER stress and apoptosis by endoplasmic reticulum calcium leak via translocon during unfolded protein response: involvement of GRP78. FASEB J 27(4):1600–1609. https://doi.org/10.1096/fj.12-218875

Harteneck C (2005) Function and pharmacology of TRPM cation channels. Naunyn Schmiedeberg's Arch Pharmacol 371(4):307–314. https://doi.org/10.1007/s00210-005-1034-x

Hassan S, Eldeeb K, Millns PJ, Bennett AJ, Alexander SP, Kendall DA (2014) Cannabidiol enhances microglial phagocytosis via transient receptor potential (TRP) channel activation. Br J Pharmacol 171(9):2426–2439. https://doi.org/10.1111/bph.12615

Heng TS, Painter MW, Immunological Genome Project C (2008) The Immunological Genome Project: networks of gene expression in immune cells. Nat Immunol 9(10):1091–1094. https://doi.org/10.1038/ni1008-1091

Heo DK, Lim HM, Nam JH, Lee MG, Kim JY (2015) Regulation of phagocytosis and cytokine secretion by store-operated calcium entry in primary isolated murine microglia. Cell Signal 27(1):177–186. https://doi.org/10.1016/j.cellsig.2014.11.003

Huang P, Zou Y, Zhong XZ, Cao Q, Zhao K, Zhu MX, Murrell-Lagnado R, Dong XP (2014) P2X4 forms functional ATP-activated cation channels on lysosomal membranes regulated by luminal pH. J Biol Chem 289(25):17658–17667. https://doi.org/10.1074/jbc.M114.552158

Hurwitz S (1996) Homeostatic control of plasma calcium concentration. Crit Rev Biochem Mol Biol 31(1):41–100. https://doi.org/10.3109/10409239609110575

Jardin I, Gomez LJ, Salido GM, Rosado JA (2009) Dynamic interaction of hTRPC6 with the Orai1-STIM1 complex or hTRPC3 mediates its role in capacitative or non-capacitative Ca(2+) entry pathways. Biochem J 420(2):267–276. https://doi.org/10.1042/BJ20082179

Kashio M, Sokabe T, Shintaku K, Uematsu T, Fukuta N, Kobayashi N, Mori Y, Tominaga M (2012) Redox signal-mediated sensitization of transient receptor potential melastatin 2 (TRPM2) to temperature affects macrophage functions. Proc Natl Acad Sci U S A 109(17):6745–6750. https://doi.org/10.1073/pnas.1114193109

Kerkhoff C, Nacken W, Benedyk M, Dagher MC, Sopalla C, Doussiere J (2005) The arachidonic acid-binding protein S100A8/A9 promotes NADPH oxidase activation by interaction with p67phox and Rac-2. FASEB J 19(3):467–469. https://doi.org/10.1096/fj.04-2377fje

Khalil M, Babes A, Lakra R, Forsch S, Reeh PW, Wirtz S, Becker C, Neurath MF, Engel MA (2016) Transient receptor potential melastatin 8 ion channel in macrophages modulates colitis through a balance-shift in TNF-alpha and interleukin-10 production. Mucosal Immunol 9(6):1500–1513. https://doi.org/10.1038/mi.2016.16

Kilpatrick BS, Yates E, Grimm C, Schapira AH, Patel S (2016) Endo-lysosomal TRP mucolipin-1 channels trigger global ER Ca2+ release and Ca2+ influx. J Cell Sci 129(20):3859–3867. https://doi.org/10.1242/jcs.190322

Kilpatrick BS, Eden ER, Hockey LN, Yates E, Futter CE, Patel S (2017) An endosomal NAADP-sensitive two-pore Ca(2+) channel regulates ER-endosome membrane contact sites to control growth factor signaling. Cell Rep 18(7):1636–1645. https://doi.org/10.1016/j.celrep.2017.01.052

Kim JY, Zeng W, Kiselyov K, Yuan JP, Dehoff MH, Mikoshiba K, Worley PF, Muallem S (2006) Homer 1 mediates store- and inositol 1,4,5-trisphosphate receptor-dependent translocation and retrieval of TRPC3 to the plasma membrane. J Biol Chem 281(43):32540–32549. https://doi.org/10.1074/jbc.M602496200

Kiselyov K, Xu X, Mozhayeva G, Kuo T, Pessah I, Mignery G, Zhu X, Birnbaumer L, Muallem S (1998) Functional interaction between InsP3 receptors and store-operated Htrp3 channels. Nature 396(6710):478–482. https://doi.org/10.1038/24890

Knowles H, Heizer JW, Li Y, Chapman K, Ogden CA, Andreasen K, Shapland E, Kucera G, Mogan J, Humann J, Lenz LL, Morrison AD, Perraud AL (2011) Transient Receptor Potential Melastatin 2 (TRPM2) ion channel is required for innate immunity against Listeria mono-cytogenes. Proc Natl Acad Sci U S A 108(28):11578–11583. https://doi.org/10.1073/pnas.1010678108

Koulen P, Cai Y, Geng L, Maeda Y, Nishimura S, Witzgall R, Ehrlich BE, Somlo S (2002) Polycystin-2 is an intracellular calcium release channel. Nat Cell Biol 4(3):191–197. https://doi.org/10.1038/ncb754

Kusner DJ, Adams J (2000) ATP-induced killing of virulent Mycobacterium tuberculosis within human macrophages requires phospholipase D. J Immunol 164(1):379–388. https://doi.org/10.4049/jimmunol.164.1.379

Kusner DJ, Barton JA (2001) ATP stimulates human macrophages to kill intracellular virulent Mycobacterium tuberculosis via calcium-dependent phagosome-lysosome fusion. J Immunol 167(6):3308–3315. https://doi.org/10.4049/jimmunol.167.6.3308

Lammas DA, Stober C, Harvey CJ, Kendrick N, Panchalingam S, Kumararatne DS (1997) ATP-induced killing of mycobacteria by human macrophages is mediated by purinergic P2Z(P2X7) receptors. Immunity 7(3):433–444. https://doi.org/10.1016/s1074-7613(00)80364-7

Lang S, Erdmann F, Jung M, Wagner R, Cavalie A, Zimmermann R (2011) Sec61 complexes form ubiquitous ER Ca2+ leak channels. Channels (Austin) 5(3):228–235. https://doi.org/10.4161/chan.5.3.15314

Larson L, Arnaudeau S, Gibson B, Li W, Krause R, Hao B, Bamburg JR, Lew DP, Demaurex N, Southwick F (2005) Gelsolin mediates calcium-dependent disassembly of Listeria actin tails. Proc Natl Acad Sci USA 102(6):1921–1926. https://doi.org/10.1073/pnas.0409062102

Lee HC, Zhao YJ (2019) Resolving the topological enigma in Ca2+−signaling by cyclic ADP-ribose and NAADP. J Biol Chem. https://doi.org/10.1074/jbc.REV119.009635

Leveque M, Penna A, Le Trionnaire S, Belleguic C, Desrues B, Brinchault G, Jouneau S, Lagadic-Gossmann D, Martin-Chouly C (2018) Phagocytosis depends on TRPV2-mediated calcium influx and requires TRPV2 in lipids rafts: alteration in macrophages from patients with cystic fibrosis. Sci Rep 8(1):4310. https://doi.org/10.1038/s41598-018-22558-5

Lim HM, Woon H, Han JW, Baba Y, Kurosaki T, Lee MG, Kim JY (2017) UDP-induced phagocytosis and ATP-stimulated chemotactic migration are impaired in STIM1(−/−) microglia in vitro and in vivo. Mediat Inflamm 2017:8158514. https://doi.org/10.1155/2017/8158514

Linden J, Koch-Nolte F, Dahl G (2019) Purine release, metabolism, and signaling in the inflammatory response. Annu Rev Immunol 37:325–347. https://doi.org/10.1146/annurev-immunol-051116-052406

Lindmark IM, Karlsson A, Serrander L, Francois P, Lew D, Rasmusson B, Stendahl O, Nusse O (2002) Synaptotagmin II could confer Ca(2+) sensitivity to phagocytosis in human neutrophils. Biochim Biophys Acta 1590(1–3):159–166. https://doi.org/10.1016/s0167-4889(02)00209-4

Link TM, Park U, Vonakis BM, Raben DM, Soloski MJ, Caterina MJ (2010) TRPV2 has a pivotal role in macrophage particle binding and phagocytosis. Nat Immunol 11(3):232–239. https://doi.org/10.1038/ni.1842

Lisak D, Schacht T, Gawlitza A, Albrecht P, Aktas O, Koop B, Gliem M, Hofstetter HH, Zanger K, Bultynck G, Parys JB, De Smedt H, Kindler T, Adams-Quack P, Hahn M, Waisman A, Reed JC, Hovelmeyer N, Methner A (2016) BAX inhibitor-1 is a Ca(2+) channel critically important for immune cell function and survival. Cell Death Differ 23(2):358–368. https://doi.org/10.1038/cdd.2015.115

Lischke T, Heesch K, Schumacher V, Schneider M, Haag F, Koch-Nolte F, Mittrcker HW (2013) CD38 Controls the innate immune response against listeria monocytogenes. Infect Immun 81(11):4091–4099. https://doi.org/10.1128/IAI.00340-13

Liu Q (2017) TMBIM-mediated Ca(2+) homeostasis and cell death. Biochim Biophys Acta, Mol Cell Res 1864(6):850–857. https://doi.org/10.1016/j.bbamcr.2016.12.023

Lloyd-Evans E, Waller-Evans H (2019) Lysosomal Ca(2+) homeostasis and signaling in health and disease. Cold Spring Harb Perspect Biol. https://doi.org/10.1101/cshperspect.a035311

Luik RM, Wu MM, Buchanan J, Lewis RS (2006) The elementary unit of store-operated Ca2+ entry: local activation of CRAC channels by STIM1 at ER-plasma membrane junctions. J Cell Biol 174(6):815–825. https://doi.org/10.1083/jcb.200604015

Maschalidi S, Nunes-Hasler P, Nascimento CR, Sallent I, Lannoy V, Garfa-Traore M, Cagnard N, Sepulveda FE, Vargas P, Lennon-Dumenil AM, van Endert P, Capiod T, Demaurex N, Darrasse-Jeze G, Manoury B (2017) UNC93B1 interacts with the calcium sensor STIM1 for efficient antigen cross-presentation in dendritic cells. Nat Commun 8(1):1640. https://doi.org/10.1038/s41467-017-01601-5

Masubuchi H, Ueno M, Maeno T, Yamaguchi K, Hara K, Sunaga H, Matsui H, Nagasawa M, Kojima I, Iwata Y, Wakabayashi S, Kurabayashi M (2019) Reduced transient receptor potential vanilloid 2 expression in alveolar macrophages causes COPD in mice through impaired phagocytic activity. BMC Pulm Med 19(1):70. https://doi.org/10.1186/s12890-019-0821-y

Mederos YSM, Gudermann T, Storch U (2018) Emerging roles of diacylglycerol-sensitive TRPC4/5 channels. Cell 7(11). https://doi.org/10.3390/cells7110218

Mery L, Magnino F, Schmidt K, Krause KH, Dufour JF (2001) Alternative splice variants of hTrp4 differentially interact with the C-terminal portion of the inositol 1,4,5-trisphosphate receptors. FEBS Lett 487(3):377–383. https://doi.org/10.1016/s0014-5793(00)02362-0

Michaelis M, Nieswandt B, Stegner D, Eilers J, Kraft R (2015) STIM1, STIM2, and Orai1 regulate store-operated calcium entry and purinergic activation of microglia. Glia 63(4):652–663. https://doi.org/10.1002/glia.22775

Miller BA (2006) The role of TRP channels in oxidative stress-induced cell death. J Membr Biol 209(1):31–41. https://doi.org/10.1007/s00232-005-0839-3

Mizoguchi Y, Kato TA, Seki Y, Ohgidani M, Sagata N, Horikawa H, Yamauchi Y, Sato-Kasai M, Hayakawa K, Inoue R, Kanba S, Monji A (2014) Brain-derived neurotrophic factor (BDNF) induces sustained intracellular Ca2+ elevation through the up-regulation of surface transient receptor potential 3 (TRPC3) channels in rodent microglia. J Biol Chem 289(26):18549–18555. https://doi.org/10.1074/jbc.M114.555334

Morgan AJ, Platt FM, Lloyd-Evans E, Galione A (2011) Molecular mechanisms of endolysosomal Ca2+ signalling in health and disease. Biochem J 439(3):349–374. https://doi.org/10.1042/BJ20110949

Morgan AJ, Davis LC, Galione A (2015) Imaging approaches to measuring lysosomal calcium. Methods Cell Biol 126:159–195. https://doi.org/10.1016/bs.mcb.2014.10.031

Muller C, Morales P, Reggio PH (2018) Cannabinoid ligands targeting TRP channels. Front Mol Neurosci 11:487. https://doi.org/10.3389/fnmol.2018.00487

Murrell-Lagnado RD, Frick M (2019) P2X4 and lysosome fusion. Curr Opin Pharmacol 47:126–132. https://doi.org/10.1016/j.coph.2019.03.002

Naert R, Lopez-Requena A, Voets T, Talavera K, Alpizar YA (2019) Expression and functional role of TRPV4 in bone marrow-derived CD11c(+) cells. Int J Mol Sci 20(14). https://doi.org/10.3390/ijms20143378

North RA (2016) P2X receptors. Philos Trans R Soc Lond Ser B Biol Sci 371(1700). https://doi.org/10.1098/rstb.2015.0427

Nunes P, Demaurex N (2010) The role of calcium signaling in phagocytosis. J Leukoc Biol 88(1):57–68. https://doi.org/10.1189/jlb.0110028

Nunes P, Cornut D, Bochet V, Hasler U, Oh-Hora M, Waldburger JM, Demaurex N (2012) STIM1 juxtaposes ER to phagosomes, generating Ca(2)(+) hotspots that boost phagocytosis. Curr Biol 22(21):1990–1997. https://doi.org/10.1016/j.cub.2012.08.049

Nunes P, Demaurex N, Dinauer MC (2013) Regulation of the NADPH oxidase and associated ion fluxes during phagocytosis. Traffic 14(11):1118–1131. https://doi.org/10.1111/tra.12115

Nunes-Hasler P, Demaurex N (2017) The ER phagosome connection in the era of membrane contact sites. Biochim Biophys Acta, Mol Cell Res 1864(9):1513–1524. https://doi.org/10.1016/j.bbamcr.2017.04.007

Nunes-Hasler P, Maschalidi S, Lippens C, Castelbou C, Bouvet S, Guido D, Bermont F, Bassoy EY, Page N, Merkler D, Hugues S, Martinvalet D, Manoury B, Demaurex N (2017) STIM1 promotes migration, phagosomal maturation and antigen cross-presentation in dendritic cells. Nat Commun 8(1):1852. https://doi.org/10.1038/s41467-017-01600-6

Olson DK, Frohlich F, Farese RV Jr, Walther TC (2016) Taming the sphinx: mechanisms of cellular sphingolipid homeostasis. Biochim Biophys Acta 1861(8 Pt B):784–792. https://doi.org/10.1016/j.bbalip.2015.12.021

Ong HL, Ambudkar IS (2019) The endoplasmic reticulum-plasma membrane junction: a hub for agonist regulation of Ca(2+) entry. Cold Spring Harb Perspect Biol. https://doi.org/10.1101/cshperspect.a035253

Otsuguro K, Tang J, Tang Y, Xiao R, Freichel M, Tsvilovskyy V, Ito S, Flockerzi V, Zhu MX, Zholos AV (2008) Isoform-specific inhibition of TRPC4 channel by phosphatidylinositol 4,5-bisphosphate. J Biol Chem 283(15):10026–10036. https://doi.org/10.1074/jbc.M707306200

Ousingsawat J, Wanitchakool P, Kmit A, Romao AM, Jantarajit W, Schreiber R, Kunzelmann K (2015) Anoctamin 6 mediates effects essential for innate immunity downstream of P2X7 receptors in macrophages. Nat Commun 6:6245. https://doi.org/10.1038/ncomms7245

Partida-Sanchez S, Randall TD, Lund FE (2003) Innate immunity is regulated by CD38, an ecto-enzyme with ADP-ribosyl cyclase activity. Microbes Infect 5(1):49–58. https://doi.org/10.1016/s1286-4579(02)00055-2

Patel S (2018) Danger-associated molecular patterns (DAMPs): the derivatives and triggers of inflammation. Curr Allergy Asthma Rep 18(11):63. https://doi.org/10.1007/s11882-018-0817-3

Patel S, Cai X (2015) Evolution of acidic Ca(2)(+) stores and their resident Ca(2)(+)-permeable channels. Cell Calcium 57(3):222–230. https://doi.org/10.1016/j.ceca.2014.12.005

Penna A, Juvin V, Chemin J, Compan V, Monet M, Rassendren FA (2006) PI3-kinase promotes TRPV2 activity independently of channel translocation to the plasma membrane. Cell Calcium 39(6):495–507. https://doi.org/10.1016/j.ceca.2006.01.009

Penny CJ, Kilpatrick BS, Eden ER, Patel S (2015) Coupling acidic organelles with the ER through Ca(2)(+) microdomains at membrane contact sites. Cell Calcium 58(4):387–396. https://doi.org/10.1016/j.ceca.2015.03.006

Pereira DMS, Mendes SJF, Alawi K, Thakore P, Aubdool A, Sousa NCF, da Silva JFR, Castro JA (2018) Transient receptor potential canonical channels 4 and 5 mediate Escherichia coli-derived thioredoxin effects in lipopolysaccharide-injected mice. Oxidative Med Cell Longev 2018:4904696. https://doi.org/10.1155/2018/4904696

Perez-Flores G, Hernandez-Silva C, Gutierrez-Escobedo G, De Las PA, Castano I, Arreola J, Perez-Cornejo P (2016) P2X7 from j774 murine macrophages acts as a scavenger receptor for bacteria but not yeast. Biochem Biophys Res Commun 481(1–2):19–24. https://doi.org/10.1016/j.bbrc.2016.11.027

Pizzo P, Lissandron V, Capitanio P, Pozzan T (2011) Ca(2+) signalling in the Golgi apparatus. Cell Calcium 50(2):184–192. https://doi.org/10.1016/j.ceca.2011.01.006

Prakriya M, Lewis RS (2015) Store-operated calcium channels. Physiol Rev 95(4):1383–1436. https://doi.org/10.1152/physrev.00020.2014

Prinz WA, Toulmay A, Balla T (2019) The functional universe of membrane contact sites. Nat Rev Mol Cell Biol. https://doi.org/10.1038/s41580-019-0180-9

Qian X, Numata T, Zhang K, Li C, Hou J, Mori Y, Fang X (2014) Transient receptor potential melastatin 2 protects mice against polymicrobial sepsis by enhancing bacterial clearance. Anesthesiology 121(2):336–351. https://doi.org/10.1097/ALN.0000000000000275

Qiu R, Lewis RS (2019) Structural features of STIM and Orai underlying store-operated calcium entry. Curr Opin Cell Biol 57:90–98. https://doi.org/10.1016/j.ceb.2018.12.012

Quarona V, Zaccarello G, Chillemi A, Brunetti E, Singh VK, Ferrero E, Funaro A, Horenstein AL, Malavasi F (2013) CD38 and CD157: a long journey from activation markers to multifunctional molecules. Cytometry B Clin Cytom 84(4):207–217. https://doi.org/10.1002/cyto.b.21092

Qureshi OS, Paramasivam A, Yu JC, Murrell-Lagnado RD (2007) Regulation of P2X4 receptors by lysosomal targeting, glycan protection and exocytosis. J Cell Sci 120(Pt 21):3838–3849. https://doi.org/10.1242/jcs.010348

Riazanski V, Gabdoulkhakova AG, Boynton LS, Eguchi RR, Deriy LV, Hogarth DK, Loaec N, Oumata N, Galons H, Brown ME, Shevchenko P, Gallan AJ, Yoo SG, Naren AP, Villereal ML, Beacham DW, Bindokas VP, Birnbaumer L, Meijer L, Nelson DJ (2015) TRPC6 channel translocation into phagosomal membrane augments phagosomal function. Proc Natl Acad Sci USA 112(47):E6486–E6495. https://doi.org/10.1073/pnas.1518966112

Sakaguchi R, Mori Y (2019) Transient receptor potential (TRP) channels: biosensors for redox environmental stimuli and cellular status. Free Radic Biol Med. https://doi.org/10.1016/j.freeradbiomed.2019.10.415

Samanta A, Hughes TET, Moiseenkova-Bell VY (2018) Transient receptor potential (TRP) channels. Subcell Biochem 87:141–165. https://doi.org/10.1007/978-981-10-7757-9_6

Samie M, Wang X, Zhang X, Goschka A, Li X, Cheng X, Gregg E, Azar M, Zhuo Y, Garrity AG, Gao Q, Slaugenhaupt S, Pickel J, Zolov SN, Weisman LS, Lenk GM, Titus S, Bryant-Genevier M, Southall N, Juan M, Ferrer M, Xu H (2013) A TRP channel in the lysosome regulates large particle phagocytosis via focal exocytosis. Dev Cell 26(5):511–524. https://doi.org/10.1016/j.devcel.2013.08.003

Santoni G, Morelli MB, Amantini C, Santoni M, Nabissi M, Marinelli O, Santoni A (2018) "Immuno-transient receptor potential ion channels": the role in monocyte- and macrophage-mediated inflammatory responses. Front Immunol 9:1273. https://doi.org/10.3389/fimmu.2018.01273

Santulli G, Nakashima R, Yuan Q, Marks AR (2017) Intracellular calcium release channels: an update. J Physiol 595(10):3041–3051. https://doi.org/10.1113/JP272781

Sawyer DW, Sullivan JA, Mandell GL (1985) Intracellular free calcium localization in neutrophils during phagocytosis. Science 230(4726):663–666. https://doi.org/10.1126/science.4048951

Scheraga RG, Abraham S, Niese KA, Southern BD, Grove LM, Hite RD, McDonald C, Hamilton TA, Olman MA (2016) TRPV4 mechanosensitive ion channel regulates lipopolysaccharide-stimulated macrophage phagocytosis. J Immunol 196(1):428–436. https://doi.org/10.4049/jimmunol.1501688

Serrander L, Skarman P, Rasmussen B, Witke W, Lew DP, Krause KH, Stendahl O, Nusse O (2000) Selective inhibition of IgG-mediated phagocytosis in gelsolin-deficient murine neutrophils. J Immunol 165(5):2451–2457. https://doi.org/10.4049/jimmunol.165.5.2451

Shibasaki K (2016) Physiological significance of TRPV2 as a mechanosensor, thermosensor and lipid sensor. J Physiol Sci 66(5):359–365. https://doi.org/10.1007/s12576-016-0434-7

Sogkas G, Stegner D, Syed SN, Vogtle T, Rau E, Gewecke B, Schmidt RE, Nieswandt B, Gessner JE (2015a) Cooperative and alternate functions for STIM1 and STIM2 in macrophage activation and in the context of inflammation. Immun Inflamm Dis 3(3):154–170. https://doi.org/10.1002/iid3.56

Sogkas G, Vogtle T, Rau E, Gewecke B, Stegner D, Schmidt RE, Nieswandt B, Gessner JE (2015b) Orai1 controls C5a-induced neutrophil recruitment in inflammation. Eur J Immunol 45(7):2143–2153. https://doi.org/10.1002/eji.201445337

Soyombo AA, Tjon-Kon-Sang S, Rbaibi Y, Bashllari E, Bisceglia J, Muallem S, Kiselyov K (2006) TRP-ML1 regulates lysosomal pH and acidic lysosomal lipid hydrolytic activity. J Biol Chem 281(11):7294–7301. https://doi.org/10.1074/jbc.M508211200

Srikanth S, Ribalet B, Gwack Y (2013) Regulation of CRAC channels by protein interactions and post-translational modification. Channels (Austin) 7(5):354–363. https://doi.org/10.4161/chan.23801

Stamboulian S, Moutin MJ, Treves S, Pochon N, Grunwald D, Zorzato F, De Waard M, Ronjat M, Arnoult C (2005) Junctate, an inositol 1,4,5-triphosphate receptor associated protein, is present in rodent sperm and binds TRPC2 and TRPC5 but not TRPC1 channels. Dev Biol 286(1):326–337. https://doi.org/10.1016/j.ydbio.2005.08.006

Steinckwich N, Schenten V, Melchior C, Brechard S, Tschirhart EJ (2011) An essential role of STIM1, Orai1, and S100A8-A9 proteins for Ca2+ signaling and FcgammaR-mediated phagosomal oxidative activity. J Immunol 186(4):2182–2191. https://doi.org/10.4049/jimmunol.1001338

Stober CB, Lammas DA, Li CM, Kumararatne DS, Lightman SL, McArdle CA (2001) ATP-mediated killing of Mycobacterium bovis bacille Calmette-Guerin within human macrophages is calcium dependent and associated with the acidification of mycobacteria-containing phagosomes. J Immunol 166(10):6276–6286. https://doi.org/10.4049/jimmunol.166.10.6276

Stokes L, Surprenant A (2009) Dynamic regulation of the P2X4 receptor in alveolar macrophages by phagocytosis and classical activation. Eur J Immunol 39(4):986–995. https://doi.org/10.1002/eji.200838818

Suurvali J, Boudinot P, Kanellopoulos J, Ruutel Boudinot S (2017) P2X4: a fast and sensitive purinergic receptor. Biom J 40(5):245–256. https://doi.org/10.1016/j.bj.2017.06.010

Syed Mortadza SA, Wang L, Li D, Jiang LH (2015) TRPM2 channel-mediated ROS-sensitive Ca(2+) signaling mechanisms in immune cells. Front Immunol 6:407. https://doi.org/10.3389/fimmu.2015.00407

Taha TA, Hannun YA, Obeid LM (2006) Sphingosine kinase: biochemical and cellular regulation and role in disease. J Biochem Mol Biol 39:113–131. https://doi.org/10.5483/bmbrep.2006.39.2.113

Tano JY, Smedlund K, Lee R, Abramowitz J, Birnbaumer L, Vazquez G (2011) Impairment of survival signaling and efferocytosis in TRPC3-deficient macrophages. Biochem Biophys Res Commun 410(3):643–647. https://doi.org/10.1016/j.bbrc.2011.06.045

Tran Van Nhieu G, Kai Liu B, Zhang J, Pierre F, Prigent S, Sansonetti P, Erneux C, Kuk Kim J, Suh PG, Dupont G, Combettes L (2013) Actin-based confinement of calcium responses during Shigella invasion. Nat Commun 4:1567. https://doi.org/10.1038/ncomms2561

Trebak M, Putney JW Jr (2017) ORAI calcium channels. Physiology (Bethesda) 32(4):332–342. https://doi.org/10.1152/physiol.00011.2017

Trebak M, Lemonnier L, DeHaven WI, Wedel BJ, Bird GS, Putney JW Jr (2009) Complex functions of phosphatidylinositol 4,5-bisphosphate in regulation of TRPC5 cation channels. Pflugers Arch 457(4):757–769. https://doi.org/10.1007/s00424-008-0550-1

Treves S, Franzini-Armstrong C, Moccagatta L, Arnoult C, Grasso C, Schrum A, Ducreux S, Zhu MX, Mikoshiba K, Girard T, Smida-Rezgui S, Ronjat M, Zorzato F (2004) Junctate is a key element in calcium entry induced by activation of InsP3 receptors and/or calcium store depletion. J Cell Biol 166(4):537–548. https://doi.org/10.1083/jcb.200404079

Treves S, Vukcevic M, Griesser J, Armstrong CF, Zhu MX, Zorzato F (2010) Agonist-activated Ca2+ influx occurs at stable plasma membrane and endoplasmic reticulum junctions. J Cell Sci 123(Pt 23):4170–4181. https://doi.org/10.1242/jcs.068387

Vaeth M, Zee I, Concepcion A R, Maus M, Shaw P, Portal-Celhay C, Zahra A, Kozhaya L, Weidinger C, Philips J, Unutmaz D, Feske S (2015) Ca2+ signaling but not store-operated Ca2+ entry is required for the function of macrophages and dendritic cells. J Immunol 195: 1202–1217. https://doi.org/10.4049/jimmunol.1403013

Valdebenito S, Barreto A, Eugenin EA (2018) The role of connexin and pannexin containing channels in the innate and acquired immune response. Biochim Biophys Acta Biomembr 1860(1):154–165. https://doi.org/10.1016/j.bbamem.2017.05.015

Van Goethem E, Silva EA, Xiao H, Franc NC (2012) The Drosophila TRPP cation channel, PKD2 and Dmel/Ced-12 act in genetically distinct pathways during apoptotic cell clearance. PLoS One 7(2):e31488. https://doi.org/10.1371/journal.pone.0031488

Venkatachalam K, Montell C (2007) TRP channels. Annu Rev Biochem 76:387–417. https://doi.org/10.1146/annurev.biochem.75.103004.142819

Venkatachalam K, Wong CO, Zhu MX (2015) The role of TRPMLs in endolysosomal trafficking and function. Cell Calcium 58(1):48–56. https://doi.org/10.1016/j.ceca.2014.10.008

Verma P, Kumar A, Goswami C (2010) TRPV4-mediated channelopathies. Channels (Austin) 4(4):319–328. https://doi.org/10.4161/chan.4.4.12905

Vieira OV, Botelho RJ, Grinstein S (2002) Phagosome maturation: aging gracefully. Biochem J 366(Pt 3):689–704. https://doi.org/10.1042/BJ20020691

Vinet AF, Fukuda M, Descoteaux A (2008) The exocytosis regulator synaptotagmin V controls phagocytosis in macrophages. J Immunol 181(8):5289–5295. https://doi.org/10.4049/jimmunol.181.8.5289

Weigert A, Olesch C, Brune B (2019) Sphingosine-1-phosphate and macrophage biology-how the sphinx tames the big eater. Front Immunol 10:1706. https://doi.org/10.3389/fimmu.2019.01706

Westman J, Grinstein S, Maxson ME (2019) Revisiting the role of calcium in phagosome formation and maturation. J Leukoc Biol 106(4):837–851. https://doi.org/10.1002/JLB.MR1118-444R

Wong LH, Copic A, Levine TP (2017) Advances on the transfer of lipids by lipid transfer proteins. Trends Biochem Sci 42(7):516–530. https://doi.org/10.1016/j.tibs.2017.05.001

Worley PF, Zeng W, Huang GN, Yuan JP, Kim JY, Lee MG, Muallem S (2007) TRPC channels as STIM1-regulated store-operated channels. Cell Calcium 42(2):205–211. https://doi.org/10.1016/j.ceca.2007.03.004

Xu SZ, Muraki K, Zeng F, Li J, Sukumar P, Shah S, Dedman AM, Flemming PK, McHugh D, Naylor J, Cheong A, Bateson AN, Munsch CM, Porter KE, Beech DJ (2006) A sphingosine-1-phosphate-activated calcium channel controlling vascular smooth muscle cell motility. Circ Res 98(11):1381–1389. https://doi.org/10.1161/01.RES.0000225284.36490.a2

Yang J, Zhao Z, Gu M, Feng X, Xu H (2019) Release and uptake mechanisms of vesicular Ca(2+) stores. Protein Cell 10(1):8–19. https://doi.org/10.1007/s13238-018-0523-x

Yona S, Heinsbroek SE, Peiser L, Gordon S, Perretti M, Flower RJ (2006) Impaired phagocytic mechanism in annexin 1 null macrophages. Br J Pharmacol 148(4):469–477. https://doi.org/10.1038/sj.bjp.0706730

Yu C, Munoz LE, Mallavarapu M, Herrmann M, Finnemann SC (2019) Annexin A5 regulates surface alphav-beta5 integrin for retinal clearance phagocytosis. J Cell Sci 132(20). https://doi.org/10.1242/jcs.232439

Zhang H, Clemens RA, Liu F, Hu Y, Baba Y, Theodore P, Kurosaki T, Lowell CA (2014) STIM1 calcium sensor is required for activation of the phagocyte oxidase during inflammation and host defense. Blood 123(14):2238–2249. https://doi.org/10.1182/blood-2012-08-450403

Zhang X, Gueguinou M, Trebak M (2018) Store-independent Orai channels regulated by STIM. In: Kozak JA, Putney JW, Jr. (eds) Calcium entry channels in non-excitable cells. Boca Raton, CRC Press pp 197–214. doi:https://doi.org/10.1201/9781315152592-11

Zhao JF, Shyue SK, Lee TS (2016) Excess nitric oxide activates TRPV1-Ca(2+)-calpain signaling and promotes PEST-dependent degradation of liver X receptor alpha. Int J Biol Sci 12(1):18–29. https://doi.org/10.7150/ijbs.13549

Zholos AV (2014) Trpc5. Handb Exp Pharmacol 222:129–156. https://doi.org/10.1007/978-3-642-54215-2_6

Zhong XZ, Yang Y, Sun X, Dong XP (2017) Methods for monitoring Ca(2+) and ion channels in the lysosome. Cell Calcium 64:20–28. https://doi.org/10.1016/j.ceca.2016.12.001

Zumerle S, Cali B, Munari F, Angioni R, Di Virgilio F, Molon B, Viola A (2019) Intercellular calcium signaling induced by ATP potentiates macrophage phagocytosis. Cell Rep 27(1):1–10. e14. https://doi.org/10.1016/j.celrep.2019.03.011

Calpain Activation by Ca^{2+} and Its Role in Phagocytosis

8

Sharon Dewitt and Maurice B. Hallett

Abstract

Although the cytosolic Ca^{2+} signalling event in phagocytosis is well established, and the mechanism for generating such signals also understood, the target for the Ca^{2+} signal and how this relates to the phagocytic outcome is less clear. In this chapter, we present the evidence for a role of the Ca^{2+} activated protease, calpain, in phagocytosis. The abundant evidence for Ca^{2+} changes and calpain activation during cell shape changes is extended to include the specific cell shape change which accompanies phagocytosis. The discussion therefore includes a brief description of the domain structure of calpain and their functions. Also the mechanism by which calpain activation is limited at the cell periphery subdomains, and how this would allow phagocytic pseudopodia to form locally.

Keywords

Ca^{2+} domains · Localised signalling · Intrawrinkle Ca^{2+} · Protease activity

S. Dewitt
School of Dentistry, Cardiff University, Cardiff, UK

M. B. Hallett (✉)
School of Medicine, Cardiff University, Cardiff, UK
e-mail: hallettmb@cf.ac.uk

Introduction

In the chapter, the role of the Ca^{2+} activated protease, calpain, in phagocytosis will be discussed. This discussion is intimately involved with both Ca^{2+} and ezrin. On the subject of Ca^{2+} in phagocytosis, the reader is recommended to re-read Chap. 7. Similarly, the role of ezrin in phagocytosis is the subject of Chap. 6.

In Fig. 8.1, the interplay of the roles of Ca^{2+}, calpain and ezrin are shown in relationship to changes in the plasma membrane which are essential for phagocytosis, This figure begins with a reminder of the way in which ezrin connects the plasma membrane to the underlying cortical actin network (Fig. 8.1a). When cytosolic Ca^{2+} is elevated, calpain is activated (Fig. 8.1b) and binds to ezrin and cleaves it (Fig. 8.1c). As a result of ezrin cleavage by Ca^{2+} activated calpain, the plasma membrane is no longer held tightly by the cortical actin network and it position can fluctuate, allowing the Brownian ratchet to operate to add actin monomers to the growing tip of actin polymers, and thus push out the membrane (Fig. 8.1d). As the pushing proceeds, actin branch points may be added until the tension in membrane exceeds that which allows further Brownian fluctuations in the position of the plasma membrane. It is thought that these molecular events underlie pseudopod formation and hence phagocytosis. This model

© Springer Nature Switzerland AG 2020
M. B. Hallett (ed.), *Molecular and Cellular Biology of Phagocytosis*, Advances in Experimental
Medicine and Biology 1246, https://doi.org/10.1007/978-3-030-40406-2_8

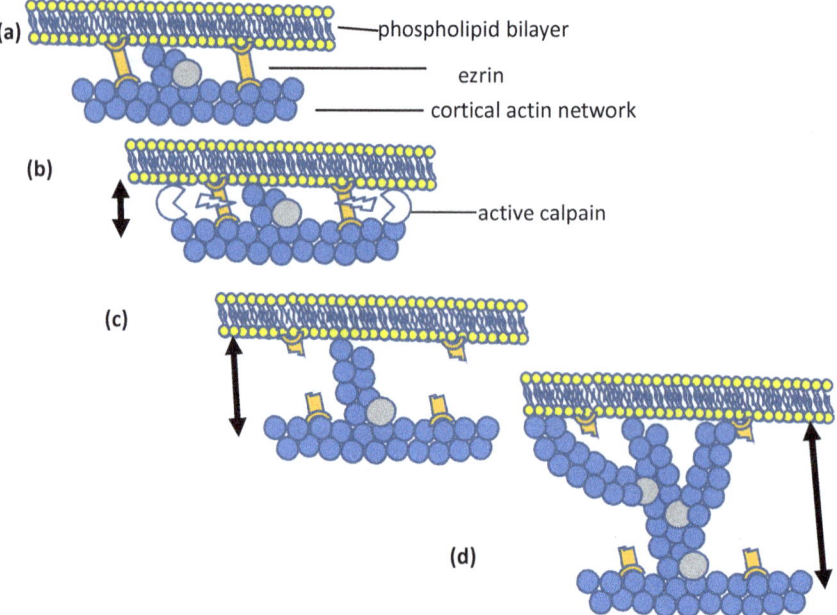

Labels in figure:
- phospholipid bilayer
- ezrin
- cortical actin network
- active calpain

(a) (b) (c) (d)

Fig. 8.1 Proposed molecular mechanism underlying the role of calpain in phagocytosis. (**a**) The configuration of the cortical cytoplasm is shown with the phospholipid plasma membrane, the cortical actin network and the crosslinking molecular ezrin holding the membrane tight to the actin cortex. A WASP generated branch point in the actin filament is shown, but the actin polymer cannot grow because the proximity of the plasma membrane prevent the addition of further actin monomers. (**b**) Activated calpain interferes with the original configuration by cleaving ezrin its substrate. (**c**) Without the restraint of ezrin, the space between the plasma membrane and the underlying actin network increases, allowing actin polymer to grow and push out the plasma membrane. (**d**) As the distance between the cortical actin network and the plasma membrane grows, addition WASP branch points are added and multiple actin polymers push out the membrane further. (This figure was originally published in Intl. J. Mol. Sci 20: Article Number: 1383 (2019))

places Ca^{2+}, ezrin and, of course, calpain, the subject of this chapter, in central roles.

Although Ca^{2+} has many cytosolic targets, surprisingly pharmacological inhibition of calmodulin or PKC has no effect on phagocytosis by neutrophils. However, there were reports that the calmodulin-activated phosphatase, calcineurin, was crucial for cell motility and pseudopodia formation in neutrophils (Hendey et al. 1992; Lawson and Maxfield 1995). In these studies, specific inhibition of calcinerin was achieved by the introduction of specific peptide inhibitors into neutrophils by a technique which involves the uptake of soluble material into pinosomes under hyperosmotic conditions and then inducing their intracellular lysis (and thereby releasing the inhibitors into the cytosol) by hypo-osmotic shock. This technique showed that calcineurin inhibitors prevented cell motility

and pseudopodia extensions. In would be interesting to repeat these experiments using direct microinjection, which can be achieved in neutrophils by modified injection techniques SLAM (simple lipid assisted microinjection) or electro-microinjection (Campbell and Hallett 2015; Laffafian and Hallett 1998, 2000; Lewis et al. 2014). Both techniques have surprising little effect of subsequent Ca^{2+} signalling or the ability of cells to undergo cell shape changes (Lewis et al. 2014). Ultimately, the inhibitory effects of calcineurin inhibition on motility were shown to be due to an inability of the treated cells to detach from their substrate (Lawson and Maxfield 1995; Hendey et al. 1996), rather than an effect of pseudopodia formation.

In newt eosinophils, which are giant motile cells amenable to microinjection (Pettit and Fay 1998), a suppressive role was reported for PKC on pseudopodia formation. However, it seemed that the effect of PCK was via a reduction in of Ca^{2+} signalling (Gilbert et al. 1994). Also in newt eosinophils, photolytic uncaging of microinjected caged inhibitory peptide, which interfered with calcium-calmodulin activation of myosin light chain kinase (MLCK) was reported to blocked newt eosinophil pseudopodia formation and loco-motion (Walker et al. 1998). It is not clear whether this also applies to phagocytosis. However, there is a strong case for another Ca^{2+} activated pro-tein, calpain, being the target of elevated Ca^{2+} in phagocytosis.

This chapter will, therefore, discuss the ev-idence for the model shown in Fig. 8.1 at the cell biological level, first in the roles of Ca^{2+} in phagocytosis, and then of Ca^{2+} and calpain cell shape changes, followed by the evidence for the specific role of calpain in phagocytosis. Then the molecular structure of calpain will be briefly de-scribed. Finally, the mechanism by which calpain activation is limited at the cell periphery to locally form pseudopodia as required for phagocytosis will be discussed.

Ca^{2+} in Phagocytosis and Cell Shape Change

It has been known, since the work of Hamburger (Hamburger 1910, 1915, 1916), that extracellular Ca^{2+} is important for phagocytosis. He found that when neutrophils were deprived of Ca^{2+} for a period, their phagocytic power was reduced to zero (see Chap. 2). However, the phagocytic ca-pacity could be restored by the addition of extra-cellular Ca^{2+} but not strontium, barium of other ions (Hamburger and de Haan 1910). Since this ground-breaking discovery the change in cytoso-lic Ca^{2+} during phagocytosis has been observed using a number of fluorescent Ca^{2+} indicator probes. In 1985, Sawyer et al. reported that a wave of Ca^{2+} spread through the neutrophil dur-ing phagocytosis (Sawyer et al. 1985). Similarly, Marks and Maxfield (1990) reported a localized cytosolic Ca^{2+} increase in the periphagosomal region of neutrophils, with some phagocytic stim-uli. There were also a report of a Ca^{2+} wave from

the phagosome (Schwab et al. 1992), a ring of high Ca^{2+} around the phagosome in monocytes (Kim et al. 1992) and of Ca^{2+} stores localising near the phagosome (Stendahl et al. 1994; Theler et al. 1995). Others have failed to detect such localised Ca^{2+} changes (Dewitt et al. 2003) but this may reflect the morphology of the phagocy-tosing cell and opsonisation (type of opsonin or none) of the target (Murata et al. 1987). However during cell spreading on a surface which engaged integrins, which can be regarded as "frustrated phagocytosis", contact of the cell with the surface triggers a cytosolic Ca^{2+} signal in neutrophils and results in cell spreading (Pettit and Hallett 1996). This Ca^{2+} signal was resolved transversely across the contact points using z-scanning confocal mi-croscopy of Ca^{2+} -dependent fluo3 intensity. It was found that the global elevation of cytoso-lic free Ca^{2+} was preceded by localized Ca^{2+} changes at the cell surface adhesion points (Pettit and Hallett 1996). These may be similar to the hotspots of Ca^{2+} near the phagosome reported by (Nunes et al. 2012).

There is agreement that the Ca^{2+} signal precedes completion of phagocytosis (Dewitt and Hallett 2002; Dewitt et al. 2003; Francis and Heinrich 2018; Kruskal and Maxfield 1987; Marks and Maxfield 1990; Meagher et al. 1991; Theler et al. 1995). It is necessary for effective phagocytosis by accelerating the step from phagocytic cup formation to phagosome closure (Dewitt and Hallett 2002). However, the identity of the molecular target of this cytosolic Ca^{2+} signal in phagocytosis, until recently, has been less clear.

Calpain Activation in Shape Change

The role of calpain in cell shape change has long been recognised in a number of cell types. Inhi-bition of calpain activity reduces cell movement as a results of inhibition of cell spreading (Hut-tenlocher et al. 1997; Kulkarni et al. 1999; Potter et al. 1998; Stewart et al. 1998) and actin based cell shape changes (Dourdin et al. 2001; Franco et al. 2004a, b; Glading et al. 2001; Huttenlocher et al. 1997). Calpain-deficient fibroblasts (lacking the small subunit: Capn 4−/−) have reduced motility and pseudopod formation (Dourdin et al. 2001). Activation of μ-calpain by an elevation of cytosolic Ca^{2+} has therefore been proposed as the mechanism for neutrophil membrane expansion and cell spreading.

The similarities between rapid cell spreading and frustrated phagocytosis have yielded additional information. As with phagocytosis, there is a large Ca^{2+} signal before cell spreading occurs (Kruskal et al. 1986). Evidence has been presented for the involvement of calcineurin, a Ca^{2+} activated phosphatase in this step (Hendey et al. 1992; Lawson and Maxfield 1995). However, since Huttenlocher et al. (1997) reported that calpain was involved in cell migration of CHO cells, (Chinese hamster ovary cells, an epithelial cell line), there has been an accumulation of evidence for a role for the Ca^{2+} activated cytosolic, calpain, in cell shape change and in phagocytosis. For example, the spreading of platelets (Croce et al. 1999) and lymphocytes (Rock et al. 2000; Stewart et al. 1998) is prevented by inhibition of calpain. Also platelets from calpain-1 null mice also had defective spreading (Azam et al. 2001). Surprisingly in some cell types not usually associated with cell spreading (pancreatic cells, vascular smooth muscle cells, CHO cells and osteoclast), their spreading in cell culture is also inhibited by calpain inhibitors (Kulkarni et al. 2002; Marzia et al. 2002; Parnaud et al. 2005; Paulhe et al. 2001).

Pharmacological inhibition of calpain also prevents cell spreading of neutrophils both in tightly controlled in vitro conditions (Anderson et al. 2000; Dewitt and Hallett 2002; Wiemer et al. 2010) and on endothelium (Noble et al. 1998). Neutrophils from calpain1 null mice also have defective cell spreading (Ishak and Hallett 2018). There have also been reports that pharmacological inhibition of calpain also activates neutrophil random movement and shape change (Katsube et al. 2008; Lokuta et al. 2003; Noma et al. 2009). However, care must be taken in interpreting these effects as many calpain inhibitors are also agonist of the formylated peptide receptor (Fujita et al. 2011), a notoriously promiscuous receptor (Migeotte

et al. 2006) that induces chemokinesis (and other responses) in neutrophils. This was reported by Kitagawa's group (Fujita et al. 2011), who had previously reported the stimulatory effect of calpain inhibitors in two papers as being the result of calpain inactivation (Fujita et al. 2011). It is well known that activation of these receptors on neutrophils, will activate random migration and shape change. It is therefore important that either pharmacological inhibition of calpain is used carefully or that non-pharmacological interventions are employed.

Circulating neutrophils isolated from calpain-1 null mice, when tested *ex vivo*, showed a clear defect in the ability to undergo neutrophil shape change (Ishak and Hallett 2018). In order to exclude the possibility that the defect was a result of signalling up stream of Ca^{2+} (ie receptor engagement, G-proteins etc), the cells were loaded with caged IP_3, so that the Ca^{2+} signal could be controlled in amount of Ca^{2+} and its timing. Under these conditions, calpain-1 null neutrophils failed to spread when IP_3 was uncaged, under conditions that triggered the wild type neutrophils to spread fully within 130 s (Ishak and Hallett 2018).

Although not a cell-type which undergoes rapid shape change (or phagocytosis), NIH 3T3 fibroblasts are amenable to genetic manipulation and have thus provided strong evidence for a role of calpain in cell shape change. Following sedimentation of fibroblasts in suspension onto a surface, the cells undergo a transition and slowly spread on the substrate. Cell permeant inhibitors of calpain, including calpeptin, produce an immediate cessation of spreading by these cells (Potter et al. 1998). Knocking out the common sub unit of calpain1 and calpain 2 produced fibroblast which when spread had unusual morphology (Franco et al. 2004a, b). In particular, protrusion of lamellipodia at the leading edge did not properly extend. By also "silencing" calpain2 large subunit expression, this effect was attributed specifically to calpain 2, which also reduced proteolysis of the cytoskeletal membrane cross-linking protein, talin, as it inhibited cell spreading (Franco et al. 2004a, b). Thus it was concluded that in this cell type,

Fig. 8.2 "Map" of calpain domains. The large and small subunits of calpain1 and 2 are shown as a "cartoon". The domains are labelled by the usual convention and the main feature of each domain indicated. The site of inhibition by calpastatin is indicated and Ca^{2+} ions are shown interacting witht he EF hands in domains IV and VI. This is the regulatory domain and the site at which the mercaptoacrylate inhibitors of calpain activation bind

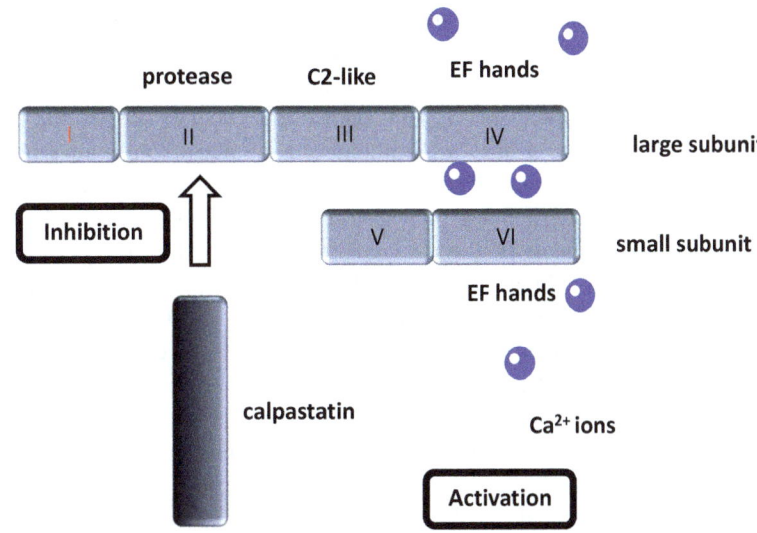

calpain 2 activity was an obligatory step in cell spreading. This conclusion re-affirmed an earlier study that did not involve knocking out calpain, but instead inhibited its activity by expressing calpastatin, the endogenous inhibitor of calpain. This molecule binds to the proteolytic site on calpain (Fig. 8.2) which is exposure in the Ca^{2+} activated form of calpain (Hanna et al. 2008). Expression of calpastatin may be a physiological mechanism of preventing potentially dangerous calpain activity in some cells. Potter et al. (1998) expressed calpastatin in NIH-3 T3 fibroblasts and selected high calpastatin expressing clones (cell lines) for experimentation. They found that all of the high calpastatin expressing cell lines failed to spread, and attributed this to a failure to extend lamellipodia (Potter et al. 1998). They also found that the actin-membrane crosslinking protein, ezrin, but not its related ERM family members (radixin or moesin), was increased in the high calpastatin cell lines. This suggested that ezrin was the substrate for calpain, and that its concentration rose because with calpain inhibition (ie in the absence of calpain proteolysis), there was an accumulation of uncleaved substrate. This suggested a role for this substrate (ezrin) in calpain-mediated cell spreading. The evidence for an involvement of ezrin in phagocytosis and membrane tension is discussed in this book in Chap. 6.

Calpain in Phagocytosis

Cell shape change during spreading and the formation of cell protrusions is probably easier to study than phagocytosis. The bulk of evidence for the role of calpain is, thus, in relation to generalised cell shape change. Obviously, the cell shape change that occurs during phagocytosis is a special case, with localised and unique features. However, it is part of the bigger set of cell shape changes, for which the evidence of an involvement of calpain is strong. Cell shape changes, and particularly phagocytosis, are dependent and influenced by the cell membrane tension (see Chap. 6), which is controlled by the calpain substrate ezrin (Liu et al. 2012; Rouven Brückner et al. 2015; see Chap. 6). Together these findings suggest for a role for calpain in cell shape changes. However, the difficulties in studying the role in phagocytosis has limited the investigation of calpain specifically in phagocytosis to the use of inhibitors of calpain. There are, however, some "technical" considerations that should not be forgotten when using calpain inhibitors. It is important that the two distinct classes of calpain inhibitor, calpain proteolytic inhibitors and calpain activation inhibitors, are recognised and used appropriately. The largest class of calpain inhibitors, which include the commonly used inhibitors ALLN and calpeptin, act on the proteolytic site of calpains ie domain II (Fig. 8.2). As domain II shares a number of features with other cysteine protease, these inhibitors cannot have selective specificity for calpains. However, they are very useful as they will inhibit active calpain (in addition to other proteases). Expression of calpastain, it perhaps the most specific way to inhibit domain II, the calpain protease activity. Calpastain binds to domain II in the active configuration (Fig. 8.2) but does not prevent the initial activation step. The second group of inhibitors are the mercaptoacrylates, such as PD150606 (Wang et al. 1996) and some newer more potent analogues (Adams et al. 2012, 2015). These act on the Ca^{2+} regulatory site ie domain IV and VI (Adams et al. 2014; Lin et al. 1997). These inhibitors therefore prevent the activation of calpain Ca^{2+} This gives them selectivity for Ca^{2+} activation of calpain1 and 2, since they can only inhibit enzymes that use the same Ca^{2+} activation mechanism of calpain dVI (Adams et al. 2014; Lin et al. 1997). However, after calpain has been activated by Ca^{2+}, there often follows an auto-proteolytic step separating the proteolytic domain from its regulatory constraint, thereby perpetuating the proteolytic activity of calpains (Cong et al. 1989). Therefore adding a mercaptoacrylate calpain inhibitor after the calpain has been activated, does not inhibit the active enzyme. After a cell has experienced a Ca^{2+} influx step, these inhibitors can have no effect on calpain activity and therefore have no cellular effect. These are important considerations, when deciding on which inhibitors to use, and in interpreting the results of calpain inhibition.

There are also "technical" issues surround the seemingly easy measurement of phagocytosis. A simple method often used is to allow fluorescent particles to sediment onto (or be centrifuged onto) cells and then measuring the number of particles per cell, by microscopic observation or by flow cytometry. However, this, and many other methods do not distinguish particles which are associated with cells, eg by adherence, in phagocytic cups etc. from particles that have been phagocytosed (ie within the cell). This is not a trivial distinction, as inhibitors may have an effect at any one of the 5 or 6 steps of phagocytosis. If binding is inhibited, then obviously all subsequent steps

cannot occur. It would be incorrect to conclude that such an experimental procedure inhibited pseudopod extension etc. This is especially relevant to calpain inhibition, since the role of calpain activation is after phagocytic cup formation, so that quite firm attachment between phagocytic target and the cell can occur, giving a positive count by flow cytometry and quick microscopic inspection, but without the completion of phagocytosis. At present, these subcellular events, can only be studied and understood microscopically at the single cell level.

Micropipette delivery of the phagocytic stimulus to the cell enables the complete phagocytic event, from first contact to phagosome formation, to be observed and dissected (Dewitt and Hallett 2002; Francis and Heinrich 2018). Figure 8.3a shows a typical experiment, with the three steps of contact, phagocytic cup formation and completion of phagocytosis by closure of the phagosome (Fig. 8.3a). In the cell shown, cytosolic free Ca^{2+} was also monitored by ratiometric imaging/fluorimetry of cytosolic fura2. It is seen that contact, binding and phagocytic cup formation occurs over the first 53 s without a significant change in cytosolic Ca^{2+} (Fig. 8.3b). Then there is a quick rise in free Ca^{2+}, over the subsequent 3 s (Fig. 8.3b) during which phagocytic pseudopodia extend around the target (compare image for 53 s with that for 54 s and 55 s). By 89 s, phagocytosis is complete and the cell begins to round up as cytosolic Ca^{2+} falls to resting levels by 200 s (Fig. 8.3b). After this point, the cell will respond in a similar manner to a second (and third) particle (Dewitt and Hallett 2002, Dewitt et al. 2003). When the Ca^{2+} signal is prevented by removal of extracellular Ca^{2+} or by inhibition of Ca^{2+} influx by the blocking ion Ni^{2+}, the rate of phagocytosis is also significantly reduced (Fig. 8.4a). A similar reduction is also achieved by pre-treating the cells with a mercaptoacrylate calpain activation inhibitor such as PD150606, (Fig. 8.4a: Dewitt and Hallett 2002) or its analogues (Adams et al. 2012, 2015), suggesting a link between Ca^{2+} elevation and calpain activation. However, experiments at the single cell level revealed the step in phagocytosis from which the inhibition originated. In the non-calpain inhibited

(a) contact >>>> phagocytic cup >>>>> closure

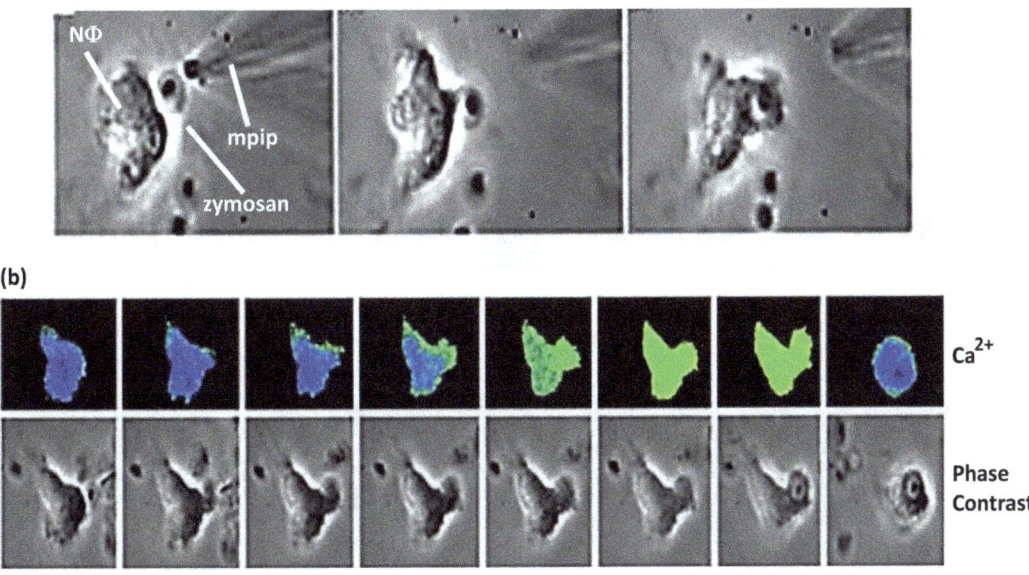

Ca^{2+}

Phase
Contrast

0s   36s   46s   53s   54s   56s   80s   197s

Fig. 8.3 Ca^{2+} and phagocytosis. (**a**) The sequence of images show the phagocytic stimulus, C3bi-opsonised zymosan particle (labelled "zymosan"), being delivered by micropipette (labelled "mpip") to a human neutrophil (labelled "NΦ"). The first image shows the point of contact between the particle and the cell. The second image shows the formation of the phagocytic cup. The final image shows the closure of the phagosome and completion of phagocytosis. (**b**) Shows a similar experiment with simultaneous imaging of cytosolic free Ca^{2+}, pseudocoloured to show the rise in Ca^{2+}, where blues and cold colours represent low Ca^{2+} and greens, yellows and warmer colour represent raised Ca^{2+}. (The reader should refer to the web version for the full colour rendition of this figure. © 2002 Dewitt SD and Hallett MB. Originally published in J. Cell Biol. https://doi.org/10.1083/jcb.200206089)

cells (absence of calpain inhibitor), the phagocytic pseudopodia extension from phagocytic cup accelerates to enclose the particle within 100 s of contact (Fig. 8.5: see Chap. 6, Fig. 6.4; Dewitt and Hallett 2002). The acceleration begins from the point at which the Ca^{2+} signal fires, when a critical phagocytic cup size (possibly when a critical number of opsonisation receptors have engaged) is reached. Inhibition of calpain activation, does not affect the ability of the cell to contact and bind to the particle (Fig. 8.4bi, ii). A phagocytic cup also forms sufficiently to generate a normal Ca^{2+} signal (Fig. 8.4biii). However, phagocytosis halts at this stage (Fig. 8.4biv). The open phagocytic cup can be seen in Fig. 8.4biv with the target zymosan particle drifting away from the aborted phagocytic cup (Fig. 8.4bv). These experiments link the cytosolic Ca^{2+} signal of phagocytosis with calpain activation at the pseudopodia extension step of phagocytosis,

and after phagocytic cup formation. It was thus concluded that Ca^{2+} activation of calpain was the accelerator step (Fig. 8.5a). Without the accelerator, the time to complete phagocytosis would be prolonged considerable (Fig. 8.5b). The creep of the phagocytic cup, leading to half the phagocytic target (ie before the acceleration) has been measured and modelled by Irmscher et al. (2013). They report that "the speed of cup progression is remarkably well conserved" at about 20 nm s^{-1}. Before the phagocytic Ca^{2+} signal in neutrophils, the slow phase of cup progression occurred at a similar rate (Dewitt and Hallett 2002) ie about 18 nms^{-1}. These values suggest a pseudopodia progression against a restraint, as compared to the fastest modelled rate of 120 nms^{-1} (Irmscher et al. 2013). The speed of pseudopodia progression in neutrophils after the Ca^{2+} signal nearly reached this maximum speed, reaching about 100 nms^{-1} (Dewitt and Hallett 2002). If the

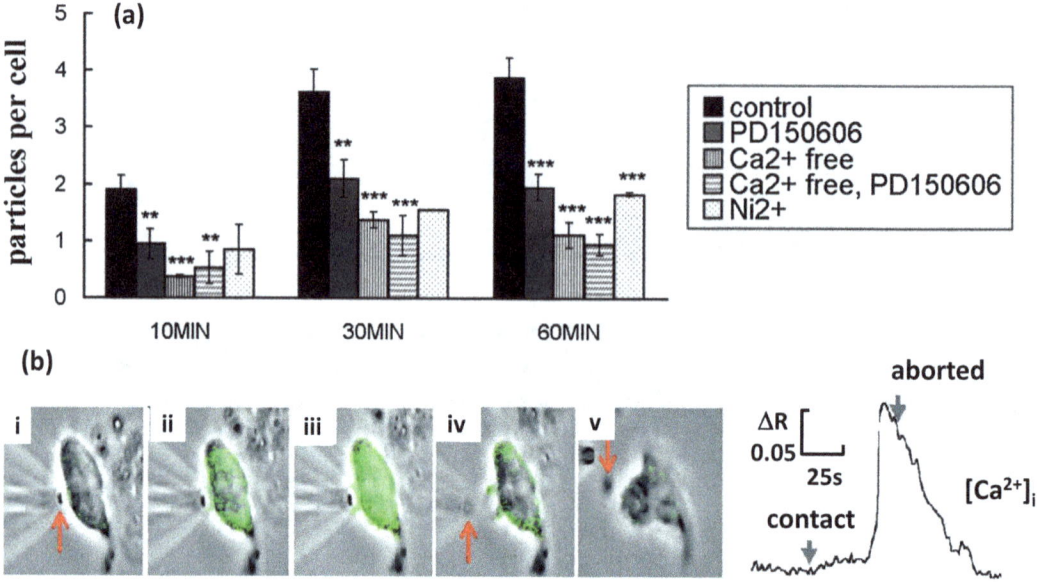

Fig. 8.4 Calpain and phagocytosis. (**a**) The uptake of zymosan particles (in populations of cells) in the presence of inhibitors at the times shown. The key on the right indicates "control" ie DMSO solvent alone: PD150606, calpain activation inhibitor: the absence of extracellular Ca^{2+} ("Ca^{2+} free"); and with the Ca^{2+} influx blocking ion Ni^{2+}. (**b**) Show a typical experiment, where a zymosan particle (indicated by the arrow) was presented to the cell loaded with the cytosolic Ca^{2+} indicator fura2 and the phase contrast image pseudocoloured to show the timing of the Ca^{2+} signal. (i) contact of the particle with the cell; (ii) the onset of Ca^{2+} signalling; (iii) the peak Ca^{2+} signal; (iv) abortive phagocytic cup; and (v) after Ca^{2+} return to resting phagocytosis. (The reader should refer to the web version for the full colour rendition of the figure. © 2002 Dewitt SD and Hallett MB. Originally published in J. Cell Biol. https://doi.org/10.1083/jcb.200206089)

rate of the slow phase continued linearly, in the absence of calpain activation, completion would take 3.5 fold longer (Fig. 8.5b). This creep of pseudopodia around the phagocytic target, without the need for calpain activation, may be the result of the physical pulling of membrane by the particle ie the biological zipper effect (Griffin et al. 1975; Swanson and Baer 1995; see also Chap. 6) or a strong physical interaction, such as with polystyrene or latex beads. This would be possible if the contact binding between the particle and the cell were greater than that holding wrinkles in place. It can be seen in biophysical experiments that when laser tweezers can pull on bead attached to the cell surface by concanavalin A, a cell tether several microns long can be pulled out (Raucher and Sheetz 1999). Thus the binding through concanavilin A is stronger than that which holds together surface wrinkles.

It is interesting to note that although Ca^{2+} signalling accelerates phagocytosis, phagocytosis can also occur without the need for a Ca^{2+} (Di Virgilio et al. 1988; Lew et al. 1985). This confirms that unfolding of cell surface wrinkles to provide additional membrane can occur in a non-calpain dependent manner, probably by physical unwrinkling of the membrane reservoir. Unfortunately, the temporal resolution of phagocytosis in the studies by Lew et al. (1985) and Di Virgilio et al. (1988) was not sufficient (ie 5–30 min) to establish whether phagocytosis in the absence of a Ca^{2+} signal was significantly slower than in its presence. From the calpain inhibitor studies, however, the unwrinkling of the membrane reservoir is made easier and quicker after calpain activation. A similar conclusion was also reached for frustrated phagocytosis (cell spreading) by neutrophils (see preceding section), when uncaging IP_3 or uncaging Ca^{2+} within the neutrophil cytosol was the defining trigger for spreading. Cell spreading under these conditions, is blocked completely be calpain

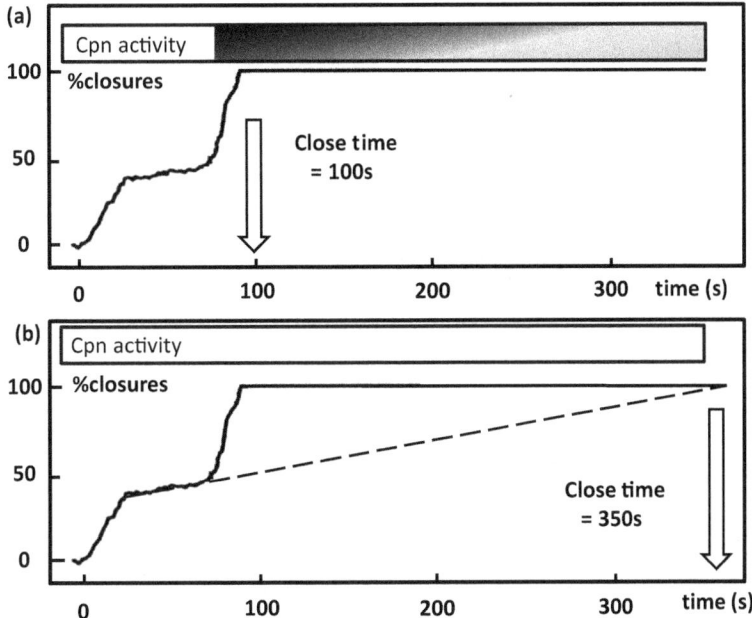

Fig. 8.5 Phagocytic pseudopodia extension and the calpain accelerator. (**a**) The progression of phagocytic pseudopodia around a target zymosan particle, is shown as percentage of the complete encirclement (ie 100%- = closure of the phagosome). The closure time of 100 s is shown, together with the onset of calpain activation (shaded bar labelled "cpn") inferred from the Ca^{2+} signal (not shown). The phagocytic pseudopodia progression and Ca^{2+} signalling data are taken from Dewitt and Hallett (2002). (**b**) The same progression of phagocytic pseudopodia, but with the anticipated continued rate (dotted line) in the absence of the accelerator effect of a Ca^{2+} signal and calpian activation. (© 2002 Dewitt SD and Hallett MB. Originally published in J. Cell Biol. https://doi.org/10.1083/jcb.200206089)

inhibitors (Dewitt et al. 2013). The requirement for additional membrane (and hence unfolding of cell surface wrinkles) is greater in cell spreading than in phagocytosis, especially for a single phagocytic event or with phagocytosis of small targets, such as bacteria. Thus, cell spreading may be expected to be more dependent on the need to signal Ca^{2+} and calpain activation.

It was therefore predicted that calpain is activated during phagocytic Ca^{2+} signalling. Confirmation of this required a methodology for monitoring calpain activity in individual neutrophils undergoing phagocytosis. To this end, a microinjection technique that works well with neutrophils has been developed (Lewis et al. 2014; Campbell and Hallett 2015). The technique involves a localised electroporation of the phagocyte and the synchronised iontophoretic release of a pulse of injectate (Haas et al. 2001). This is a no-touch injection (point and shoot) technique which is surprisingly "gentle". Neutrophils can be injected while undergoing chemotaxis without apparently "noticing" the injection (Lewis et al. 2014) and non-spread neutrophils, injected by this technique, spread normally in response to a cytosolic Ca^{2+} elevating stimulus (Lewis et al. 2014). Using this approach, a cell impermeant fluorogenic peptide substrate with specificity for calpain cleavage (H-Lys(FAM)-Glu-Val-Tyr-Gly-Met-Met-Lys(Dabcyl)-OH) was introduced into neutrophils (Fig. 8.6a). The amino acid sequence is the calpain-1 cleavage site of α-spectrin. The fluorescence of the fluorescein moiety at one end of the peptide is internally quenched through fluorescence resonant energy transfer (FRET) to the Dabcyl fluorescence quencher at the other end. When cleaved by calpain between the Tyr-Gly residues, quenching is relaxed and there is a resultant enhancement

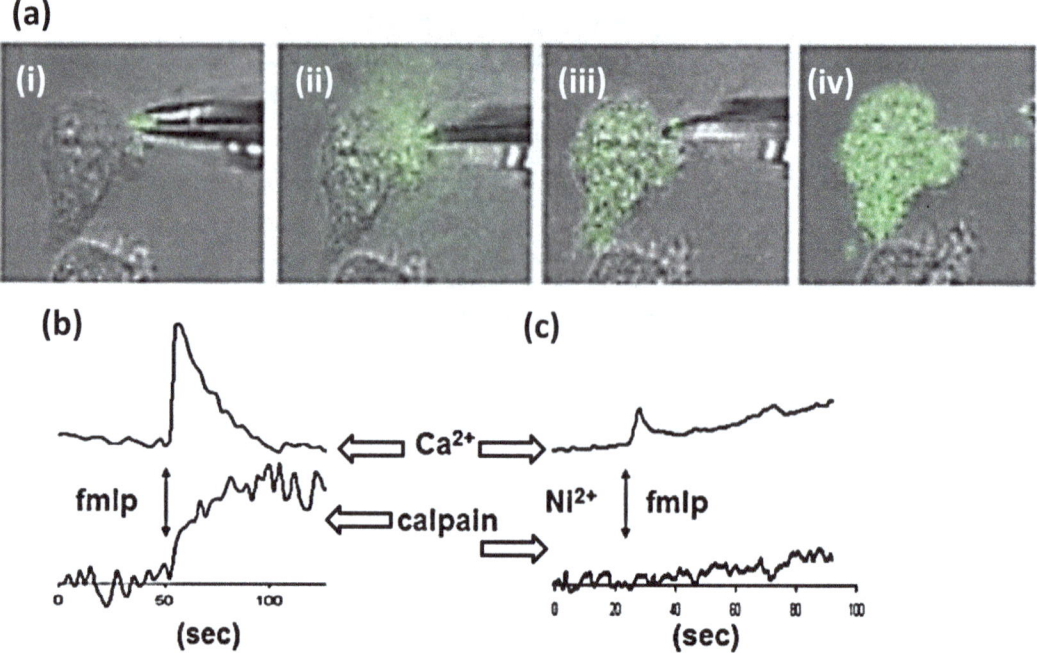

Fig. 8.6 Ca^{2+} dependent calpain activation in phago-cytes. (**a**) Electroinjection of the calpain FRET probe(H-Lys(FAM)-Glu-Val-Tyr-Gly-Met-Met-Lys(Dabcyl)-OH). The sequence of images shows (i) the micropipette close to a human neutrophil (ii) the effect of the voltage pulse ejecting the weakly fluorescent probe from the micropipette and transiently electroporating the cell, (iii) the cell after electroinjection with weakly fluorescent calpain FRET substrate, (iv) the increase in fluorescence after Ca^{2+} signalling. (**b**) Shows the time courses of the increase in FRET signal (labelled "calpain") and cytosolic Ca^{2+} concentration (labelled "$Ca2^{+}$") measured in the same cell during stimulation with the formylated peptide, f-met-leu-phe. (labelled fmlp). (**b**) Shows the time courses of the increase in FRET signal (labelled "calpain") and cytosolic Ca^{2+} concentration (labelled "$Ca2^{+}$") measured in the same cell during stimulation with the formylated peptide, f-met-leu-phe. (labelled fmlp) in the presence of the Ca^{2+} influx blocking ion, Ni^{2+}. (The reader should refer to the web version for the full colour rendition of the figure. Reprinted from Biochimica et Biophysica Acta (BBA) – Molecular Cell Research, vol 1843, Kimberley J. Lewis, Benjamin Masterman, Iraj Laffafian, Sharon Dewitt, Jennie S. Campbell, Maurice B. Hallett, Minimal impact electro-injection of cells undergoing dynamic shape change reveals calpain activation, pp. 1182–1187, © 2014, with permission from Elsevier)

of fluorescence (Mittoo et al. 2003). This approach successfully reported calpain activation in neutrophils stimulated by the cytosolic Ca^{2+} elevating formylated peptide, f-met-leu-phe (Fig. 8.6b). It was shown that there was a requirement for Ca^{2+} influx as calpain activation was inhibited by preventing Ca^{2+} influx with the blocking ion Ni^{2+} (Fig. 8.6b). Interestingly, the release of Ca^{2+} from Ca^{2+} stores within the neutrophil was not sufficient (or more likely, in the right location) to activate calpain (Fig. 8.6b). The problem with microinjecting the peptide substrate was a single Ca^{2+} signal (as from f-met-leu-phe) consumed nearly all the

peptide substrate within 50 s (Fig. 8.6b) making it difficult to study phagocytosis. Also, there was often a "resting" rate of consumption of peptide consumed substrate before the experiment could be undertaken (Fig. 8.7). Lokuta et al. (2003) also reported a basal "calpain" activity using the peptidase substrate CMAC, t-BOC-LM as an indicator. Although, the consumption of FRET substrate was reduced by ALLN and calpeptin (Fig. 8.7d), the consumption may not have been due to genuine calpain activity, but instead may be due to other non-specific protease activity. Calpeptin and ALLN are not specific inhibitors of calpain proteolysis (see earlier) and

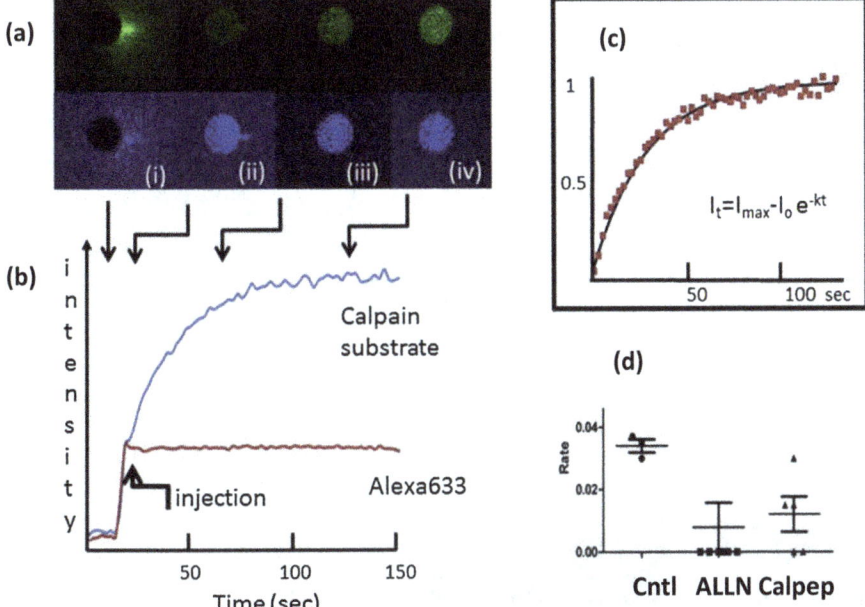

Fig. 8.7 "Spontaneous calpain activation" in phagocytes. (**a**) A sequence of paired images showing the simultaneous electroinjection of calpain FRET substrate (upper sequence) and Alexa 663 (lower sequence) in a solution containing EGTA (10 mM), which lowered the resting cytosolic free Ca^{2+} to below 50 nM during the time of electroinjection. (i) The voltage pulse ejecting the two fluors. Note that at this stage the cell is seen only as a dark silhouette. (ii) Immediately after electroinjection. (iii and iv) show the situation at the times indicated by the arrow on the graph (**b**) beneath. Note that the FRET signal from the calpain substrate increases while the Alexa 663 control signal does not. (**b**) Shows the complete time course of the data in (**a**). (**c**) Shows a comparison of the FRET/Alexa ratio first order reaction kinetics (shown **b**) plotted as dots with the theoretical first order reaction kinetics (shown as line). (**d**) Shows the effect of pretreatment of the neutrophils with ALLN or calpeptin on the FRET consumption rate constant. (The reader should refer to the web version for the full colour rendition of the figure. Reprinted from Biochem. Biophys. Res. Commun. vol 457: Campbell and Hallett, pp. 341–346,© 2015 with permission from Elsevier)

the consumption of the FRET substrate was not prevented by microinjections of substrate with 10 mM EGTA, which lowered the cytosolic free Ca^{2+} (Campbell and Hallett 2015). However, despite this high background rate, by including a second fluor (Alexa 633) in the injectate to act as a control for cell shape and thickness (Fig. 8.7a, b), it was possible to measure enzyme activity (Fig. 8.7c) in neutrophils undergoing phagocytosis and compare this with non-phagocytic neutrophils. The substrate-cleaving enzyme activity in "resting" non-phagocytic neutrophils had a mean rate of about 0.021 s^{-1}, whereas cells which were undergoing phagocytosis had an increased enzyme rate of 0.032 s^{-1} (sig diff at p = 0.057). This small increase (approx. 50% increase) and the inability to measuring the consumption

rate before and after phagocytosis in the same cells, made further study difficult (Campbell and Hallett 2015). The problem was that the amount of substrate in the cell that could be microinjected was restricted by the cell volume and solubility of the substrate, which ultimately limited the usefulness of this approach. To overcome this, a membrane-permeant fluorogenic substrate was used, which was present throughout the experiment and thus provided a vast supply of substrate for the cell. Diffusion of substrate into the cell from the extracellular medium would replenish intracellularly cleaved substrate during the experiment. Gitler and Spira (1998) have successfully used this approach in neurones with (CBZ-Ala Ala)$_2$ R110, a non-fluorescent calpain substrate derivative of rhodamine 110

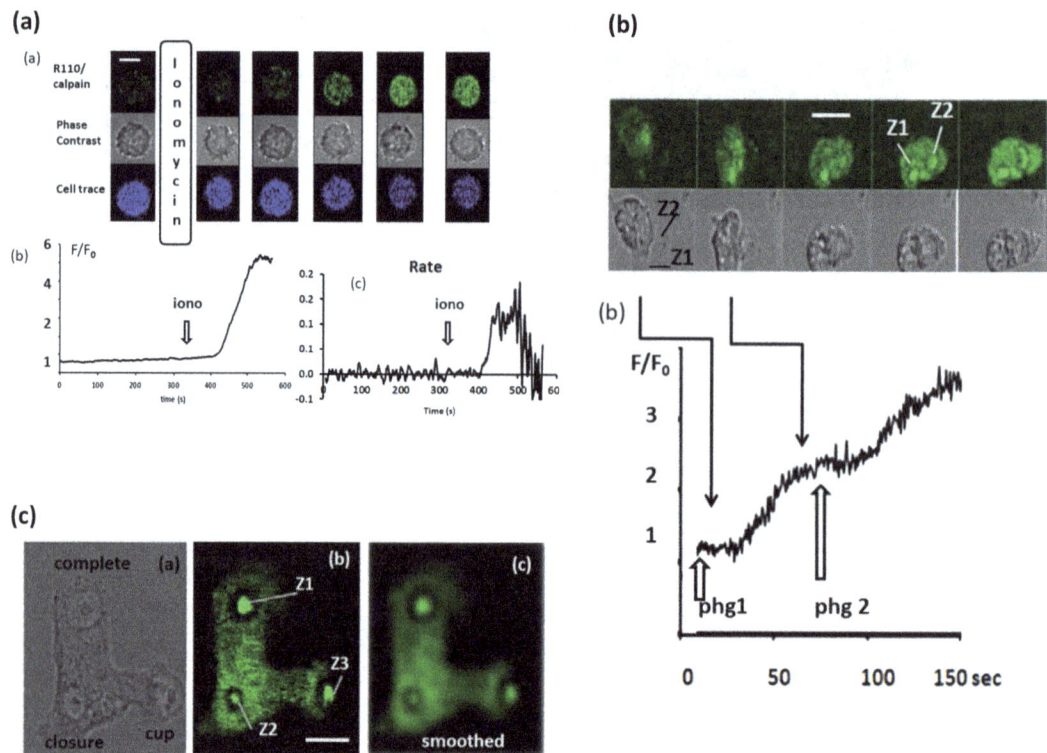

Fig. 8.8 Calpain "bursts" during phagocytosis. (**a**) Ionomycin-induced calpain activation monitored in a single neutrophil using the cell permeant calpain fluorogenic substrate (CBZ-Ala Ala)$_2$ R110. (i) The upper sequence of images shows the time course of R110 fluorescence (as a marker of calpain activity); The middle sequence of images, show the corresponding phase contrast and the lower sequence, the inert control Celltrace (inert control), as indicated. The scale bar indicates 5 μm on all images. (ii) The time course of R110 fluorescence normalised for Cell trace intensity (as F/F0) and (iii) the rate of change of fluorescence, proportional to the calpain activity. (**b**) Calpain activity and cytosolic Ca^{2+} during phagocytosis. (i) A sequence of images showing the time course of R110 fluorescence (upper sequence) and phase contrast (lower sequence) images during consecutive phagocytoses of two C3bi-opsonised zymosan particles (Z1 and Z2). Note each

zymosan particle has an autofluorescent central region. (ii) The time course of normalised R110 intensity (as F/F0) during the double phagocytosis with the arrows indicating the two phagocytotic events (phag 1and phag 2). The arrow show where in the time sequence, the images were taken. The scale bar of 10 μm refers to all images insection b. (**c**) Localisation of calpain activity. Images (**a**), (**b**) and (**c**) show a neutrophil undergoing multiple phagocytoses of C3bi-opsonised zymosans. (i) shows the phase contrast image with the three phagocytic vents labelled as "complete" ie a closed phagosome; "closure" a recently closed phagosome; and "cup" a phagocytic cup. In (ii), the R110 fluorescence image these are labelled Z1, Z2, Z3. Image (iii) is the result of noise reduction "smoothing". The scale bar indicates 5 μm. (Reprinted from Biochem Biophys Res Commun. 515, S. Jennings, M.B. Hallett, pp. 163–168, copyright (2019) with permission from Elsevier)

(R110). This molecule releases R110, a bright green emitting fluor when cleaved (Leytus et al. 1983). Although (CBZ-Ala Ala)$_2$ R110 is also a substrate for the neutrophil granule enzyme, elastase (Johnson et al. 1993), there was little background consumption of the probe and there was no increase in fluorescence within neutrophils granules (Fig. 8.8a–c). The

increase in (CBZ-Ala Ala)$_2$ R110 signal triggered by elevating cytosolic Ca^{2+} using ionomycin was about 6 times the background (Fig. 8.8a; Jennings and Hallett 2019), suggesting that the probe would be sensitive enough to report calpain activity dynamically in real time. In 36/42 phagocytic events recorded, there was a detectable "burst" of calpain activity co-incident with the phagocytosis. Multiple phagocytic events triggered in the same cell were always accompanied with bursts of calpain activity (Fig. 8.8b: Jennings and Hallett 2019). Each calpain burst also corresponded to the phagocytic Ca^{2+} signal. Thus, detectable calpain activity was triggered during phagocytosis. The location of calpain activity could be identified by R110 subcellular distribution as R110 diffuses away from it generation site. It was estimated that the diffusion constant for R110 in neutrophil cytoplasm was about 17.5 μm^2/s (Jennings and Hallett 2019). The mean displacement of 5 μm from its source will therefore take about 230–400 ms (depending on the geometry). However, by fast image acquisition (sacrificing spatial resolution), images showing the source of the elevated calpain activity established that the origin was the early phagosome. For example, in Fig. 8.8c, there is little R110 fluorescence around an old phagosome (Z_1); a significant amount of R110 diffusing from the site of a recently formed phagosome (Z_2); and R110 fluorescence restricted to the phagocytic cup (Z_3) of an ongoing phagocytosis (Jennings and Hallett 2019). Together, these data therefore provides good evidence of the activation of calpain during phagocytosis.

In calpain-1/2 knock out mice, there is also a clear defect in neutrophil behaviour (Kumar et al. 2014). Transgenic mice expressing Cre recombinase in innate immune cells was used to generate conditional capns1 (the common small subunit) knockout mice, ie deficient in calpain 1 and 2 activity. When the knockout mice were challenged by an intraperitoneal injection of bacterial, there was significantly impaired bacterial killing within the peritoneal cavity and the animal developed bacteraemia (Kumar et al. 2014). As with many whole animal studies, there was more than one underlying cause for the effect. Firstly, adherence of neutrophils to the endothelial cells lining the blood vessels, flattening (spreading) and then chemotaxis are each calpain-dependent. Not surprisingly, then, the numbers of neutrophils in the peritoneal cavity of knockout mice was lower than in the wild type mice, contributing to poor bacterial clearance. However, Kumar et al. (2014) also found that killing of bacterial and activation

of the neutrophil oxidase were also reduced in the knockout cells. These are both events that follow phagocytosis and therefore was consistent with a role of calpain for phagocytosis. However, the authors concluded that the mouse macrophages were phagocytically competent over long periods. However, Figure 4A of their paper shows that there were significantly more extracellular viable bacteria in the *in vitro* knockout macrophages assay, especially at the earliest time, 15 min, suggesting a reduced uptake of bacteria by the calpain null macrophages. This leaves open the question of whether the accelerator of phagocytosis was impaired by knockout of calpain1/2. Neutrophils isolated from the circulation of calpain-1 knockout mice, which were shown to have a severe defect in cell spreading (Ishak and Hallett 2018), were also phagocytically "competent" (ie they could internalise particles). However, the time between presentation of a zymosan particle and completion of phagocytosis was significantly extended in the calpain-1 deficient mice, from 114 s for wild-type neutrophils to 325 s for calpain null neutrophils (Ishak 2012). This was consistent with the calpain-dependent accelerator effect on phagocytosis. However, the delay in phagocytosis by calpain1 null neutrophils was mainly due the time taken for calpain-1 null neutrophils to form a large enough phagocytic cup to signal a global Ca^{2+} change (WT = 69 s; KO = 195 s) or be observed microscopically (WT = 61 s: KO = 260 s). This may suggest that the phagocytic defect in calpain-1 null neutrophils was before phagocytic cup formation. However, there was also an extended delay in calpain-1 null neutrophils from the triggering of the Ca^{2+} signal to closure of the phagosome (WT = 45 s; K0 = 129 s), with some calpain-1 null neutrophils signalling Ca^{2+} but failing to complete phagocytosis.

Structure of Calpain 1 and 2

The calpain family has at least 15 known members which are papain-like thiol or cysteine proteases. Probably the two most important calpains for phagocytosis and pseudopodia extension are μ-calpain (also known as calpain-

1) and m-calpain (also known as calpain-2). Calpain 1 and 2 are non-lysosomal proteolytic enzymes that are in the cytosol, where they are inactive until activated by Ca^{2+}. There is no unique or consensus sequence at which calpain cleaves, but there are several algorithms and programs aimed at predicting calpain cleavage sites (eg DuVerle and Mamitsuka 2019; DuVerle et al. 2011; Fan et al. 2019; Liu et al. 2011). As a consequent of the promiscuous proteolytic activity of calpain, there is a broad spectrum of proteins identified as calpain substrates, either *in vitro* or *in vivo* (Storr et al. 2011). More than 100 substrates for μ-calpain have been identified under experimental (often *in vitro*) conditions. These can be broadly categorised into three main groups, (i) cytoskeletal and membrane-associated proteins, (ii) transcription factors, and (iii) kinases/phosphatases. There is a clear significance of cross-linker proteins, ie those linking the cortical actin network to the plasma membrane, including members of the Ezrin-Radixin-Moesin (ERM) family (see Chap. 6), being calpain substrates for phagocytosis.

Calpain1 and calpain 2 proteins share 62% sequence similarity. Calpain 1 and 2 are both heterodimeric protein composed of a large 80 kDa subunit (CAPN1 and CAPN2 respectively), encoded on chromosome 1; and a smaller 28 kDa regulatory subunit (CAPNS1), encoded on chromosome 19 in humans. The small regulatory subunit (also called the common subunit and in older literature calpain 4) and the large subunit of calpains 1 and 2 can be divided into domains. Domains V on the large subunit and domain VI on the small subunit both have a calmodulin-like region, consisting of five EF hand motifs which bind Ca^{2+}. The binding of Ca^{2+} to these EF hands is key to calpain activation (Blanchard et al. 1997). Domain III on the large subunit is a C2-like domain, and may have a regulatory function by facilitating the translocate of calpain to the plasma membrane. C2 domains bind to the plasma membrane by engaging the inner leaflet of phospholipid phosphatidylserine in a Ca^{2+}-dependent manner, thus causing translocation to the membrane in regions with locally elevated Ca^{2+} concentration.

Domain II on the large subunit provides the proteolytic activity of the calpain. It results from a catalytic triad of histidine, asparagine and cysteine which together attack a carbonyl group in the target substrate to hydrolyse it. This domain is only accessible by substrates after conformational change to the calpain molecule after Ca^{2+} binding to the EF hands in the large and small subunits (Blanchard et al. 1997; Moldoveanu et al. 2002). Domain II is also the region that binds the endogenous calpain inhibitor, calpastatin, preventing activity of the catalytic site. Calpastatin therefore only binds to Ca^{2+} activated calpain (Hanna et al. 2008). Domain I of the larger subunit does not have any sequence homology to any other known protein. A domain cartoon map of calpain 1 or calpain 2 is shown in Fig. 8.2.

Calpain-1 and 2 are also identified by suffixes of μ- and m-calpain respectively which indicate the difference in Ca^{2+} concentration required for activation, namely "micromolar" 20–60 μM Ca^{2+} for μ-calpain and "millimolar" 0.3–1.3 mM Ca^{2+} for m-calpain respectively (Goll et al. 2003). Both Ca^{2+} concentrations exceed the global cytosolic free Ca^{2+} concentrations reported in neutrophils, which peaks at near 1 μM in non-muscle cells. However, the important cellular location for active calpain is the cell periphery where the cytsoskeletal-membrane interaction is important for phagocytosis. μ-Calpain translocates from the cytosol to the cell periphery via its C2-like domains, in response to elevated Ca^{2+} influx (Gil-Parrado et al. 2003). Also calpain accumulates within cell pseudopodia, being 2.5 times more concentrated than in the bulk cytosol (Jia et al. 2005). Both locations are where Ca^{2+} is expected to be considerably higher than the cytosol and probably reaches the tens of micromolar.

Physiological Activation of Calpain by Ca^{2+} in Cells

Most of the cytosolic enzymes which are physiologically activated by a change in Ca^{2+}, have dissociation constants (kd) near 0.3–

0.6 μM. Thus at the resting cytosolic free Ca^{2+} of 0.1 μM they will be inactive, but when the Ca^{2+} rises to 1 μM they will become active. Thus the physiological rise in Ca^{2+} acts as a switch for these enzymes. Physiologically bulk cytosolic Ca^{2+} in non-muscle cells rarely rises more than 1–2 μM. So it is perhaps surprising to find that the calpain-1 is activated by Ca^{2+} at 20–60 μM Ca^{2+} and calpain 2 at 0.3–1.3 mM (Goll et al. 2003). These Ca^{2+} concentration requirements have given these two calpains their alternative names of μ-calpain and m-calpain where μ indicates μmolar Ca^{2+} (albeit high μmolar) and m indicates millilmolar Ca^{2+} (albeit low millimolar). With such Ca^{2+} activation concentrations, which calpains in the bulk cytosol will never experience physiologically, they are safely held in an inactive state. As calpain, like other proteases, has a wide spectrum of substrates (although they are mainly cytoskeletal, see previous section). Such an active protease in the cytosol would cause unpredictable molecular havoc. When cells are dying, the membrane leaks or ATP levels fall below that required for effective Ca^{2+} pumping, cytosolic Ca^{2+} can rise to high levels (ultimately millimolar Ca^{2+}), and calpain can be activated in cell death. This has led to ideas that calpains are enzymes of cell death (whether by necrosis or apoptosis) whose function is to assist in the demise of the cell and clear up the aftermath by randomly destroying the proteinateous cell contents. This pathological activation of calpains may explain the blebbing which often accompanies cell death (Majno and Joris 1995), as calpain preferentially cleaves cytoskeletal proteins, thus destroying plasma membrane/cortical cytoskeleton links, and so allowing blebs to form (Larsen et al. 2008).

However, calpain is a specialist enzyme whose activity, while prevented in the bulk cytosol (and so can be safely controlled), can only be activated at special locations within the cell where cytosolic Ca^{2+} reaches high activating levels. Such localised Ca^{2+} hotspots are known to occur in cells and are usually associated with Ca^{2+} influx channels, whether at the plasma membrane or intracellular Ca^{2+} storage sites (Giacomello et al. 2010; Parker et al. 1996). At the inner mouth of the open Ca^{2+} channel on the plasma membrane, the Ca^{2+} concentration will be the similar to the concentration of Ca^{2+} in the extracellular medium (ie 1–2 mM). However, diffusion into the highly Ca^{2+} buffered cytosol lowers this to sub-μ molar level within a micrometre of the channel (Fig. 8.9b). Even so, clusters of Ca^{2+} channels such as IP$_3$ receptors on the endoplasmic reticulum generate distinct Ca^{2+} hotspots (Giacomello et al. 2010; Parker et al. 1996). It is not clear whether these hot spot function simply as a prelude to cell-wide Ca^{2+} signals and are "ignition points" (Berridge et al. 2000), or whether some physiological processes require localised higher levels of Ca^{2+}. For example, in neutrophils, internal perfusion of the cell with Ca^{2+} triggered exocytosis measured by the increase in membrane capacitance (ie the increase in membrane surface area) but required 50–100 μM (Nusse and Lindau 1988). A similarly high concentration of Ca^{2+}, 25 μM, was required to stimulate exocytosis in permeabilised neutrophils (Smolen et al. 1986).

Neutrophils on a non-stimulating surface remained non-spreading until a Ca^{2+} signal is experimentally induced by either photolytic uncaging of Ca^{2+} or IP$_3$, when neutrophil spreading was immediately triggered (Dewitt et al. 2013; Pettit and Hallett 1998). However, it was found that that whereas cell spreading occurred at physiological levels of bulk Ca^{2+} when the signal was triggered by caged IP$_3$ (Dewitt et al. 2013), unquantifiably high levels of cytosolic Ca^{2+} were required to induce cell spreading if the source were from caged Ca^{2+} (Pettit and Hallett 1998). This differential would be explained by the locations of the Ca^{2+} signal, which is purely cytosolic for caged Ca^{2+}. IP$_3$ initially signal release of Ca^{2+} from stores, which in turn signals the opening of physiological Ca^{2+} channels on the plasma membrane and hence Ca^{2+} influx (a process called "store operated Ca^{2+} influx"). This suggests that close to the plasma membrane Ca^{2+} influx, generates very high Ca^{2+} concentrations (as high as the caged Ca^{2+}) at sites driven by Ca^{2+} influx channels. This explanation also points to the involvement of an unusually low Ca^{2+} affinity protein at the downstream mediator of the spreading response.

This, of course, is consistent with a role for calpain.

In suspensions of neutrophils, attempts have been made to measure near plasma membrane Ca^{2+} levels. Using a lipophilic and low Ca^{2+} affinity version of the fluorescent Ca^{2+} probes fura-2 (FFP-18) which remains within the plasma, during stimulated Ca^{2+} influx near the plasma membrane Ca^{2+} has been estimated to rise to a level of at least 30 μM (Davies and Hallett 1996, 1998). This is a high enough near the plasma membrane Ca^{2+} concentration to locally activate calpain-1. Total internal reflection fluorescence (TIRF) microscopy has also been used to report near plasma membrane Ca^{2+} during stimulated Ca^{2+} influx in neutrophils. TIRF generates an evanescent wave of excitation 100 nm from the reflection surface and into the cell (Steyer and Almers 2001), requiring the membrane of the neutrophil to also be within 100 nm of the imaging surface. The Ca^{2+} measurements were therefore only be made on neutrophils which were closely opposed to the imaging surface ie fully spread on the imaging surface, or if more loosely attached with wrinkles in place within the wrinkles. TIRF reported the ratio of fluo3 fluorescence at the membrane to be only 35% higher than in the bulk cytosol (Omann and Axelrod 1996). In is unclear how this equates to Ca^{2+}, but as fluo3 has a Kd 580 nM Ca^{2+}, the peak cytosolic signal of about 0.8–1 μM Ca^{2+} would result in about 70–75% saturated of fluo3 with Ca^{2+}. Therefore a 35% increase in fluorescence signal would be nearing full saturate fluo3, and so report a near membrane Ca^{2+} of greater than 5–6 μM Ca^{2+}. If the cells chosen for measurement were spherical (rather than flattened) and attached by the tips of the wrinkles (with valleys between wrinkles being greater than 100 nm), then the relative increase in near-membrane Ca^{2+} would be localized into "hot spots" (100 nm tips of wrinkles). The Ca^{2+} change in the wrinkles would be greater (eg 50% coverage means the Ca^{2+} concentration within the wrinkle-tips would be >10 μM). The discrepancy between the two ways of measuring near membrane Ca^{2+} thus suggests that the high near-membrane Ca^{2+} measured by the two

methods was the result in a feature present in neutrophils in suspension which was absent or reduced in the flattened (spread) neutrophils. From the discussion of cell surface wrinkles in Chap. 6, an obvious candidate for the location of high Ca^{2+} would be within the wrinkles and microridges, which are present in neutrophils in suspension but absent at the cell spreading contact surface.

Mathematical modelling of the Ca^{2+} hot spots also points to wrinkles as a site of high Ca^{2+} hotspots (Brasen et al. 2010, 2011). Essentially, the wrinkles represents a small compartment with a high surface area to volume ratio. The number of Ca^{2+} influx channels on the wrinkles and microridges (distributed evenly over the plasma membrane) "feed" a small volume of cytosol (Fig. 8.9a). When the rate of Ca^{2+} influx into the wrinkle exceeds the rate of diffusion of Ca^{2+} out of the wrinkle at its base and into the bulk cytosol, Ca^{2+} will accumulate in the wrinkle (Brasen et al. 2010). This, by itself, would elevate Ca^{2+} locally within the wrinkle (Fig. 8.9a). However, the concentration of mobile Ca^{2+} buffers which exist in the cytosol (Ca^{2+} binding proteins and small molecules) is high, around 0.76 mM with an average K_d of 0.5 μM (Von Tscharner et al. 1986). This gives a measured buffering capacity of around 1000–3000:1 (Von Tscharner et al. 1986; Al-Mohanna and Hallett 1988). Thus for every 1000–3000 Ca^{2+} ions entering the cytosol, only 1 will be a free Ca^{2+} ion. This effectively buffers Ca^{2+} fluctuations. In the wrinkles, the mathematical model, shows that this buffering system will be locally overwhelmed (Brasen et al. 2010) to be replaced by diffusion of buffer from the cytosol through the narrow "mouth" of the wrinkle (Fig. 8.9b) into the wrinkle cytosol. with a diffusion constant of 13 μm^2/s (Allbritton et al. 1992). This bottle neck allows the free Ca^{2+} ion concentration in the wrinkles to rise without much restraint (Fig. 8.9c, d), giving a "run away" situation (like the sharp fall in pH of pH-buffered solution on adding acid, once the pH is below the buffering range, near the pKa of the pH buffer). The mathematical model shows that intra-wrinkle Ca^{2+} can reach 80 μM, depending on the wrinkle shape and length (Brasen

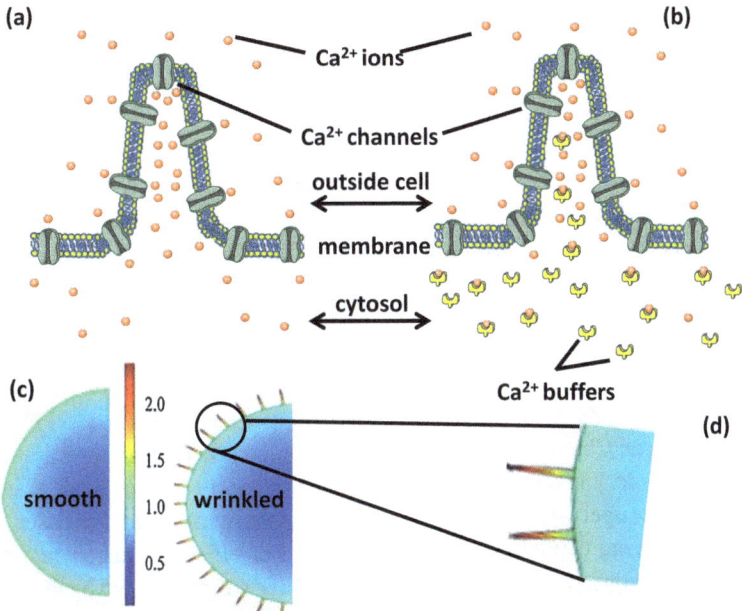

Fig. 8.9 Localised activation of calpain by Ca^{2+} hotspots in the cell surface wrinkles. (**a**) Shows a "wrinkle" of plasma membrane with open Ca^{2+} channels allowing Ca^{2+} influx. Although the Ca^{2+} channels are evenly spaced, Ca^{2+} ions tend to accumulate within the wrinkle as a result of the increased surface are to volume ratio of this region. (**b**) Mobile Ca^{2+} buffers are added to the model shown in (**a**) and show the effect on the bulk cytosol of effectively buffering changes in cytosolic free Ca^{2+}. These buffers must diffuse into the wrinkles to have the same effect and if overwhelmed, free Ca^{2+} rises without restraint. (**d**) Shows the result of mathematical modelling the effect of a wrinkled cell surface on the near membrane Ca^{2+} concentration and compares smooth and wrinkled surfaces pseudocoloured according to the scale shown. (**d**) Shows an enlargement of the wrinkles, in which the gradient within the wrinkles can be seen. (The reader should refer to the web version for the full colour rendition of the figure. The graphics were compiled using elements provided by Servier Medical Art (https://smart.servier.com/category/cellular-biology/intracellular-components/))

et al. 2010). Recently, measurements of Ca^{2+} concentration within wrinkles have been made using a low affinity Ca^{2+} indicator attached to ezrin (which localises to cell surface wrinkles) and reports that intra-wrinkle Ca^{2+} may reach 20–70 µM (Roberts 2017; Roberts et al. 2017). Thus, the wrinkled membrane encloses a space where Ca^{2+} rises high enough to activate calpain when physiological Ca^{2+} channel open, and so will be the location for physiological calpain activation. As Ca^{2+} influx during the phagocytosis is not restricted but occurs around the cell (Dewitt et al. 2003), this mechanism would release wrinkles around the cell, and so provide the forming phagocytic pseudopodia with addition a membrane. It would also allow phagocytic cups at other locations to form more easily. This is often seen experimental, but surprisingly, when con-fronted with two phagocytic cups, usually only one at a time progresses (eg Dewitt et al. 2006). This effect has been called a mechanical "bottleneck effect" (van Zon et al. 2009) and modelling suggests that when the phagocytic cup is half-way up its target, there may be a competition between the individual particles. Tension in the membrane limits success to only one phagocytic cup at a time, the others having to await the outcome (van Zon et al. 2009). The geometry of the phagocytic cup is a thin veil of cytoplasm, like an extended (and circular) microridge. When the Ca^{2+} signal fires, the effect on calpain activity will be immediately manifest in the phagocytic cup and may explain the preferential extension of these pseudopodia at the phagocytic cup. Perhaps as no two phagocytic cups are identical, the phagocytic cup that wins the tension struggle for additional membrane, wins the additional membrane available for it to progress (ie a "membrane supply and demand" situation).

That Ca^{2+} influx is the key part of signalling for phagocytosis was inferred by the inhibition of phagocytosis by Ca^{2+} influx blocking ions (eg Ni^{2+}) and from theoretical conditions of the target for high Ca^{2+} and the locations at which high Ca^{2+} may occur. There is however, a question as to whether these Ca^{2+} influx channels are connected to the store operated Ca^{2+} entry (SOCE) system which involves the joining of STIM-1 and STIM2 to Orai1 to form a Ca^{2+} influx channel at the plasma membrane (see Chap. 8). Vaeth et al. (2015) generated a double knockout of STIM1 and STIM2. The store operated Ca^{2+} influx (triggered by store emptying using thapsigargin) was totally inhibited in mouse macrophages and bone marrow–derived dendritic cells. However, phagocytosis by these cells was unimpaired, despite the lack of a working SOCE system. Surprisingly, this phagocytosis was dependent on extracellular Ca^{2+} (inhibited by the Ca^{2+} chelator EGTA) and inhibited by cytosolic Ca^{2+} chelator, BAPTA (Introduced as BAPTA-AM). This showed that phagocytosis that was being observed was dependent on a cytosolic free Ca^{2+} signal, but that it was not triggered by a SOCE mechanism. There may be a number of Ca^{2+} ion influx channels which do not require a SOCE trigger (discussed in Chap. 8). For example, Vaeth et al. (2015) showed that Ca^{2+} signalling by ATP, acting mainly on the P2X7 receptor channel, was unimparied in the STIM1/2 deficient cells. These data seem to point strongly to a role for Ca^{2+} influx (via a non SOCE channel) which is essential for phagocytosis. The mathematical model for calpain activation within wrinkles (Fig. 8.9) also suggests that Ca^{2+} influx (ratherthan simply a global Ca^{2+} signal) is also crucial.

Conclusions

Taken together, the vast amount of evidence given in this chapter pointing to a role for calpain activity in cell shape changes and in phagocytosis is persuasive. There are many other papers which support this view, but were not mentioned here. There is the possibility that phagocytosis can occur by a physical pulling apart of the wrinkles as an artificial target binds strongly to a lipid membrane, as with biophysical experiments. However, these are probably non-physiological situations and the result is a slow phagocytosis. It is well known that, opsonisation of particles that are poorly internalised, results in a rapid phagocytosis, often at increased rates of 200–300%. For efficient phagocytosis, a Ca^{2+} signal is required (often induced by the opsonin-receptor binding). The evidence presented in this chapter, points to the localised activation of calpain as Ca^{2+} in to wrinkles exceeds that required for calpain activation. This results in the cleavage of actin network-plasma membrane crosslinking proteins (including ezrin) and the release of the wrinkles at the plasma membrane for use in forming pseudopodia and phagosomes. At the molecular level, the model given at the start of this chapter (Fig. 8.1) can be now be extended to show how a surface area which initially formed the "valley" between wrinkles (redundant cell surface area) can be released to form hills ie phagocytic pseudopodia (Fig. 8.10). The initial situation is familiar (Fig. 8.10a) with the plasma membrane held in folds and wrinkles by ezrin linking the plasma membrane to the cortical actin network. This maintains tension in the membrane and a reservoir of membrane. At the phagocytic Ca^{2+} signal, Ca^{2+} ions enter the wrinkles and reach a concentration high enough to activate calpain, and cleave ezrin (Fig. 8.10b).This relaxes tension

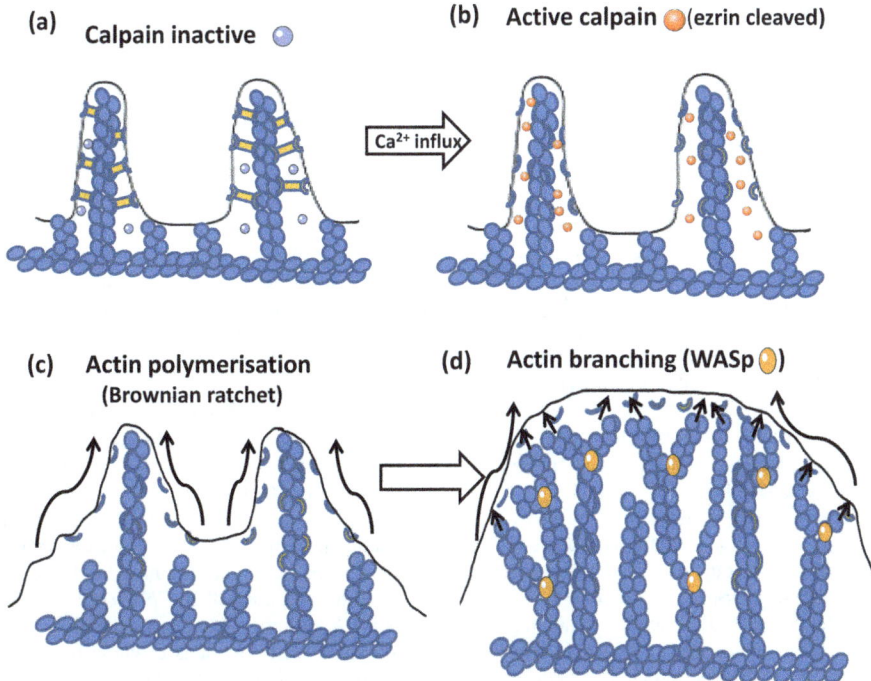

(a)   Calpain inactive ○

(b)   Active calpain ○ **(ezrin cleaved)**

Ca²⁺ influx

(c)   Actin polymerisation
(Brownian ratchet)

(d)   Actin branching (WASp ○ **)**

Fig. 8.10 Proposed molecular mechanism of cell surface unwrinkling in phagocytosis. (**a**) The configuration of the cortical actin network and the crosslinking molecular ezrin within two cell surface wrinkles is shown. Without calpain activation, the membrane held tightly to the actin cortex. (**b**) After Ca²⁺ influx into the wrinkles, calpain within the wrinkles is activated and ezrin is cleaved. (**c**) Without the restraint of ezrin, the plasma membrane no longer held in the membrane, so that it is subject to Brownian tightly and the Brownian ratchet of actin polymerisation can operate and actin pushes the plasma membrane in the "valley" between wrinkles as well as the wrinkle membrane itself. (**d**) The actin network encroaches into the space between the wrinkles as g actin polymers grow, pushing out the plasma membrane and additional WASP branch points are added. (This figure is pmodifiedfrom one originally published in Intl. J. Mol. Sci 20: Article Number: 1383 (2019))

in the membrane, so that it is subject to Brownian fluctuations in position and permits actin to continue to polymerise and so push out the membrane (Fig. 8.10c). The Brownian ratchet is explained in more detail in Chap. 6. As the actin polymer grows, proteins Wiskott Aldrich syndrome proteins (WASp; Snapper and Rosen 1999) are included in the long chains of actin to make branch points and give a firmer footing for the "actin push" and also increase the number of growing actin tips (Fig. 8.10d). This process transforms the two wrinkles in Fig. 8.10a into a single larger protrusion. Further growth of the protrusion to make a phagocytic pseudopodium requires adjacent wrinkles to undergo similar transformations. In this model, the trigger for the entire process is thus the Ca²⁺ activation of calpain.

References

Adams SE, Parr C, Miller DJ, Allemann RK, Hallett MB (2012) Potent inhibition of Ca²⁺-dependent activation of calpain-1 by novel mercaptoacrylates. Med Chem Commun 3:566–570

Adams SE, Rizkallah PJ, Miller DJ, Robinson EJ, Hallett MB, Allemann RK (2014) The structural basis of differential inhibition of human calpain by indole and phenyl alpha-mercaptoacrylic acids. J Struct Biol 187:236–241

Adams SE, Robinson EJ, Miller DJ, Rizkallah PJ, Hallett MB, Allemann RK (2015) Conformationally restricted calpain inhibitors. Chem Sci 6:6865–6871

Allbritton NL, Meyer T, Stryer L (1992) Range of messenger action of calcium ion and inositol 1,4,5-trisphosphate. Science 258:1812–1815

Al-Mohanna FA, Hallett MB (1988) The use of fura-2 to determine the relationship between cytoplasmic free Ca²⁺ and oxidase activation in rat neutrophils. Cell Calcium 9:17–26

Anderson SI, Hotchin NA, Nash GB (2000) Role of the cytoskeleton in rapid activation of CD11b/CD18 function and its subsequent downregulation in neutrophils. J Cell Sci 113:2737–2745

Azam M, Andrabi SS, Sahr KE, Kamath L, Kuliopulos A, Chishti AH (2001) Disruption of the mouse mu-calpain gene reveals an essential role in platelet function. Mol Cell Biol 21:2213–2220

Berridge MJ, Lipp P, Bootman MD (2000) The versatility and universality of calcium signalling. Nat Rev Mol Cell Biol 1:11–21

Blanchard H, Grochulski P, Li Y, Arthur JSC, Davies PL, Elce JS, Cygler M (1997) Structure of a calpain Ca^{2+}-binding domain reveals a novel EF-hand and Ca^{2+}-induced conformational changes. Nat Struct Biol 4:532–538

Brasen JC, Olsen LF, Hallett MB (2010) Cell surface topology creates high Ca^{2+} signalling microdomains. Cell Calcium 47:339–349

Brasen JC, Jacobsen JCB, Holstein-Rathlou NH (2011) Modeling Ca^{2+} microdomains. In: Mosekilde E, Sosnovtseva O, Rostami-Hodjegan A (eds) Biosimulation in biomedical research. Springer, Vienna

Campbell JS, Hallett MB (2015) Active calpain in phagocytically competent human neutrophils: Electroinjection of fluorogenic calpain substrate. Biochem Biophys Res Commun 457:341–346

Cong J, Goll DE, Peterson AM, Kapprell HP (1989) The role of autolysis in activity of the Ca^{2+}-dependent proteinases, mu-calpain and m-calpain. J Biol Chem 264:10096–10103

Croce K, Flaumenhaft R, Rivers M, Furie B, Furie BC, Herman IM, Potter DA (1999) Inhibition of calpain blocks platelet secretion, aggregation, and spreading. J Biol Chem 274:36321–36327

Davies EV, Hallett MB (1996) Near membrane Ca^{2+} changes resulting from store release in neutrophils: detection by FFP-18. Cell Calcium 19:355–362

Davies EV, Hallett MB (1998) High micromolar Ca^{2+} beneath the plasma membrane in stimulated neutrophils. Biochem Biophys Res Commun 248:679–683

Dewitt S, Hallett MB (2002) Cytosolic free Ca^{2+} changes and calpain activation are required for ß2 integrin-accelerated phagocytosis by human neutrophils. J Cell Biol 159:181–189

Dewitt S, Laffafian I, Hallett MB (2003) Phagosomal oxidative activity during ß2 integrin (CR3)-mediated phagocytosis by neutrophils is triggered by a non-restricted Ca^{2+} signal: Ca^{2+} controls time not space. J Cell Sci 116:2857–2865

Dewitt S, Tian W, Hallett MB (2006) Localised PI(3,4,5)P3 or PI (3,4)P2 at the phagocytic cup is required for both phagosome closure and Ca^{2+} signalling in HL60 neutrophils. J Cell Sci 119:443–451

Dewitt S, Francis RJ, Hallett MB (2013) Ca^{2+} and calpain control membrane expansion during the rapid cell spreading of neutrophils. J Cell Sci 126:4627–4635

Di Virgilio F, Meyer BC, Greenberg S, Silverstein SC (1988) Fc receptor-mediated phagocytosis occurs in macrophages at exceedingly low cytosolic Ca^{2+} levels. J Cell Biol 106:657–666

Dourdin N, Bhatt AK, Dutt P, Greer PA, Arthur JS, Elce JS, Huttenlocher A (2001) Reduced cell migration and disruption of the actin cytoskeleton in calpain-deficient embryonic fibroblasts. J Biol Chem 276:48382–48388

DuVerle DA, Mamitsuka H (2019) CalCleaveMKL: a tool for calpain cleavage prediction. Methods Mol Biol 1915:121–147. https://doi.org/10.1007/978-1-4939-8988-1_11

DuVerle DA, Ono Y, Sorimachi H, Mamitsuka H (2011) Calpain cleavage prediction using multiple kernel learning. PLoS One 3:e19035. https://doi.org/10.1371/journal.pone.0019035

Fan YX, Pan X, Zhang Y, Shen HB (2019) LabCaS for ranking potential calpain substrate cleavage sites from amino acid sequence. Methods Mol Biol 1915:111–120. https://doi.org/10.1007/978-1-4939-8988-1_10

Francis EA, Heinrich V (2018) Mechanistic understanding of single-cell behavior is essential for transformative advances in biomedicine. Yale J Biol Med 91:279–289

Franco S, Perrin B, Huttenlocher A (2004a) Isoform specific function of calpain 2 in regulating membrane protrusion. Exp Cell Res 299:179–187

Franco SJ, Rodgers MA, Perrin BJ, Han JW, Bennin DA, Critchley DR, Huttenlocher A (2004b) Calpain-mediated proteolysis of talin regulates adhesion dynamics. Nat Cell Biol 6:977–983

Fujita H, Kato T, Watanabe N, Takahashi T, Kitagawa S (2011) Calpain inhibitors stimulate phagocyte functions via activation of human formyl peptide receptors. Arch Biochem Biophys 513:51–60

Giacomello M, Drago L, Bortolozz M, Scorzeto M, Gianelle A, Pizzo P, Pozzan T (2010) Ca^{2+} hot spots on the mitochondrial surface are generated by Ca^{2+} mobilization from stores, but not by activation of store-operated Ca^{2+} channels. Mol Cell 38:280–290

Gilbert SH, Perry K, Fay FS (1994) Mediation of chemoattractant-induced changes in [Ca^{2+}]i and cell shape, polarity and locomotion by InsP(3), DAG and protein kinase-c in newt eosinophils. J Cell Biol 127:489–503

Gil-Parrado S, Popp O, Knoch TA, Zahler S, Bestvater F, Felgentrager M, Holloschi A, Fernandez-Montalvan A, Auerswald EA, Fritz H (2003) Subcellular localization and in vivo subunit interactions of ubiquitous mu-calpain. J Biol Chem 278:16336–16346

Gitler D, Spira ME (1998) Real time imaging of calcium-induced localized proteolytic activity after axotomy and its relation to growth cone formation. Neuron 20:1123–1135

Glading A, Uberall F, Keyse SM, Lauffenburger DA, Wells A (2001) Membrane proximal ERK signaling is required for M-calpain activation downstream of epidermal growth factor receptor signalling. J Biol Chem 276:23341–23348

Goll DE, Thompson VF, Li HQ, Wei W, Cong JY (2003) The calpain system. Physiol Rev 83:731–801

Griffin FM, Griffin JA, Leider JE, Silverstein SC (1975) Studies on mechanism of phagocytosis 1. Requirement for circumfential atthacjment of particle-bound ligands to specific receptors on macrophage plasma membrane. J Exp Med 142:1263–1282

Haas K, Sin WC, Javaherian A, Li Z, Cline HT (2001) Single-cell electroporation for gene transfer in vivo. Neuron 29:583–591

Hamburger HJ (1910) The influence of small amounts of calcium on the motion of phagocytes. Roy Neth Acad Arts Sci (KNAW) Amst Proc 13:66–79. https://www.dwc.knaw.nl/DL/publications/PU00013276.pdf

Hamburger HJ (1915) Researches on phagocytosis. Nature 96:19–23

Hamburger HJ (1916) Researches on phagocytosis. Br Med J 1:37–41

Hamburger HJ, de Haan J (1910) The biology of phagocytes. VI. The effect of alkaline earth salts on phagocytosis (Ca, Ba, Sr, Mg). Biochem Z 24:470–477

Hanna RA, Campbell RL, Davies PL (2008) Calcium-bound structure of calpain and its mechanism of inhibition by calpastatin. Nature 456:409–412

Hendey B, Klee CB, Maxfield FR (1992) Inhibition of neutrophil chemokinesis on vitronectin by inhibitors of calcineurin. Science 258:296–299

Hendey B, Lawson M, Marcantonio EE, Maxfield FR (1996) Intracellular calcium and calcineurin regulate neutrophil motility on vitronectin through a receptor identified by antibodies to integrins alpha v and beta 3. Blood 87:2038–2048

Huttenlocher A, Palecek SP, Lu Q, Zhang W, Mellgren RL, Lauffenburger MH, Ginsberg DA, Horwitz AF (1997) Regulation of cell migration by the calcium-dependent protease calpain. J Biol Chem 272:32719–32722

Irmscher M, de Jong AM, Kress H, Prins MWJ (2013) A method for time-resolved measurements of the mechanics of phagocytic cups. J R Soc Interface 10:20121048. https://doi.org/10.1098/rsif.2012.1048

Ishak R (2012) Calpain-1: investigating its role in murine neutrophils. PhD thesis, Cardiff University. http://orca.cf.ac.uk/37448/1/2012IshakRPhD.pdf

Ishak R, Hallett MB (2018) Defective rapid cell shape and transendothelial migration by calpain-1 null neutrophils. Biochem Biophys Res Commun 506:1065–1070

Jennings S, Hallett MB (2019) Single cell measurement of calpain activity in neutrophils reveals link to cytosolic Ca^{2+} elevation and individual phagocytotic events. Biochem Biophys Res Commun 515:163–168

Jia ZJ, Barbier L, Stuart H, Amraei M, Pelech S, Dennis JW, Metalnikov P, O'Donnell P, Nabi IR (2005) Tumor cell pseudopodial protrusions. J Biol Chem 280:30564–30573

Johnson AF, Struthers MD, Pierson KB, Mangel WF, Smith LM (1993) Nonisotopic DNA detection system employing elastase and a fluorogenic rhodamine substrate. Anal Chem 65:2352–2359

Katsube M, Kato Y, Kitagawa M, Noma H, Fujita H, Kitagawa S (2008) Calpain-mediated regulation of the distinct signaling pathways and cell migration in human neutrophils. J Leukoc Biol 84:255–253

Kim E, Enelow RI, Sullivan GW, Mandell GR (1992) Regional and generalized changes in cytosolic free calcium in monocytes during phagocytosis. Infect Immun 60:1244–1248

Kruskal BA, Maxfield FR (1987) Cytosolic free calcium increases before and oscillates during frustrated phagocytosis in macrophages. J Cell Biol 105:2685–2693

Kruskal BA, Shak S, Maxfield FR (1986) Spreading of human neutrophils is immediately preceded by a large increase in cytoplasmic free calcium. Proc Natl Acad Sci U S A 83:2919–2923

Kulkarni S, Saido TC, Suzuki K, Fox JE (1999) Calpain mediates integrin-induced signaling at a point upstream of Rho family members. J Biol Chem 274:21265–21275

Kulkarni S, Goll DE, Fox JEB (2002) Calpain cleaves RhoA generating a dominant-negative form that inhibits integrin-induced actin filament assembly and cell spreading. J Biol Chem 277:24435–24441

Kumar V, Everingham S, Hall C, Greer PA, Craig AWB (2014) Calpains promote neutrophil recruitment and bacterial clearance in an acute bacterial peritonitis mode. Eur J Immunol 44:831–841

Laffafian I, Hallett MB (1998) Lipid-assisted microinjection: introducing material into the cytosol and membranes of small cells. Biophys J 75:2558–2563

Laffafian I, Hallett MB (2000) Gentle microinjection for myeloid cells using SLAM. Blood 95:3270–3271

Larsen AK, Lametsch R, Elce JS (2008) Genetic disruption of calpain correlates with loss of membrane blebbing and differential expression of RhoGDI-1, cofilin and tropomyosin. Biochem J 411:657–668

Lawson MA, Maxfield FR (1995) Ca^{2+} and calcineurin-dependent recycling of integrin to the front of migrating neutrophils. Nature 377:75–79

Lew DP, Andersson T, DiVirgilio F, Pozzan T, Stendahl O (1985) Ca^{2+} dependent and Ca^{2+} independent phagocytosis in human neutrophils. Nature 315:509–511

Lewis KJ, Masterman B, Laffafian I, Hallett MB (2014) Minimal impact electro-injection of cells undergoing dynamic shape change reveals calpain activation. Biochim Biophys Acta 1843:1182–1187

Leytus SP, Patterson WL, Mangel WF (1983) New class of sensitive and selective fluorogenic substrates for serine proteinases. Amino acid and dipeptide derivatives of rhodamine. Biochem J 215:253–260

Lin GD, Chattopadhyay D, Maki M, Wang KKW, Carson M, Jin L, Yuen P, Takano E, Hatanaka M, DeLucas LJ et al (1997) Crystal structure of calcium bound domain VI of calpain at 1.9 angstrom resolution and its role in enzyme assembly, regulation, and inhibitor binding. Nat Struct Biol 4:539–547

Liu Z, Cao J, Gao X, Ma Q, Ren J, Xue Y (2011) GPS-CCD: a novel computational program for the prediction of calpain cleavage sites. PLoS One 6:e19001

Liu Y, Belkina NV, Park C, Nambiar R, Loughhead SM, Patino-Lopez G, Ben-Aissa K, Hao J-J, Kruhlak MJ, Qi H, von Andrian UH, Kehrl JH, Tyska MJ, Shaw S (2012) Constitutively active ezrin increases membrane tension, slows migration, and impedes endothelial transmigration of lymphocytes in vivo in mice. Blood 119:445–453

Lokuta MA, Nuzzi PA, Huttenlocher A (2003) Calpain regulates neutrophil chemotaxis. Proc Natl Acad Sci U S A 100:4006–4011

Majno G, Joris I (1995) Apoptosis, oncosis and necrosis-an overview of cell death. Am J Pathol 146:3–15

Marks PW, Maxfield FR (1990) Local and global changes in cytosolic free calcium in neutrophils during chemotaxis and phagocytosis. Cell Calcium 11:181–190

Marzia M, Neff L, Baron R (2002) Calpain contributes to the regulation of osteoclast attachment and spreading. J Bone Miner Res 17(Suppl 1):S252–S252

Meagher LC, Moonga BS, Haslett C, Huang CL-H, Zaidi M (1991) Single pulses of cytoplasmic calcium associated with phagocytosis of individual zymosan particles by macrophages. Biochem Biophys Res Commun 177:460–465

Migeotte I, Communi D, Parmentier M (2006) Formyl peptide receptors: a promiscuous subfamily of G protein-coupled receptors controlling immune responses. Cytokine Growth Factor Rev 17:501–519

Mittoo S, Sundstrom LE, Bradley M (2003) Synthesis and evaluation of fluorescent probes for the detection of calpain activity. Anal Biochem 319:234–238

Moldoveanu T, Hosfield CM, Lim D, Elce JS, Jia Z, Davies PL (2002) A Ca^{2+} switch aligns the active site of calpain. Cell 108:649–660

Murata T, Sullivan JA, Sawyer DW, Mandell GL (1987) Influence of type and opsonization of injested particle on intracellular free calcium distribution and superoxide production by human neutrophils. Infect Immun 55:1784–1791

Noble KE, Yong K, Khwaja A (1998) Neutrophil transendothelial migration is regulated by the calcium dependent protease calpain. Blood 92(Suppl 1):535A–535A

Noma H, Kato Y, Fujita H, Kitagawa M, Yamano T, Kitagawa S (2009) Calpain inhibition induces activation of the distinct signalling pathways and cell migration in human monocytes. Immunology 128:e487–e496

Nunes P, Cornut D, Bochet V, Hasler U, Oh-Hora M, Waldburger JM, Demaurex N (2012) STIM1 juxtaposes ER to phagosomes, generating Ca^{2+} hotspots that boost phagocytosis. Curr Biol 22:1990–1997

Nusse O, Lindau M (1988) The dynamics of exocytosis in human neutrophils. J Cell Biol 107:2117–2123

Omann GM, Axelrod D (1996) Membrane-proximal calcium transients in stimulated neutrophils detected by total internal reflection fluorescence. Biophys J 71:2885–2891

Parker I, Choi J, Yao Y (1996) Elementary events of InsP(3)-induced Ca^{2+} liberation in Xenopus oocytes: hot spots, puffs and blips. Cell Calcium 20:105–121

Parnaud G, Hammar E, Rouiller DG, Bosco D (2005) Inhibition of calpain blocks pancreatic beta-cell spreading and insulin secretion. Am J Physiol Endocrinol Metab 289:E313–E321

Paulhe F, Bogyo A, Chap H, Perret B, Racaud-Sultan C (2001) Vascular smooth muscle cell spreading onto fibrinogen is regulated by calpains and phospholipase C. Biochem Biophys Res Commun 288:875–881

Pettit EJ, Fay FS (1998) Cytosolic free calcium and the cytoskeleton in the control of leukocyte chemotaxis. Physiol Rev 78:949–967

Pettit EJ, Hallett MB (1996) Localised and global cytosolic Ca^{2+} changes in neutrophils during engagement of CD11b/CD18 integrin visualised using confocal laser scanning reconstruction. J Cell Sci 109:1689–1694

Pettit EJ, Hallett MB (1998) Release of "caged" cytosolic Ca^{2+} triggers rapid spreading of human neutrophils adherent via integrin engagement. J Cell Sci 111:2209–2215

Potter DA, Tirnauer JS, Janssen R, Croall DE, Hughes CN, Fiacco KA, Mier JW, Maki M, Herman IM (1998) Calpain regulates actin remoding during cell spreading. J Cell Biol 141:647–662

Raucher D, Sheetz MP (1999) Characteristics of a membrane reservoir buffering membrane tension. Biophys J 77:1992–2002

Roberts RE (2017) The μ-calpain-ezrin axis: a potential target for therapy in inflammatory disease. PhD thesis, Cardiff University. http://orca.cf.ac.uk/108477/1/2018RobertsR_PhD.pdf

Roberts RE, Vervliet T, Bultynck G, Parys JB, Hallett MB (2017) Dynamics of ezrin location at the plasma membrane: relevance to neutrophil spreading. Eur J Clin Investig 47(Suppl 1):148–148

Rock MT, Dix AR, Brooks WH, Roszman TL (2000) Beta 1 integrin-mediated T cell adhesion and cell spreading are regulated by calpain. Exp Cell Res 261:260–270

Rouven Brückner B, Pietuch A, Nehls S, Rother J, Janshoff A (2015) Ezrin is a major regulator of membrane tension in epithelial cells. Sci Rep 5:14700. https://doi.org/10.1038/srep14700

Sawyer DW, Sullivan JA, Mandell GL (1985) Intracellular free calcium localization in neutrophils during phagocytosis. Science 230:633–666

Schwab JC, Leong DA, Mandell G (1992) A wave of elevated intracellular calcium spreads through human neutrophils during phagocytosis of zymosan. J Leukoc Biol 51:437–443

Smolen JE, Stoehr SJ, Boxer LA (1986) Human neutrophils permeabilized with digitonin respond with lysosomal enzyme release when exposed to micromolar levels of free calcium. Biochim Biophys Acta 886:1–17

Snapper SB, Rosen FS (1999) The Wiskott-Aldrich syndrome protein (WASp): roles in signaling and cytoskeletal organization. Annu Rev Immunol 17:905–929

Stendahl O, Krause KH, Krishers J et al (1994) Redistribution of intracellular Ca^{2+} stores during phagocytosis in human neutrophils. Science 265:1439–1441

Stewart MP, McDowall A, Hogg N (1998) LFA-1-mediated adhesion is regulated by cytoskeletal restraint and by a Ca^{2+}-dependent protease, calpain. J Cell Biol 140:699–707

Steyer JA, Almers W (2001) A real-time view of life within 100 nm of the plasma membrane. Nat Rev Mol Cell Biol 2:268–275

Storr SJ, Carragher NO, Frame MC, Parr T, Martin SG (2011) The calpain system and cancer. Nat Rev Cancer 11:364–374

Swanson JA, Baer SC (1995) Phagocytosis zipper and triggers. Trends Cell Biol 5:89–93

Theler JM, Lew DP, Jaconi ME, Krause KH, Wollheim CB, Schlegel W (1995) Intracellular pattern of cytosolic Ca^{2+} changes during adhesion and multiple phagocytosis in human neutrophils. Dynamics of intracellular Ca^{2+} stores. Blood 85:2194–2201

Vaeth M, Zee I, Concepcion AR, Maus M, Shaw P, Portal-Celhay C, Zahra A, Kozhaya L, Weidinger C, Philips J, Unutmaz D, Feske S (2015) Ca^{2+} signaling but not store-operated Ca^{2+} entry is required for the function of macrophages and dendritic cells. J Immunol 195:1202–1217

van Zon JS, Tzircotis G, Caron E, Howard M (2009) A mechanical bottleneck explains the variation in cup growth during FcgammaR phagocytosis. Mol Syst Biol 5:298. https://doi.org/10.1038/msb.2009.59

Walker JW, Gilbert SH, Drummond RM, Yamada M, Sreekumar R, Carraway RE, Ikebe M, Fay FS (1998) Signaling pathways underlying eosinophil cell motility revealed by using caged peptides. Proc Natl Acad Sci U S A 95:1568–1573

Von Tscharner, Deranleau DA, Baggiolini M (1986) Calcium fluxes and calcium buffering in human neutrophils. J Biol Chem 261:163–168

Wang KKW, Nath R, Posner A, Raser KJ, BurokerKilgore M, Hajimohammadreza I, Probert AW, Marcoux FW, Ye QH, Takano E et al (1996) An alpha-mercaptoacrylic acid derivative is a selective nonpeptide cell-permeable calpain inhibitor and is neuroprotective. Proc Natl Acad Sci U S A 93:6687–6692

Wiemer AJ, Lokuta MA, Surfus JC, Wernimont SA, Huttenlocher A (2010) Calpain inhibition impairs TNF-alpha-mediated neutrophil adhesion, arrest and oxidative burst. Mol Immunol 47:894–902

The NADPH Oxidase and the Phagosome

9

Hana Valenta, Marie Erard, Sophie Dupré-Crochet,
and Oliver Nüße

Abstract

The key purpose of phagocytosis is the destruction of pathogenic microorganisms. The phagocytes exert a wide array of killing mechanisms that allow mastering the vast majority of pathogens. One of these mechanisms consists in the production of reactive oxygen species inside the phagosome by a specific enzyme, the phagocyte NADPH oxidase. This enzyme is composed of 6 proteins that need to assemble to form a complex on the phagosomal membrane. Multiple signaling pathways tightly regulate the assembly. We briefly summarize key features of the enzyme and its regulation. We then focus on several related topics that address the activity of the NADPH oxidase during phagocytosis. Novel fluorescence microscopy techniques combined with fluorescent protein labeling of NADPH oxidase subunits opened the view on the structure and dynamics of these proteins in living cells. This combination revealed details of the role of anionic phospholipids in the control of phagosomal ROS production. It also added critical information to propose a 3D model of the complex between the cytosolicsubunits prior

to activation, in complement to other structural data on the oxidase.

Keywords

Phagocytosis · NADPH oxidase · ROS ·
Neutrophil · Fluorescence microscopy ·
Phospholipid dynamics

Introduction

Once the phagocyte has captured its prey and enclosed it into a newly formed phagosome the microorganism needs to be killed and degraded. Coevolution of phagocytes and bacterial or fungal microorganisms has led to an arms race. A growing number of microbicidal mechanisms spurred the selection of resistant microorganisms. Like modern multidrug therapy of infectious diseases, the phagocyte employs multiple mechanisms of destruction thereby limiting the chances of the microorganism to resist to all of them. Acidification of the phagosome, delivery of numerous lytic enzymes and the production of reactive oxygen species (ROS) appear to be the main mechanism for killing inside the phagosome (Nauseef 2007). The relative importance of these mechanisms differs between pathogens and also between phagocytes. Here we will focus on the ROS production by neutrophils, which represent the most abundant and the most aggressive phago-

H. Valenta · M. Erard · S. Dupré-Crochet · O. Nüße (✉)
Université Paris-Saclay, Orsay, France

CNRS U8000, ICP, Orsay, France
e-mail: oliver.nusse@universite-paris-saclay.fr

© Springer Nature Switzerland AG 2020
M. B. Hallett (ed.), *Molecular and Cellular Biology of Phagocytosis*, Advances in Experimental
Medicine and Biology 1246, https://doi.org/10.1007/978-3-030-40406-2_9

cyte population. We will present the responsible enzyme, the phagocyte NADPH oxidase, its composition, structure and regulation. Although much is known about this enzyme, numerous aspects remain poorly understood. In fact, the investigation of the NADPH oxidase faces technical difficulties and we will discuss some of them.

The absence of a functional phagocyte NADPH oxidase leads to Chronic Granulomateous Disease, CGD. Patients suffer from numerous, severe infections, which demonstrates the importance of the NADPH oxidase for our antimicrobial defense (Roos 2019; Stasia and Li 2008). *Staphylococcus aureus* is the most frequent pathogen in CGD patients suggesting that ROS-production is particularly important for the immune defense against this widespread bacterium (Buvelot et al. 2017). Surprisingly, *Staphylococcus aureus* expresses a wide array of antioxidant mechanisms besides numerous other means to survive the phagocyte attack (Greenlee-Wacker et al. 2015). It is not evident, why killing of this well protected bacterium is still strongly ROS-dependent. One hypothesis is that the metabolic cost of producing the complete set of defense mechanisms is too high for the bacterium. Thus in the absence of a functional NADPH oxidase, *Staphylococcus aureus* may concentrate its efforts to protect itself against ROS-independent mechanisms of killing.

The Phagocyte NADPH Oxidase – A Dynamic Protein Complex

Since the early 2000, seven isoforms of the catalytic core of the NADPH oxidases have been identified, NOX1, NOX2, NOX3, NOX4, NOX5, DUOX1 and DUOX2 in different tissues (Buvelot et al. 2019). Those enzymes are dedicated to ROS production as opposed to other enzymes where ROS appear to be side products. NOX1-3 require cytosolic subunits to be activated whereas NOX4 is constitutively active. NOX5, DUOX 1 and 2 contain Ca^{2+}-binding EF hand domains. The activity of NOX1-3 and 5 is detected as superoxide anion production. Hydrogen peroxide is mainly detected for NOX4 activity

and exclusively detected for DUOX 1 and 2. Depending on their concentration, ROS can play different roles in cells. At high concentration ROS are responsible for pathogen killing, on the other hand, at low concentrations they contribute to the regulation of physiological pathways. The roles of the seven isoforms are not fully understood. The first discovered isoform and the best-known one is NOX2, the phagocyte NADPH oxidase. It oxidizes cytosolic NADPH and transfers electrons across the membrane to molecular oxygen present in the phagosome giving rise to superoxide anion. The latter then dismutates to H_2O_2 and undergoes further transformations to other oxygen species such as HOCl. Although ROS have several beneficial effects on mammalian cells, they are clearly toxic above certain concentrations and, in the phagosome, the large quantities of ROS create an army ready to kill pathogens. Superoxide anion production was estimated at 5–10 nmol/min for 10^6 neutrophils (Masoud et al. 2017; Schrenzel et al. 1998). Therefore, the NADPH oxidase activity needs to be strictly controlled. Several decades of research on this enzyme have revealed numerous levels of regulation and we may not know all of them yet.

The phagocyte NADPH oxidase is composed of two membrane proteins, NOX2 (also called gp91^{phox}) and p22^{phox}, three cytosolic subunits (p40^{phox}, p47^{phox}, p67^{phox}) and the small GTPase Rac (Sumimoto 2008) (Fig. 9.1). Other proteins are likely to interact with these proteins and influence enzyme activity such as the calcium-binding proteins S100A8/A9 (Bréchard et al. 2013). In the resting state, the cytosolic subunits and Rac are separated from the membrane subunits. This physical separation is a major security measure to prevent inappropriate ROS production. In fact, activation of the NADPH oxidase requires phosphorylation of the cytosolic subunits and their translocation to the membrane. Phosphorylation induces structural modifications that allow the subunits to interact on one hand with NOX2 and p22^{phox} and on the other hand with lipid head groups in the target membrane. The NADPH oxidase actively produces ROS only when these interactions have been established. We will discuss the signaling events that control NADPH oxi-

Fig. 9.1 Multiple signals link phagocytosis and NADPH oxidase activation. (**a**): Membrane receptors such as FcγRs recognize pathogens and initiate phagocytosis. At the beginning, the NADPH oxidase is not assembled, the cytosolic subunits p40, p47 and p67 as well as Rac are separated from the membrane subunits NOX2 and p22phox. (**b**): Activation of Syk protein kinase is central to the stimulation of multiple signaling pathways, including phosphoinositide3kinase, PI3K, phospholipase D, PLD, phospholipase C, PLC, a rise in intracellular free calcium, Ca^{2+}, and protein kinase C, PKC. These pathways will converge to assemble the NADPH oxidase on the phagosome membrane and initiate ROS production

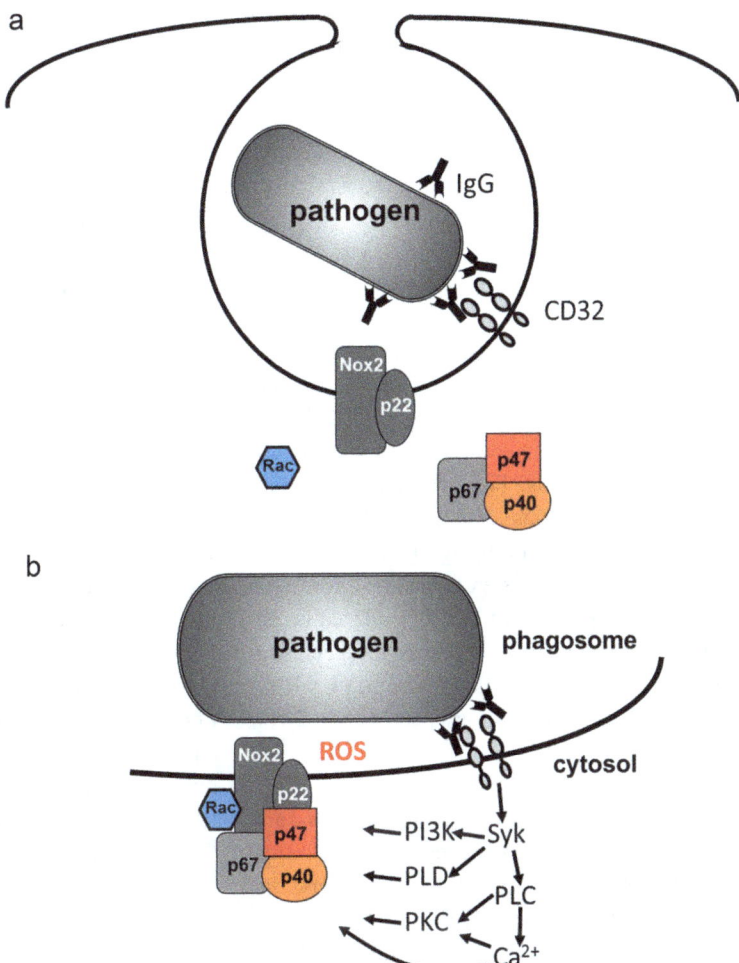

dase activation in more detail later. Most likely, disassembly of the NADPH oxidase complex, in particular dissociation of p67phox from NOX2, terminates ROS production.

Imaging NADPH Oxidase in Live Cells

To appreciate the dynamic behavior of a protein, live cell imaging is a method of choice. If we want to visualize proteins in living cells, tagging with fluorescent proteins (FP) provides interesting opportunities. Over the last 15 years, we and others have FP-tagged all cytosolic subunits as well as Rac and investigated their dynamic behavior during phagocytosis. FPs are genetically

encoded fluorophores of 25–28 kDa composed of a beta-barrel containing an alpha-helix that bears a chromophore. It results from the autocatalytic cyclisation of three amino-acids. Nowadays, there is a very wide variety of FPs available for imaging applications with color ranging from blue to far red. A database gathering main properties of FPs has been recently developed: https://www.fpbase. org/ (Lambert 2019). It contains useful pieces of information in order to choose the right FP or combination of FP.

The use of FPs requires several experimental steps:

- The first one is the fusion of the gene coding the targeted NADPH oxidase subunit to the one of the FP. A sequence coding a linker of a few amino acids may be added at this step to increase the distance between the subunit and the FP. The length of the linker may influence the functionality of the tagged subunit. The choice of the C- or the N-terminus as tagging site depends on the position of the functional domains on the subunit.

- The second step includes the choice of the cell model. Due to the fact, that neutrophils are very difficult to transfect, several myeloid cell lines that can be differentiated in neutrophils or macrophages-like cells have been developed. Those cell lines, such as HL60, PLB-985 or RAW264 cells, can be transfected using established protocols. They also express constitutively endogenous subunits of the NADPH oxidase that may compete in the molecular complexes with transient or stable expression of the FP-tagged ones. In PLB-985 cells, expression of endogenous proteins has been suppressed or reduced by gene disruption, siRNA or CRISPR-cas technology (Song et al. 2017; Wrona et al. 2017; Zhen et al. 1993). Another option is to use non phagocytic cells. Several cell lines have been also successfully used for heterologous expression of wild type or mutants of the NADPH oxidase subunits. For example, COS7 cell line stably expressing the transgenes for NOX2, $p22^{phox}$, $p47^{phox}$ and $p67^{phox}$ are referred as COS^{phox} (Price et al. 2002). For stimulation by phagocytosis of IgG-opsonized beads, FcγRII receptors need to be introduced as well. The NADPH oxidase activity at the phagosome was observed only if $p40^{phox}$ was also expressed (Suh et al. 2006).
- The last step is the imaging of the dynamic behavior of the subunits during phagocytosis. Several imaging strategies can be proposed from the conventional video-microscopy to more advanced techniques (Erard et al. 2018). The phagosome is a highly mobile cell compartment and the quantification of the fluorescence around it necessitates optical sectioning using confocal or spinning disk microscopy. At each time point, a z-stack should be recorded in order to identify, *a posteriori* during data processing, the optimal focal plane for the phagosome. Due to its

technology of illuminating many spots in parallel, a spinning disk microscope is faster than a conventional confocal microscope. Its illumination mode also reduces the photobleaching of fluorescent proteins, which is an advantage when one wants to quantify the evolution of the fluorescence intensities around the phagosome during several minutes. The fluorescence of a soluble cytosolic dye can be used as a reference for ratiometric approaches (Tlili et al. 2012).

Using a Citrine labeled $p67^{phox}$ expressed in the PLB-985 cell line, we showed that $p67^{phox}$–Citrine accumulates at the phagosome during several tens of minutes after the phagocytosis of yeast particles (Tlili et al. 2012). If the Citrine accumulated at the phagosome is photobleached intentionally with a high laser power during a very short time period, the monitoring of the fluorescence recovery (Fluorescence Recovery After Photobleaching, FRAP) is an indication of the exchange between the phagosomal and the cytosolic pools of $p67^{phox}$-Citrine. In our experiment, no fluorescence recovery was detected showing that there is no exchange with the cytosolic pool. On the other hand, the local photobleaching of half of the phagosomal membrane was followed by a fluorescence recovery from the non-bleached half (Fluorescence Loss Induced by Photobleaching, FLIP) indicating a free diffusion around the phagosome. In addition, the duration of the presence of $p67^{phox}$-Citrine at the phagosome correlates with the duration of ROS production. Altogether, these results suggest that once assembled, $p67^{phox}$ is stable in the active complex at the phagosomal membrane (Tlili et al. 2012). A very similar imaging strategy with FP-labeled $p47^{phox}$ and Rac2 showed that the accumulation of $p47^{phox}$ and Rac2 at the phagosome is only transient: in less than 3 min after the phagosome closure, both proteins detach (Faure et al. 2013). The accumulation of labeled $p40^{phox}$ at the phagosome was also observed. In this case, it is tightly linked with the presence of phosphoinositol 3-phosphate (PI(3)P) in the membrane as explained below (Song et al. 2017). The imaging strategies mentioned above allowed also deciphering more precisely the role of specific sequences or domains of the NADPH oxidase subunits in their assembly at the resting or active state. Several examples will be detailed in the paragraph "Structure of the NADPH oxidase" below.

The cytosolic p47^{phox} and p67^{phox} are highly flexible and their structure is difficult to investigate with conventional techniques such as X-ray crystallography, NMR, or cryo-EM. Using FP-labeled subunits, we proposed recently an imaging workflow for structural and quantitative studies of the interactions within the NADPH oxidase cytosolic complex in the resting state in live cells (Ziegler et al. 2019). We monitored the interactions between the FP-labeled subunits in COS7 cells by Förster Resonance Energy Transfer detected by Fluorescence Lifetime Imaging (FRET-FLIM) and Fluorescence Cross-Correlation Spectroscopy (FCCS). This approach revealed that the stoichiometry of the three cytosolic subunits in the complex is 1:1:1 and that nearly 100% of them are present in these complexes in living cells. We also created a 3D model of the cytosolic complex by blending FRET data, published crystal structures of isolated domains and small-angle X-ray scattering (SAXS) models of the subunits. The model will be detailed in the next section.

Fluorescent proteins were also engineered to lead to a large variety of biosensors (Greenwald et al. 2018) including for chemical parameters relevant for phagocytosis: there are pH (Martynov et al. 2018) and redox sensitive fluorescent proteins (Bilan and Belousov 2018; Roma et al. 2018). The FP-based biosensor can be fused directly to a NADPH oxidase subunit in order to target the biosensor at the site of ROS production. RoGFP, a redox sensitive FP, was fused to p47^{phox} and expressed in different cell lines and in vivo for a localized NADPH oxidase activity detection (Henríquez-Olguín et al. 2019; Pal et al. 2013).

NOX2 has also been tagged although the position of the tag is critical for the location and functionality of the protein (Ambasta et al. 2004; Casbon et al. 2009; Murillo and Henderson 2005; van Manen et al. 2008). Examples of p22^{phox} labeling were also reported (Ambasta et al. 2004; Casbon et al. 2009). FP-NOX2 and p22^{phox}-FP shall provide new opportunities to investigate the

activation process of the NADPH oxidase. One of them consists in detecting active oxidase by FRET between NOX2 and one of the cytosolic subunits. One of the difficulties is that only a limited portion of the cytosolic subunits engages with NOX2 at any given time. Therefore, in an imaging experiment, one has to identify this limited portion amidst the majority that is not engaged.

The FP-tag may interfere with the localization or the function of the protein of interest, for example by hindering interaction with membranes or other potential binding partners. Ideally, one would like to confirm results from FP-tagged proteins with native, endogenous proteins. However, the obvious method to analyze the localization of endogenous proteins, immunofluorescence, requires fixing the cells and thus prevents any dynamic follow-up on the same cells. Cells can be fixed and immune-stained at different time-points after phagocytosis, however, time resolution is low in this case and it is not possible to look at the same cell twice at different time points.

Super-resolution microscopy techniques push the limits of detection to the range of 50 nm. For example, these techniques reveal details of cytoskeleton organization and receptor proteins in the membrane and also the distribution of NOX2 in the phagocyte membrane (Baranov et al. 2019) and (Joly et al. unpublished).

Structure of the NADPH Oxidase, Recent Progress and Many Open Questions

Subunits and Their Interactions

The NADPH oxidase is composed of six subunits that need to assemble to create the active complex. The structure of the protein complex of the NADPH oxidase is only partially known. In this paragraph, individual subunits, their interactions and structure will be described (Fig. 9.2).

Membrane Subunits

The membrane part of the NADPH oxidase is the flavocytochrome b$_{558}$, which is a heterodimer composed of a glycosylated subunit, NOX2, and

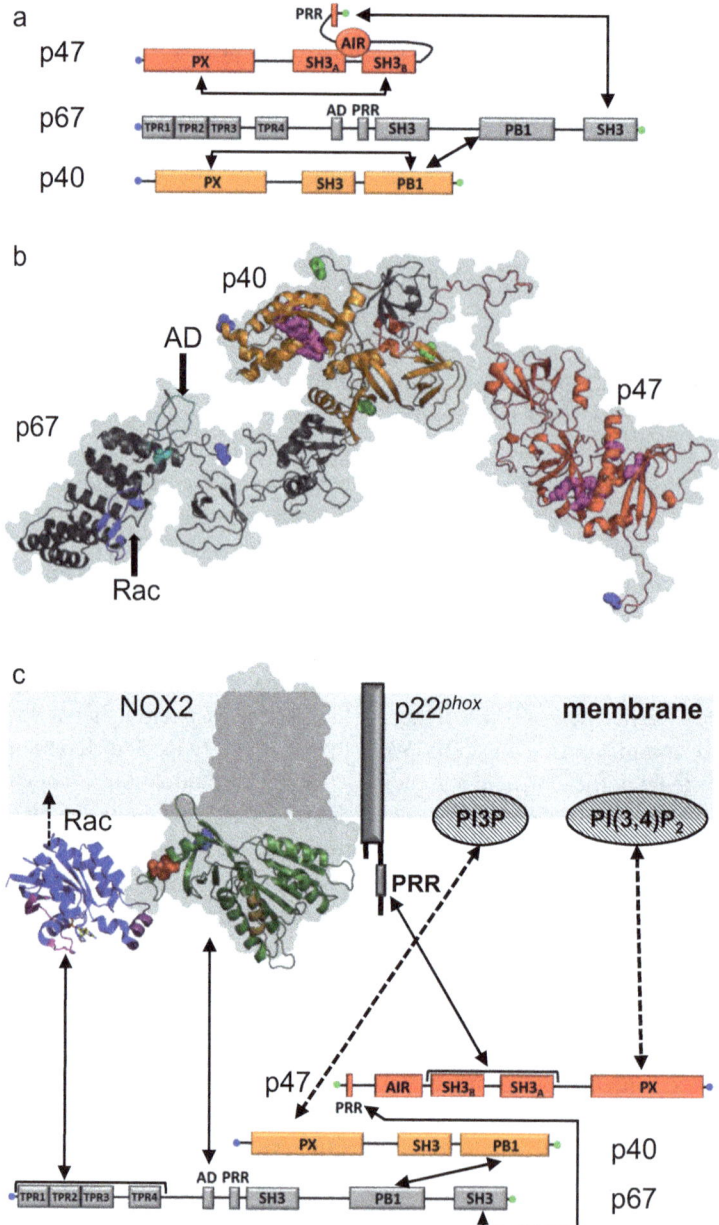

Fig. 9.2 Interactions in the resting and the active state and 3D models. (**a**) The 3 cytosolic proteins form a ternary complex in the cytosol of resting cells. The domains are represented by boxes whose size is proportional to the number of residues involved. The arrows symbolized the intra- and inter-molecular interactions. (**b**) 3-D model of the complete cytosolic heterotrimer (Ziegler et al. 2019). Cartoon representations of p67^{phox}, p40^{phox} and p47^{phox} The N-termini are labeled in blue, and the C-termini are labeled in green. In p40^{phox} and p47^{phox}, the pink spheres represent the residues that bind the phosphoinositides in the active state. They are hidden in the resting state. In p67^{phox}, blue and cyan segments represent the Rac-binding beta-hairpin insertion and the activation domains, respectively. (**c**) Schematic representation of the active state with the inter-molecular interactions represented by arrows. Interactions with lipids are represented with dashed arrows. Rac (PDB entry 1E96) is anchored at the membrane in the active state. The pink and the purple segments in Rac represent the effector and insertion domains of Rac respectively. The structural model of NOX5 is represented as its surface (Magnani et al. 2017). The homology model of the cytosolic domain of NOX2 is in cartoon in green with its N-terminus labeled in blue in the back (Beaumel et al. 2017). The NIS is in gold and the red spheres represent the 3 residues involved in the stabilization of the binding of p67^{phox} (^{369}Cys-Gly-Cys^{371}). P22^{phox} is symbolized by a box with a cytosolic C-terminus. In this figure, all structural models were drawn using PYMOL software at the same scales

a non-glycosylated subunit, p22^{phox}. The formation of the heterodimer is necessary for structural stability and the acquisition of hemes by NOX2 (DeLeo et al. 2000; Yu et al. 1998).

NOX2 is the catalytic center of phagocyte NADPH oxidase. It is a highly glycosylated protein with a molecular weight of 91 kDa. The N-terminal part forms six transmembrane α-helices, while the C-terminal end (AA 282–500), often called dehydrogenase domain, is situated in the cytosol carrying the binding sites for FAD and NADPH (Cross and Segal 2004). The NADPH is a source of electrons and its binding to NOX2 is necessary for the NADPH oxidase activity and ROS production. Upon fixation of NADPH, two electrons are transferred to FAD and then to two non-equivalent hems performing two sequential reductions of the O_2 molecule to form the superoxide anion $O_2^{\bullet-}$ (Cross and Segal 2004). The production of $O_2^{\bullet-}$ is described by the following reaction:

$$2O_2 + NADPH \rightarrow 2O_2^{\bullet-} + NADP^+ + H^+$$

The dehydrogenase domain carries an insertion sequence called NIS (NOX <u>N</u>ADPH-domain <u>I</u>nsertion <u>S</u>equence). This sequence, identified in 1993, corresponds to positions 484–504 in NOX2 (Taylor et al. 1993). The functional role of NIS, which exists only in NOX1-4 and is missing in NOX5, was recently examined (Beaumel et al. 2017). The NIS domain may act as a regulator of the NADPH oxidase activity; it is crucial for the efficiency of the electron transfer from NADPH to FAD and necessary for the translocation of the cytosolic protein to the phagosomal membrane. In addition, the phosphorylation of a specific site in NIS (Ser486) by ATM kinase limits the ROS production.

The NOX2 is complexed in 1:1 ratio with p22^{phox} subunit that has a molecular weight of 21 kDa. P22^{phox} contributes to the maturation and complete stabilization of NOX2. Its C-terminus contains a PRR (proline rich region: AA 151–160) region, which is a binding site for the cytosolic subunit p47^{phox} (Groemping and Rittinger 2005). A mutation in Pro^{156} in the PRR region of p22^{phox} was found in patients with CGD (Dinauer

et al. 1991) resulting in nonfunctional flavocytochrome b$_{558}$.

Cytosolic Subunits

The NADPH oxidase membrane part alone has no enzymatic activity. It needs to be joined by four cytosolic subunits to form a functional active complex. We will detail these cytosolic subunits in the following paragraph.

p47^{phox}

Biochemical studies have shown that the p47^{phox} subunit has an organizing function in the NADPH oxidase activation process in phagocytes, however it is not responsible for the catalytic activity (Nisimoto et al. 1999). This protein (390 AA, 44.7 kDa) is composed of several domains (from N- to C-terminus):

- A PX domain (homology of phagocyte oxidase domain)
- Two SH3 domains (SH3 N-terminal and SH3 C-terminal)
- An auto-inhibitory region (AIR)
- A PRR region (region rich in prolines) containing PxxP motif (Pro-Xaa-Xaa-Pro)

The PX domain is present in p47^{phox} and also in p40^{phox} subunit. This domain specifically recognizes phosphoinositides in the phagosome membrane, which allows anchorage of p47^{phox} after its translocation from the cytosol (Kanai et al. 2001). The PX domain can also interact with moesin, a cytoskeleton protein in neutrophils (Wientjes et al. 2001). Recently, it has been shown that proliferating cell nuclear antigen (PCNA) acts as a regulator of the NADPH oxidase activity by binding to the PX domain of p47^{phox} (Ohayon et al. 2019). P47^{phox} exists in an auto-inhibited conformation in which its SH3 domains are masked by intramolecular interactions with an auto inhibitory region (AIR; AA 292–340) in the C-terminal part of the protein. Upon activation, several sites (Ser303, Ser304, Ser328) in the AIR region are phosphorylated inducing a conformational change in p47^{phox}, which unmasks the tandem SH3 domains, thereby allowing the binding to p22^{phox} subunit and the translocation to the membrane (Sumimoto 2008).

The phosphorylation of p47^{phox} has been extensively studied (Belambri et al. 2018; El-Benna et al. 2009). It has been shown, that p47^{phox} is phosphorylated on serines located between Ser303 and Ser379, but only Ser379 is essential for oxidase activation. Moreover, priming agents such as TNFα can induce partial phosphorylation of p47^{phox} on Ser345, leading to PhosphoSer-345, which represents a binding site for the proline isomerase Pin1. The binding of Pin1 catalyzes a conformational change of p47^{phox} that facilitates the subsequent phosphorylation of p47^{phox} on other sites by protein kinase C, and consequently stimulates the activation of the NADPH oxidase (Boussetta et al. 2010; Makni-Maalej et al. 2012).

The PRR region of p47^{phox} allows interaction with p67^{phox}. Its PxxP motif attaches to the SH3 domain (at C-terminus) of p67^{phox} with high affinity ($K_d = 20$ nM) (Kami 2002). The results of Li et al. suggest that this interaction is disrupted soon after assembly of the NADPH oxidase at the phagosome (Li et al. 2009). The SH3 domain of p40^{phox} may be able to interact with the PRR region of p47^{phox} and negatively regulate the ROS production by NADPH oxidase. Nevertheless, this interaction was found to be too weak (Lapouge et al. 2002) to compete with p67^{phox} which attaches to p47^{phox} with a 250-fold higher affinity.

p67^{phox}

The largest cytosolic subunit (526 AA, 59.8 kDa) is an essential activator of the NADPH oxidase, as it interacts directly with NOX2 (Nisimoto et al. 1999). P67^{phox} is composed of the following domains (from N- to C-terminal):

- Four TPR motifs (tetratricopeptide repeat)
- An activation domain (AD)
- A small PRR stretch (10 amino acids)
- Two SH3 domains (SH3$_A$ N-terminal, SH3$_B$ C-terminal)
- A PB1 domain (between SH3$_A$ and SH3$_B$ domains)

TPR motifs are generally involved in protein-protein interactions and in the case of p67^{phox}, they are responsible for the interaction with the effector domain of the Rac protein (Koga et al. 1999; Nisimoto et al. 1997). The activation domain is defined by 11 amino acids. It has been shown by experiments *in vitro* that this domain is indispensable for the $O_2^{\bullet-}$ production (Han and Lee 2000; Han et al. 1998). Concerning the PRR region (AA 226–236), Thr233 is phosphorylated during the activation process (Belambri et al. 2018). No binding partner has been found so far for the PRR region (Groemping and Rittinger 2005). C-terminal SH3 domain has a strong affinity to the PRR region of p47^{phox} as previously described. SH3 N-terminal domain is the most conserved region in p67^{phox}. Its function remains unknown (Sumimoto et al. 2019). The PB1 domain (P for phagocyte and B1 for its homology to a yeast protein Bem1) (Ito 2001) reacts with the PB1 domain of p40^{phox} forming a heterodimer (Wilson et al. 2003). Thanks to this connection between p67^{phox} and p40^{phox}, the p67^{phox} subunit creates a 'bridge' between p47^{phox} and p40^{phox}, which allows translocation of all cytosolic subunits to the membrane during the activation process. Protein p67^{phox} interacts directly with NOX2 in the active complex, but the attachment site on NOX2 remains unclear. A recent study shows that the ^{369}Cys-Gly-Cys^{371} triad in NOX2 could form a disulfide bridge with cysteines in p67^{phox} and stabilize its binding to NOX2 (Fradin et al. 2018).

p40^{phox}

p40^{phox} subunit (339 amino acids, 39 kDa) was the last of all cytosolic subunits to be discovered by co-immunoprecipitation with p47^{phox} and p67^{phox} (Someya et al. 1993). P40^{phox} is composed of the three following domains (from N- to C-terminal), (Fig. 9.2):

- PX domain
- SH3 domain
- PB1 domain

PX domain of p40^{phox} binds with high affinity to phosphatidylinositol-3-phosphate (PI(3)P), which accumulates in phagosome membranes (Ellson et al. 2001; Song et al. 2017).

The SH3 domain of p40^{phox} shows affinity for the PRR region of p47^{phox}, but this interaction is very weak. Thus its function needs to be clarified. The interaction of p40^{phox} with p67^{phox} is performed by the PB1 domain. Another function of the PB1 domain is to prevent the PX domain from interacting with PI(3)Ps, resulting in incapability of p40^{phox} to localize to the phagosome. As a consequence, in its inactive state, p40^{phox} exists in a more 'folded' form (Honbou et al. 2007). In presence of ROS, p40^{phox} gets unfolded and acquires PI(3)P binding capabilities (Ueyama et al. 2011).

The role of p40^{phox} in the regulation of the functioning of NADPH oxidase remains controversial. It has been described as an important positive regulator in the case of NADPH oxidase activation by phagocytosis involving a Fcγ receptor (Suh et al. 2006). By binding PI(3)P at the phagosomal membrane p40^{phox} allows sustaining NADPH oxidase activation (see "The lipid connection" below) (Matute et al. 2009; Song et al. 2017). Positive regulation of ROS production by p40^{phox} has also been shown in PMA-stimulated K562 cells (Kuribayashi et al. 2002) and *in vitro* studies (Cross 2000). Some studies, however, showed a negative effect of p40^{phox} subunit on NADPH oxidase activation. It was demonstrated that p40^{phox} stabilizes the inactive state via an increased association with a cytoskeleton (Chen et al. 2007). Another study described that Thr^{154} phosphorylation of p40^{phox} leads to the inhibitory conformation that blocks the activation of NADPH oxidase (Lopes et al. 2004).

Rac Protein

Rac is a member of the Rho family of small GTPases and is a necessary subunit for the activation of the NADPH oxidase (Abo et al. 1991). Among the phagocytes, Rac2 is expressed in human neutrophils (Knaus et al. 1992), whereas monocytes and macrophages contain both, Rac1 and Rac2 (Abo et al. 1994). Rac is composed of several domains (from N- to C-terminal), (Fig. 9.2):

- Effector domain (AA 26–45)
- Insert domain (AA 124–135)

- Polybasic region (AA 182–186)

The insert domain of Rac directly associates with flavocytochrome b$_{558}$ (Freeman et al. 1996), whereas the C-terminal (polybasic) region of Rac interacts directly with the plasma membrane (Kreck et al. 1996). The effector domain participates in binding to p67^{phox} and assuring the ROS production (Nisimoto et al. 1997; Nunes et al. 2013). At the resting state, Rac is localized in the cytosol separated from the other cytosolic subunits, in a GDP-bound form that is associated with RhoGDI. Upon activation, Rac dissociates from RhoGDI by exchange of GDP for GTP and joins the active complex of the NADPH oxidase. Exchange of GDP for GTP requires the activity of guanine nucleotide exchange factors (GEFs) due to the slow dissociation rate of nucleotides. GEFs, together with GTPase-activating proteins (GAPs), contribute to GTPase regulation (Bos et al. 2007). Rac attachment to the cell membrane is achieved by the prenylation of its C-terminus and its polybasic region (see "The lipid connection" below) (Nauseef 2004). A crucial step in the active complex assembly is the interaction between p67^{phox} and active GTP-linked Rac, which occurs after both proteins have migrated to the membrane independently of each other. Rac binding is thought to induce a conformational change in p67^{phox} allowing its activation domain to interact with NOX2 (Sarfstein et al. 2004).

Conversion from Inactive to Active State of the NADPH Oxidase

The regulation of the NADPH oxidase is carried out by two mechanisms; (1) the spatial separation of membrane and cytosolic subunits, and (2) the modulation of protein-protein or protein-lipid interactions (Groemping and Rittinger 2005). Without stimulation, the cytosolic subunits exist together in the cytosol as a complex (separated from the Rac protein). Upon phagocytosis of a pathogen, p47^{phox} is phosphorylated at specific sites, which triggers migration of the ternary complex of cytosolic subunits to the phagosome membrane. There, it associates with cytochrome b$_{558}$

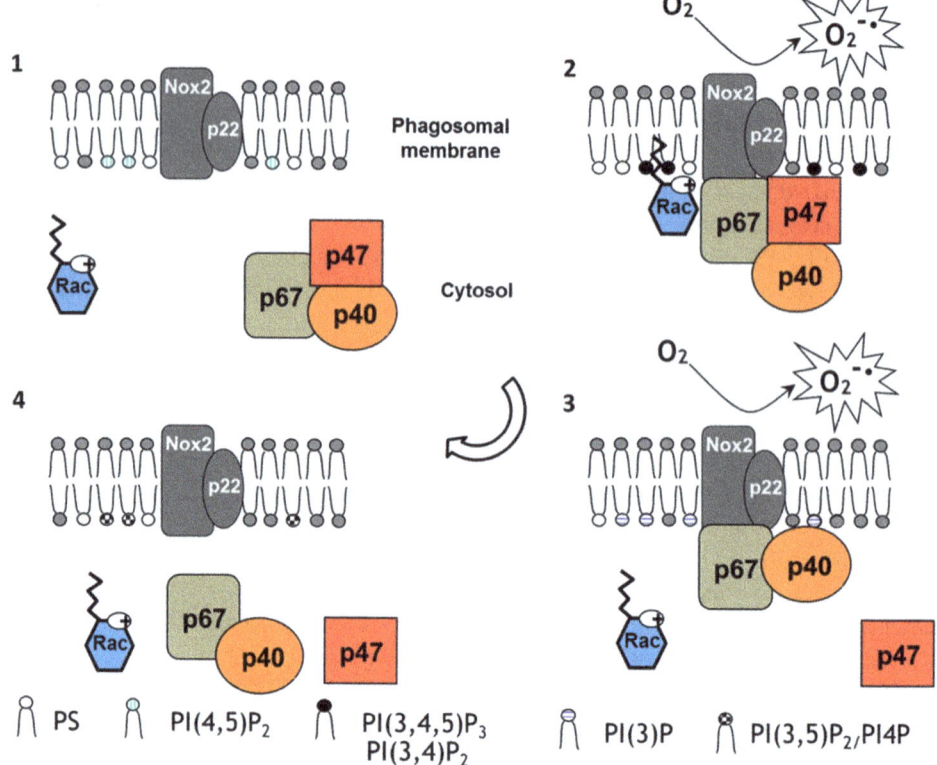

Fig. 9.3 Model for NADPH oxidase activation at the phagosome. *1.* In the resting state the NADPH oxidase subunits are separated. The membrane subunits NOX2 and p22^{phox} are located in the plasma membrane, in the secondary granules and in endosomal membranes. The heterotrimeric complex p47^{phox}/p67^{phox}/p40^{phox} and Rac are in the cytosol. *2.* Upon phagocytosis, the subunits are phosphorylated, the p47^{phox}/p67^{phox}/p40^{phox} complex and Rac accumulate at the phagosomal membrane and bind the membrane subunits. The assembly triggers NADPH oxidase activation and the ROS production. At the same time PI(3,4)P$_2$ and PI(3,4,5)P$_3$ accumulate at the phagosomal cup. PS is also present in the plasma and phagosomal membrane. PS and PI(3.4)P$_2$ may help to recruit and stabilize p47^{phox} at the phagosome. *3.* A few minutes after phagosome sealing, Rac and p47^{phox} detach from the phagosomal membrane. The detachment correlates with the drop in PI(3,4)P$_2$ and PI(3,4,5)P$_3$ levels and the accumulation of PI3P. The latter maintains p40^{phox} and p67^{phox} at the phagosomal membrane. *4.* Upon PI3P disappearance, p40^{phox} and p67^{phox} detach from the phagosoaml membrane terminating the ROS production

and forms the active complex able to produce superoxide anions. According to the different studies that assessed the dynamics of the NADPH oxidase subunits during phagocytosis (Faure et al. 2013; Li et al. 2009; Matute et al. 2009; Song et al. 2017), we propose the following model for the dynamic of Rac and the cytosolic subunits. Rac participates in the formation of an active complex, but as soon as the complex is formed and the p67^{phox} is well positioned towards NOX2, Rac2 and p47^{phox} detach from the phagosomal membrane. The detachment of p47^{phox} is concomitant to the decrease in the level of PI(3,4)P$_2$

and the accumulation of PI(3)P, in the phagosome membrane. The fixation of p40^{phox} to PI(3)P sustains the NADPH oxidase activity (see "The lipid connection" and Fig. 9.3).

Structural Models of the Inactive and Active State of the NADPH Oxidase

Numerous 2D models of the cytosolic subunits of the NADPH oxidase were proposed in several publications (Massenet et al. 2005; Nauseef 2004;

Nunes et al. 2013; Sumimoto 2008) describing the interactions in the inactive and active phase including the one we proposed recently (Fig. 9.2). Since these subunits contain several intrinsically disordered regions, information on their structure is incomplete. Recently, we combined results obtained by imaging in live cells with all the structural data collected on the individual subunits: SAXS models of p47^{phox} and p67^{phox}, 3D structure of p40^{phox} and of the interacting domains of the different subunits. The resulting 3D model of the cytosolic subunits in the resting state has a very elongated shape with the PX domain of p47^{phox} subunit pointing outward of the complex (Ziegler et al. 2019). It is perfectly located to interact with the lipids in the phagosomal membrane and the C-terminal of p22^{phox} and initiate the assembly. The non-structured C-terminus of p47^{phox} introduces a very flexible hinge in the complex. This may help to bring p67^{phox} close to NOX2 in the appropriate orientation. P40^{phox} is in the middle of the complex. This position is unfavorable for a major role during the initiation of the assembly due to the lack of accessibility of its PX domain. Nevertheless, as p47^{phox} leaves the phagosomal membrane shortly after the phagosome closure (Faure et al. 2013; Song et al. 2017), p40^{phox} could be ideally positioned to interact with the membrane and stabilize p67^{phox} for a sustained ROS production. The N-terminus of p67^{phox} is also in a favorable position, at the opposite end of the complex, to interact with Rac and NOX2 through its activation domain (AD, Fig. 9.2b).

Structural information on the transmembrane subunits is less abundant. A three dimensional homology model of the cytosolic dehydrogenase domain of NOX2 including the NIS was proposed in 2017 (Beaumel et al. 2017). At the same time, the structures of the dehydrogenase domain and the transmembrane domains of NOX5, a NOX2 homologue that does not require p22^{phox}, were solved (Magnani et al. 2017). The topology of p22^{phox} is discussed in the literature and 2–4 transmembrane segments were proposed (Campion et al. 2009; Dahan et al. 2002; Meijles et al. 2012). A homology model of NOX4/p22^{phox} with 4 transmembrane segments was proposed in 2018 (O'Neill et al. 2018). Figure 9.2c proposes a structural alignment of the NOX5 structural model represented as its surface and of the homology model of the cytosolic domain of NOX2. In this figure, different models are shown at the same scale, thus the size of the cytosolic dehydrogenase domain can be directly compared to the one of the 3D model of the cytosolic complex just above and to the one of Rac nearby. The arrows represent the identified interactions. This representation strongly suggests very high spatial constrains for an efficient assembly of the active complex in particular just after the phagosome closure when Rac, p67^{phox}, p47^{phox} and p40^{phox} are altogether at the phagosomal membrane.

Regulation of NADPH Oxidase Assembly and Activity

As ROS are toxic for both, microorganisms and mammalian hosts, the activity of the NADPH oxidase is strictly controlled in time and space. Several kinases phosphorylate oxidase subunits in at least 10 positions (Belambri et al. 2018). These kinases are activated by signaling pathways employing receptors, kinases, scaffolding proteins, and second messengers (Fig. 9.1).

Fc receptors for the immunoglobulins, mainly IgG and IgA, as well as integrins recognizing complement factor C3b are the most efficient activators of phagocytosis. These receptors recognize the phagocytic prey and then trigger the chain of events from membrane folding to actin polymerization, phagosome closure and then phagosome maturation. Multiple signaling cascades are triggered at the onset of phagocytosis (Flannagan et al. 2012; Niedergang and Grinstein 2018). Receptor activation leads to activation of the kinase Syk, which is responsible for activation of phospholipase Cγ. Cleavage of PIP2 into DAG and IP3 triggers calcium release from intracellular stores and activation of conventional PKCs. The PKCs as well as ERK and p38MAPK phosphorylate all NADPH oxidase subunits for priming and activation of the oxidase. Phosphorylation of p47^{phox} appears to be responsible for structural changes that allow this protein to bind to p22^{phox} and to specific phosphoinositides in the membrane (Belambri et al. 2018).

The activation of the NADPH oxidase starts even before the phagosome is fully closed, that is in less than 1 min. ROS production goes on for at least 30 min when cells are activated by FcR and complement receptors. There is a striking correlation between a transient rise in the cytosolic free calcium concentration and the onset of phagosomal ROS production as demonstrated on the level of individual phagocytic events (Dewitt et al. 2003). The importance of calcium signaling has been recognized since the early days of intracellular calcium measurements. Most likely, calcium acts in concert with other signaling pathways to trigger NADPH oxidase activation (Foyouzi-Youssefi et al. 1997; Pozzan et al. 1983). Calcium signaling in phagocytes primarily relies on IP3-mediated release from the endoplasmic reticulum and subsequent store operated calcium entry (SOCE) from the outside (Immler et al. 2018; Westman et al. 2019). The ER calcium sensor Stim1 and the plasma-membrane calcium channel Orai1 are key players in this process. Their inhibition by BTP2 inhibits ROS production (Steinckwich et al. 2007) and siRNA against Stim1 or Orai1 reduces phagosomal ROS production (Steinckwich et al. 2011). Studies in mouse models support the idea that SOCE contributes to the activation of NOX2 during phagocytosis (Clemens and Lowell 2019; Demaurex and Saul 2018).

None of the NADPH oxidase subunits binds calcium thus the stimulating activity of the calcium signal probably relies on intermediate signaling proteins. Conventional PKCs are activated by calcium in the presence of diacylglycerol. Therefore, PKC is a good candidate to explain the rapid onset of oxidase activity during a transient rise in intracellular calcium. It is less evident to envision how a calcium transient of less than 3 min controls the production of ROS in the phagosome for more than 30 min. Calcium hot spots in the vicinity of the phagosome could provide local calcium signals to maintain the activity of the oxidase (Nunes et al. 2012). Besides

PKC other calcium-dependent proteins are potential regulators of oxidase activity. The calcium-binding proteins S100A8/A9 appear to regulate NADPH oxidase activity by binding to several subunits (Berthier et al. 2003; Bréchard et al. 2013). During phagocytosis, S100A8/A9 translocate to the phagosome in a calcium-dependent manner and siRNA against these proteins significantly reduce phagosomal ROS production (Steinckwich et al. 2011). Most likely, calcium signals act in concert with other signaling pathways to activate the NADPH oxidase and multiple calcium-binding proteins mediate the calcium effect.

The Lipid Connection

Since translocation and attachment of the cytosolic subunits to the membrane is a critical feature, the role of the membrane lipids in the activation process deserves further attention. Indeed p47^{phox}, p40^{phox} and the small GTPase Rac are able to bind anionic phospholipids as explained above. The anionic phospholipids are mainly distributed in the inner leaflet of the cell membranes and composed of phosphatidylserine (PS), phosphatidic acid and phosphoinositides (PPI). The negative charge of the anionic phospholipids confers a negative surface charge of the cytosolic membrane (see Chap. 4). It creates an electric field equivalent to 10^5 V/cm that can attract cationic molecules (Fig. 9.4) (Yeung and Grinstein 2007). The cytosolic leaflet of the plasma membrane is the most negatively charged membrane leaflet of the cells (Yeung et al. 2008). This negative charge decreases during phagocytosis in RAW macrophages and murine neutrophils. This decrease in charge is correlated with the metabolism of anionic phospholipids (Magalhaes and Glogauer 2010; Yeung 2006). Both the modifications of the charge and the changes in phospholipids contribute to the regulation of the NADPH oxidase. We will first describe the phospholipid changes during phagocytosis and then examine the consequences for the regulation of the NADPH oxidase.

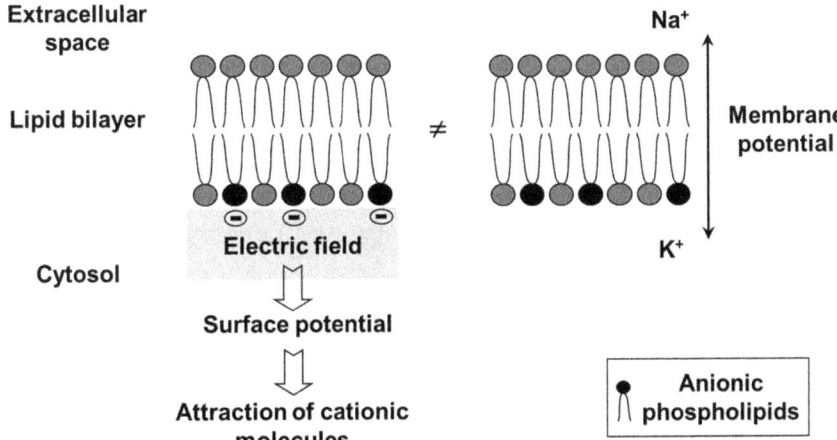

Fig. 9.4 Difference between the surface charge and the membrane potential. The negative surface charge is due to anionic phospholipides. They create an electric field that attracts cationic molecules. The membrane potential results from unequal distribution of inorganic ions on both sides of the membrane

Phospholipid Dynamics in the Cytosolic Leaflet of the Phagosomes

Phospholipid metabolism at the phagosome has been investigated by biochemical approaches and by the use of protein domains, tagged with fluorescent proteins that bind to specific phospholipids of the cytosolic leaflet. These lipid changes and especially phospholipid modifications at the phagosomal membrane have been well studied in the macrophage cell line RAW 264.7 (Levin et al. 2015). We and others have investigated several anionic phospholipid modifications in respectively the neutrophil like cells PLB-985 and human or murin neutrophils (Faure et al. 2013; Magalhaes and Glogauer 2010; Minakami et al. 2010; Song et al. 2017). In this part, we will highlight some phospholipid changes at the phagosomal membrane that are important for the NADPH oxidase regulation.

Phosphatidylserine

Phosphatidylserine (PS) represents about 15% of the lipids on the inner leaflet of the plasma membrane (Yeung and Grinstein 2007). The PS content in phagosomal and plasma membrane has been analyzed using the C2 domain of the lac-

tadherin protein tagged with fluorescent protein, FP-lactC2 (Faure et al. 2013; Yeung et al. 2009). This domain binds specifically to the PS (Shao et al. 2008). PS persists after sealing and contributes to the surface charge of the phagosomal membrane (Magalhaes and Glogauer 2010; Yeung et al. 2009). Using the C2 domain of the lactadherin tagged with the fluorescent protein mCherry, we observed a decrease in phagosomal and plasma membrane PS, 15 min after phagosome sealing (Faure et al. 2013). The reason for this decrease in phagosomal PS has not been investigated but may be due to the fusion of the neutrophil granules containing less PS with the phagosome.

Phosphoinositide Dynamics

Phosphoinositides (PPI) are generated by phosphorylation of phosphatidylinositol (PI) (Fig. 9.5). PI represents 10% of the lipids on the inner leaflet of the plasma membrane (Yeung and Grinstein 2007). Phosphatidylinositol 4-phosphate (PI4P) and PI(4,5)P$_2$ are present in resting cells on the inner leaflet of the plasma membrane. During phagocytosis, characteristic phosphoinositide changes occur. Indeed, using tagged protein domains that bind specifically to PI(4,5)P$_2$, it has been shown that its level dropped

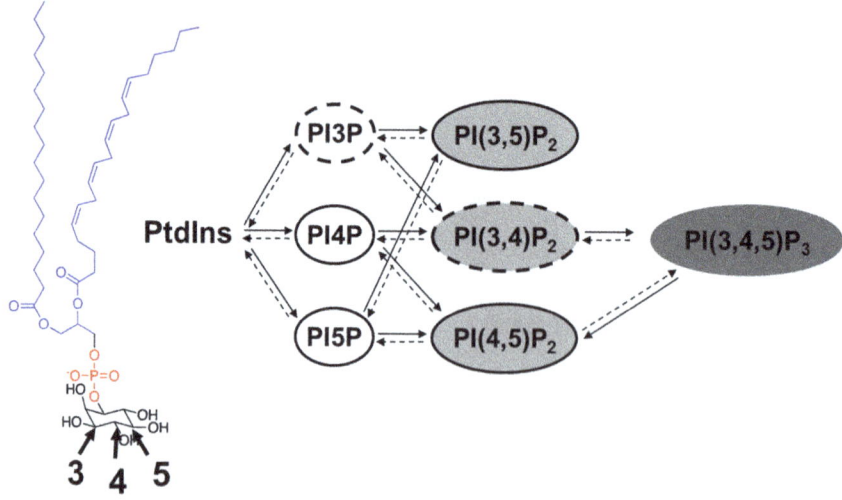

Fig. 9.5 Phosphatidylinositol and formation of phospho-inositides. The phosphatidylinositol is made of two fatty acids, a glycerol and a phosphate group links to the inositol. The inositol ring can be phosphorylated by different kinases on position 3, 4 or 5. It yields several mono, bis or tris phosphatidylinositol phosphates. The phosphoinositides that are targeted by the cytosolic subunits of the NADPH oxidase p47[phox] and p40[phox] are surrounded by a dashed line

after phagosome sealing in RAW macrophages (Levin et al. 2015). The same observation was made in human neutrophils (Minakami et al. 2010). In RAW cells, the PI(4,5)P2 level increases at the phagosomal cup before decreasing upon sealing, whereas the PI4P level rises upon sealing and then decreases 30–45 s later (Levin et al. 2017). The PI(4,5)P_2 and PI4P modifications are the consequences of activation of lipid kinases, phospholipases and phosphatases during phagocytosis. The signaling events are well known in the case of Fc receptor engagement (Levin et al. 2015; Rosales and Uribe-Querol 2017). The Class I phosphoinositides 3-kinases (PI3K) are activated and phosphorylate the 3'OH position of the inositol ring of PI(4,5)P_2 and PI4P giving rise to phosphatidylinositol (3,4,5)-trisphosphate (PIP$_3$) and phosphatidylinositol (3,4)-bisphosphate (PI(3,4)P_2) respectively (Goulden et al. 2019). The Class IA PI3Ks are comprised of 3 catalytic subunits (p110α, p110β and p110δ) and a regulatory subunit (mainly p85). The catalytic subunit p110γ belongs to the Class IB and interacts with the p101 regulatory subunit (Balla 2013). The generation of Class I PI3K products can be observed using the

pleckstrin homology (PH) domain of Akt tagged with a fluorescent protein (Fig. 9.6). PIP$_3$ can further recruit the phospholipase Cγ (PLCγ), which hydrolyses PI(4,5)P_2 into DAG and IP$_3$. Several phosphatases are also recruited to the nascent phagosomes and contribute to the drop in PI(4,5)P_2 and PI4P after sealing (Levin et al. 2017; Marion et al. 2012). Soon after phagosome closure, PIP$_3$ and PI(3,4)P_2 levels drop at the phagosome and a rise in phosphatidylinositol 3-phosphate (PI3P) is observed (Fig. 9.6) (Levin et al. 2015; Song et al. 2017). The breakdown of PIP$_3$ and PI(3,4)P_2 occurs mainly through the action of SH2 Domain-Containing Inositol 5-phosphatase (SHIP) (Bohdanowicz and Grinstein 2013) and Inositol Polyphosphatase 4A-phosphatase (INPP4A) (Nigorikawa et al. 2015). The recruitment of these phosphatases contributes to the rise in PI3P level, however an important source of PI3P is the phosphorylation of PI by the Class III PI3K (Levin et al. 2015; Song et al. 2017). This Class III PI3K is composed of a catalytic subunit Vps34 (Vacuolar protein sorting) and a regulatory one: Vps15. Using the PX domain of p40[phox], we observed an enrichment of phagosomal PI3P that lasted

15 min. The drop of PI3P thereafter is due to the recruitment of the myotubularin phosphatase, Myotubular Myopathy 1, at the phagosome (Song et al. 2017) that dephosphorylates PI3P (Begley and Dixon 2005) and to the phosphorylation of PI3P to phosphatidylinositol (3,5)-bisphosphate (Kim et al. 2014). Other proteins like Rubicon negatively regulate Vps34 and thus contributed to the drop of PI3P (Song et al. 2017).

Phospholipid Dynamics and the Regulation of the NADPH Oxidase

Changes in the level of phospholipid contents can modulate the membrane properties or affect protein-membrane interactions. The latter can occur in several ways. A protein can be recruited thanks to its specific lipid-binding domain, the recruitment may require interaction with other membrane proteins or not. Binding of the protein domain to specific lipid(s) can sometimes also induce a change in the protein conformation. The protein recruitment via anionic phospholipids can also be due to non-specific electrostatic interactions (Hammond and Balla 2015). The lipid metabolism affects NADPH oxidase activation either directly or through the regulation of other proteins. In this paragraph, we will first examine the role of phospholipase products in NADPH oxidase activation, then we will analyze the interplay between phospholipid metabolism and the cytosolic NADPH oxidase subunits bearing a lipid binding domain.

Phospholipases Are Involved in NADPH Oxidase Activation

As explained previously PLCγ generates DAG and IP3 upon receptor engagement. The level of phagosomal DAG on the cytosolic leaflet of the phagosome has been correlated with the production of the NADPH oxidase inside the phagosome (Schlam et al. 2013). DAG is involved in activation of different isoforms of the protein kinases C that phosphorylate p47^{phox}

(El-Benna et al. 2009). Upon phosphorylation p47^{phox} translocates to the plasma membrane together with p67^{phox} and p40^{phox} as described above. Other subunits, p67^{phox} (Zhao et al. 2005) and p22^{phox} (Regier et al. 1999), are phosphorylated by diverse PKC isoforms. Thus, DAG is important for NADPH oxidase assembly.

Cytosolic phospholipase A2 is also involved in NADPH oxidase activation by the generation of arachidonic acid (Dana et al. 1998). Arachidonic acid, which is a potent oxidase activator *in vitro*, has been proposed to bind the membrane subunit and increase its affinity for oxygen (Yeung and Grinstein 2007).

Rac Recruitment and Activation

Phospholipids can regulate Rac by attracting it to the membrane and by modulating the activities of Guanine Exchange Factors (GEFs) and GTPase activating proteins (GAPs). Two isoforms are found in hematopoietic cells: Rac1 and Rac2. The Rac2 isoform is the most abundant in human neutrophils (96%) whereas the macrophages express and use preferentially Rac1 (Zhao et al. 2003). The two isoforms are very similar, sharing 92% of homology. The difference concerns mainly the C-terminal polybasic domains (PB). Rac 1 bears 6 basic residues in its PB domains whereas the one of Rac2 comprises only 3 basic residues. Only the active form of Rac can be targeted to the plasma membrane as the prenylation is unmasked. However, the PB domain has been shown to determine to which membrane Rac proteins are recruited. The cytosolic leaflet of the most positively charged membrane attracts the most negatively charged domain. In the cell, the plasma membrane has the highest negative surface charge. During phagocytosis, the surface charge of the cytosolic leaflet of the phagosome decreases. The decrease was attributed to the change in PPI that occurs during phagocytosis and mainly to the drop of PI(4,5)P$_2$ (Magalhaes and Glogauer 2010; Yeung 2006; Yeung et al. 2009). As a consequence, we and others have observed that constitutively active Rac1 is

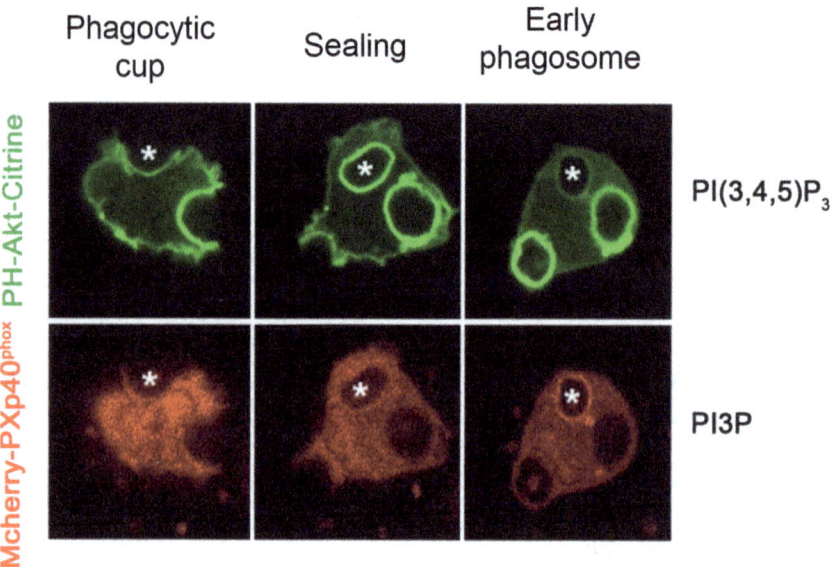

Fig. 9.6 Dynamic changes in phosphoinositide contents at the phagosome. Neutrophil-like cells PLB-985 had been transfected with plasmids encoding PH-Akt tagged with the fluorescent protein citrine and the PX domain of p40phox fused to mcherry. During phagocytosis of serum opsonised zymosan, the subcellular localisation of the two lipid probes was followed by spinning disk confocal video-microscopy. The PH domain of Akt detects PI(3,4)P2 and PI(3,4,5)P3. It accumulated at the phagosomal cup and then detached a few minutes after sealing. The p40phox PX domain, which is a PI3P biosensor, could be observed only in the early phagosome approximately 1 min after phagosome sealing

present at the plasma membrane and detaches from the phagosome upon sealing whereas active Rac2 is recruited just before phagosome sealing (Faure et al. 2013; Magalhaes and Glogauer 2010; Yeung 2006; Yeung et al. 2008). This means that, in neutrophils, active Rac2 will be able to be recruited at the phagosome. This phagosomal accumulation may be transient as we observed a detachment of constitutively active Rac2 a few minutes after sealing (Faure et al. 2013). This detachment may be due to interaction of Rac2 with other proteins. All the experiments described above used constitutively active Rac; however, Rac regulation also depends on GEF and GAP activation. Anionic phospholipids such as PIP$_3$ can modulate GEF and GAP. PIP$_3$ favors the activation of several GEFs like Vav and P-Rex (Campa et al. 2015). PIP$_3$ can also recruit GAP at the phagosomal cup (Schlam et al. 2015). Thus the role of PIP$_3$ is ambiguous and may depend of the level of PIP$_3$ or/and competition for Rac binding between GEF and GAP.

Phospholipid Dynamics and p47phox and p40phox Regulation

As seen before, upon receptor engagement the modification of membrane lipids activates signaling pathways that lead to subunit phosphorylation. PLCγ products but also the Class I PI3K pathway activate Akt, which has also been involved in p47phox phosphorylation in addition to PKC (Chen et al. 2003). As for Rac, the p47phox and p40phox bind anionic phospholipids via their PX domains. The PX domains of p47phox has two pockets: one binds to PS and phosphatidic acid and the other to PI(3,4)P$_2$ (Karathanassis et al. 2002). P40phox PX binds specifically to PI3P (Bravo et al. 2001; Ellson et al. 2001, 2006). Several studies explored the role of the PX domains of p47phox and p40phox in NADPH oxidase activation at the phagosome. Indeed, using a probe that binds and thus masks PS, FP-lactC2, we observed, in neutrophil like PLB-985 cells, a delay in FP tagged p47phox recruitment at the phagosome. We also measured a delay in ROS production during phagocytosis of opsonized yeast (Faure et al. 2013). Mutations in R90, a residue involved in PI(3,4)P2 binding (Karathanassis et al. 2002), also affects ROS production during phagocytosis in mouse neutrophils. The defect in ROS production was rather small (around 20%) and the FP tagged p47^{phox} R90A was still observed at the phagosome (Li et al. 2010). Thus, binding of p47^{phox} to phospholipids has a minor effect on NADPH oxidase activation at the phagosome. This may be explained by the fact that p47^{phox} also binds to p22^{phox} and, according to our finding, stays at the phagosome for a few minutes only. In contrast to p47^{phox}, the PX domain of p40^{phox} and its binding to PI3P is crucial for the NADPH oxidase activation. A mutation R105Q found in a CGD patient affects ROS production at the phagosome (Matute et al. 2009). Several studies with mice bearing a PX mutated p40^{phox} observed a defect in ROS production (Bagaitkar et al. 2017; Ellson et al. 2006). The time of presence at the phagocyte of a FP tagged p40^{phox} mutants (R105 Q /A), was reduced (Matute et al. 2009; Tian et al. 2008). All these studies suggest that PI3P binding of p40^{phox} is crucial for ROS production at the phagosome. We indeed showed using FP tagged p40^{phox} and p67^{phox} that both proteins stay the same time at the phagosome i.e. during all the ROS production. Therefore, the ROS production is intrinsically correlated to PI3P. Using pharmacologic and molecular approaches, we decreased or increased the PI3P level of the phagosome. Modulation of PI3P level induced a parallel modulation of the ROS production. PI3P acts as a timer for ROS production by maintaining p40^{phox}, linked to p67^{phox}, at the phagosomal membrane (Song et al. 2017). Figure 9.6 illustrates the change in anionic phospholipids at the phagosomal membrane in correlation with the dynamics of the cytosolic subunit and Rac.

What Happens Inside the Phagosome?

Since the goal of phagosomal ROS production is the destruction of a microorganism, we would like to know in more detail how ROS react inside

the phagosome. To this end, quantitative measurements of ROS inside the phagosome would be very useful. Alas, despite a wide range of ROS detection techniques, quantitative assessment remains very difficult and the choice of the redox sensor depends on the biological context (Erard et al. 2018; Kalyanaraman et al. 2017; Nault et al. 2016). ROS do not simply accumulate in the phagosome until they reach the critical concentration to kill the ingested microorganism. Instead, ROS react with many components including the microorganism but also host proteins that arrive by granule fusion. Furthermore, ROS may diffuse across membranes, in particular H_2O_2. When they diffuse out of the phagosome into the cytoplasm of the phagocyte and into the extracellular space, they may have regulatory roles that are poorly understood. The immunomodulatory role of NOX2-derived ROS attracts much attention in recent years. Patients lacking functional oxydase show signs of hyperinflammation, which may result in a positive feedback on the antimicrobial capacities of the phagocyte (Cachat et al. 2015; Nauseef 2019; Thomas 2018). Furthermore, several subunits of the oxidase are sensitive to ROS *in vitro* (Ostuni et al. 2010). Several signaling pathways inside the phagocyte may also be affected by H_2O_2 diffusing out of the phagosome. For example, some of the phosphoinositide phosphatases such as PTEN, involved in the control of oxidase assembly, are sensitive to inhibition by oxidation through H_2O_2 (Hsu and Mao 2015). SOCE is also susceptible to redox modifications again suggesting that ROS may exert a negative feedback on phagocyte calcium influx (Bhardwaj et al. 2016; Bogeski et al. 2010).

For assessing the redox-sensitivity of microorganisms, it is relatively straightforward to expose them to ROS *in vitro* and determine their survival. Inside the phagosome, conditions are changing rapidly in a complex manner. Changes of pH, liberation of granule proteins in high concentration, production, degradation and diffusion of ROS all occur simultaneously with non-linear kinetics. A side effect of these changes is the fact that any dyes we use to probe the interior of the phagosome may be altered or even destroyed by the harsh environment of the phagosome (Dupré-Crochet et al. 2013).

Probing the phagosome specifically may be achieved by coupling a specific dye or sensor onto the phagocytic prey prior to phagocytosis. For the detection of phagosomal ROS, dichlorodihydrofluorescein (DCFH2) has been used by several groups (Dewitt et al. 2003; Dupré-Crochet et al. 2019; Kamen et al. 2008; Steinckwich et al. 2011; Tlili et al. 2011). The succinimidyl-ester of DCFH2, DCFH2-SE, readily reacts with amines on the particles (beads, bacteria, fungi). The protocol is easy to adapt to standard laboratories. The detection of its green fluorescence is compatible with most fluorescence microscopes, plate readers and flow cytometers (Dupré-Crochet et al. 2019). DCFH2 is currently the only ROS-sensor commercially available with an appropriate reactive group. This dye is not specific for any particular ROS and it is subject to other potential artifacts. Nevertheless, it provides valuable information on the kinetics of phagosomal ROS production. A transient rise in intracellular calcium concentration coincides with the beginning of phagosomal ROS production (Dewitt et al. 2013). In neutrophils and neutrophil-like PLB-985 cells, the latter starts even before phagosome closure and goes on for 20–40 min, depending on the receptors that triggered phagocytosis. The duration of ROS production appears to correlate roughly with the time that p67^{phox} stays on the phagosome (Tlili et al. 2012). However, ROS production might continue for much longer at a reduced rate. Current detection methods are not sufficiently sensitive to detect low level production in late stages of the phagosome maturation. Furthermore, the ROS-sensitive dyes also saturate due to the large amount of ROS that are produced just after phagosome closure.

The concentration of phagosomal ROS has been estimated in a mathematical model providing critical insight into the level of different ROS that we can expect in the phagosome (Winterbourn et al. 2006). A key finding was that hydrogen peroxide will not accumulate in the phagosome because it reacts with phagosomal contents, or is transformed to HOCl by the myeloperoxidase, or diffuses out through the phagosome membrane. The actual role of HOCl in bacterial killing is much debated, although most researchers agree that it is a major component of the neutrophil armament (Klebanoff et al. 2013). Interestingly, the transformation of H_2O_2 to HOCl appears to focus the microbicidal action into the phagosome, because the latter does not diffuse across the membrane. In myeloperoxidase-deficient mice, Salmonella infection was still controlled, but the diffusion of H_2O_2 caused collateral damage in the surrounding tissue (Schürmann et al. 2017). Furthermore, neutrophil phagosomes displayed a considerable heterogeneity with respect to the onset of detectable HOCl-production as revealed with the rhodamine-based probe R19-S that is reportedly specific for HOCl (Albrett et al. 2018).

How does the time frame of ROS production correlate with the killing of the microorganism? Colony formation on culture plates is the standard criterion for survival. However, the method is time consuming, slow and gives an average response for a sample containing thousands of phagosomes. It does not allow identifying the critical moment of cell death and it gives no information on individual phagosomes. The microorganism may die during the period of ROS production. Alternatively, the ROS damage the microorganism but its actual cell death occurs later. Many bacteria and fungi are able to respond to ROS and other environmental changes in the phagosome. They will rapidly produce ROS scavengers and ROS degrading enzymes (Imlay 2008; Staerck et al. 2017). The timing of this response is critical. The microorganism needs to survive the initial, rapid onslaught to gain time and adapt to its environment. Many microorganisms attempt to render the phagosome less hostile, for example by inhibiting the NADPH oxidase. *Helicobacter pylori* actually diverts the oxidase towards the plasma membrane by an unknown molecular mechanism (Allen and McCaffrey 2007). *Pseudomonas aeruginosa* produces toxins, ExoS and ExoT, that are injected into the phagocyte by a type 3 secretion system and then interfere with the activation of PI3kinase and thereby inhibit ROS production in the phagosome (Vareechon et al. 2017). *Mycobacterium tuberculosis* employs at least two mechanisms to reduce phagosomal NOX2 activity in infected macrophages. The nuoG gene codes one subunit of a NADH dehydrogenase, which affects ROS production by an unknown mechanism (Miller et al. 2010). The ndkA gene codes a Nucleoside Diphosphate Kinase, which binds to and inactivates rac1 by exerting GAP activity on the small GTPase. This leads to reduced assembly of the oxidase on the mycobacterium-containing phagosome and thus lower ROS production (Sun et al. 2013). In both cases, reduced ROS production delays apoptosis of the macrophage and its subsequent uptake by other macrophages.

These examples illustrate the diversity of microbial action against phagosomal ROS production. We may expect to see the identification of other mechanisms in the future. Since the host and the pathogens employ multiple mechanisms to ensure their survival, the role of an individual virulence factor may be hidden in the complexity of the host-pathogen interaction. Elucidation of these mechanisms raises the hope of finding selective inhibitors that eventually could be used as antibiotics, not by killing directly the microorganism but by enhancing the oxidative host response.

Open Questions

Despite the large body of literature on phagosomal ROS production and its importance for microbial killing, many questions remain open. For example, we know very little about the deactivation of the NADPH oxidase (DeCoursey and Ligeti 2005). A reasonable scenario suggests that kinase activity slows down and phosphatase activity catches up dephosphorylating the NADPH oxidase subunits. The subsequent conformational changes disrupt the interactions with NOX2, $p22^{phox}$ and the membrane lipids, the cytosolic subunits detach and thus inactivate the enzyme. $P67^{phox}$ appears to be the main actor, its relocation to the cytosol stops oxidase activity. At least for $p40^{phox}$, detachment from the membrane is also a consequence of lipid phosphatases that modify inositols and thereby the affinity of

$p40^{phox}$ for the membrane. This scenario clearly needs experimental confirmation.

The anionic lipid composition of the plasma membrane is not identical to the composition of the phagosomal membrane. The relevant enzymes are not the same raising the perspective of selective intervention. Perhaps, we may target specifically the plasma membrane phosphoinositides to inhibit extracellular ROS production and preserve phagosomal ROS production.

Duration of phagosomal ROS production depends on the stimulus. When the ingested particles are opsonized by IgG and complement, they produce ROS for much longer than in response to IgG opsonization alone (Tlili et al. 2012). Does this reflect an intrinsic program of the phagocyte? Does the phagocyte sense what is going on inside the phagosome in terms of ROS concentration? How could the phagocyte "know" that it has killed the microorganism? What happens when the microorganism resists, for example with the help of its ROS degrading enzymes such as superoxide dismutase and/or catalase? Can the phagocyte react to resistance for example by fusing more granules with the phagosome? Could it switch from ROS-based killing to protease-based killing?

Answers to these questions require appropriate experimental techniques. The dynamic nature of phagocytosis and the short life of ROS complicate the task to follow phagosomes over time. Microscopy techniques improve constantly and new sensors become available to probe the interior of the phagosome and its membrane. These are good reasons to reexamine phagosomal NADPH oxidase activity with new tools.

References

Abo A, Pick E, Hall A et al (1991) Activation of the NADPH oxidase involves the small GTP-binding protein p21rac1. Nature 353:668–670. https://doi.org/10.1038/353668a0

Abo A, Webb MR, Grogan A, Segal AW (1994) Activation of NADPH oxidase involves the dissociation of p21rac from its inhibitory GDP/GTP exchange protein (rhoGDI) followed by its translocation to the plasma membrane. Biochem J 298:585–591. https://doi.org/10.1042/bj2980585

Albrett AM, Ashby LV, Dickerhof N et al (2018) Hetero-geneity of hypochlorous acid production in individual neutrophil phagosomes revealed by a rhodamine-based probe. J Biol Chem 293:15715–15724. https://doi.org/10.1074/jbc.RA118.004789

Allen LH, McCaffrey RL (2007) To activate or not to activate: distinct strategies used by Helicobacter pylori and Francisella tularensis to modulate the NADPH oxidase and survive in human neutrophils. Immunol Rev 219:103–117. https://doi.org/10.1111/j.1600-065X.2007.00544.x

Ambasta RK, Kumar P, Griendling KK et al (2004) Direct interaction of the novel nox proteins with p22phox is required for the formation of a functionally active NADPH oxidase. J Biol Chem 279:45935–45941. https://doi.org/10.1074/jbc.M406486200

Bagaitkar J, Barbu EA, Perez-Zapata LJ et al (2017) PI(3)P-p40 phox binding regulates NADPH oxidase activation in mouse macrophages and magnitude of inflammatory responses in vivo. J Leukoc Biol 101:449–457. https://doi.org/10.1189/jlb.3AB0316-139R

Balla T (2013) Phosphoinositides: tiny lipids with giant impact on cell regulation. Physiol Rev 93:1019–1137. https://doi.org/10.1152/physrev.00028.2012

Baranov MV, Olea RA, van den Bogaart G (2019) Chasing uptake: super-resolution microscopy in endocytosis and phagocytosis. Trends Cell Biol 29:727–739. https://doi.org/10.1016/j.tcb.2019.05.006

Beaumel S, Picciocchi A, Debeurme F et al (2017) Down-regulation of NOX2 activity in phagocytes mediated by ATM-kinase dependent phosphorylation. Free Radic Biol Med 113:1–15. https://doi.org/10.1016/j.freeradbiomed.2017.09.007

Begley MJ, Dixon JE (2005) The structure and regulation of myotubularin phosphatases. Curr Opin Struct Biol 15:614–620. https://doi.org/10.1016/j.sbi.2005.10.016

Belambri SA, Rolas L, Raad H et al (2018) NADPH oxidase activation in neutrophils: role of the phosphorylation of its subunits. Eur J Clin Investig 48:e12951. https://doi.org/10.1111/eci.12951

Berthier S, Paclet MH, Lerouge S et al (2003) Changing the conformation state of cytochrome b 558 initiates NADPH oxidase activation: MRP8/MRP14 regulation. J Biol Chem 278:25499–25508. https://doi.org/10.1074/jbc.M209755200

Bhardwaj R, Hediger MA, Demaurex N (2016) Redox modulation of STIM-ORAI signaling. Cell Calcium 60:142–152. https://doi.org/10.1016/j.ceca.2016.03.006

Bilan DS, Belousov VV (2018) In vivo imaging of hydrogen peroxide with HyPer probes. Antioxid Redox Signal 29:569–584. https://doi.org/10.1089/ars.2018.7540

Bogeski I, Kummerow C, Al-Ansary D et al (2010) Differential redox regulation of ORAI ion channels: a mechanism to tune cellular calcium signaling. Sci Signal 3:1–10. https://doi.org/10.1126/scisignal.2000672

Bohdanowicz M, Grinstein S (2013) Role of phospholipids in endocytosis, phagocytosis, and macropinocytosis. Physiol Rev 93:69–106. https://doi.org/10.1152/physrev.00002.2012

Bos JL, Rehmann H, Wittinghofer A (2007) GEFs and GAPs: critical elements in the control of small G proteins. Cell 129:865–877. https://doi.org/10.1016/j.cell.2007.05.018

Boussetta T, Gougerot-Pocidalo M-A, Hayem G et al (2010) The prolyl isomerase Pin1 acts as a novel molecular switch for TNF-α–induced priming of the NADPH oxidase in human neutrophils. Blood 116:5795–5802. https://doi.org/10.1182/blood-2010-03-273094

Bravo J, Karathanassis D, Pacold CM et al (2001) The crystal structure of the PX domain from p40phoxbound to phosphatidylinositol 3-phosphate. Mol Cell 8:829–839. https://doi.org/10.1016/S1097-2765(01)00372-0

Bréchard S, Plançon S, Tschirhart EJ (2013) New insights into the regulation of neutrophil NADPH oxidase activity in the phagosome: a focus on the role of lipid and Ca^{2+} signaling. Antioxid Redox Signal 18:661–676. https://doi.org/10.1089/ars.2012.4773

Buvelot H, Posfay-Barbe KM, Linder P et al (2017) Staphylococcus aureus, phagocyte NADPH oxidase and chronic granulomatous disease. FEMS Microbiol Rev 41:139–157. https://doi.org/10.1093/femsre/fuw042

Buvelot H, Jaquet V, Krause K-H (2019) Mammalian NADPH oxidases. Methods Mol Biol 1982:17–36

Cachat J, Deffert C, Hugues S, Krause K-H (2015) Phagocyte NADPH oxidase and specific immunity. Clin Sci 128:635–648. https://doi.org/10.1042/CS20140635

Campa CC, Ciraolo E, Ghigo A et al (2015) Crossroads of PI3K and Rac pathways. Small GTPases 6:71–80. https://doi.org/10.4161/21541248.2014.989789

Campion Y, Jesaitis AJ, Nguyen MVC et al (2009) New p22-phox monoclonal antibodies: identification of a conformational probe for cytochrome b558. J Innate Immun 1:556–569. https://doi.org/10.1159/000231977

Casbon A-J, Allen L-AH, Dunn KW, Dinauer MC (2009) Macrophage NADPH oxidase flavocytochrome b localizes to the plasma membrane and Rab11-positive recycling endosomes. J Immunol 182:2325–2339. https://doi.org/10.4049/jimmunol.0803476

Chen Q, Powell DW, Rane MJ et al (2003) Akt phosphorylates p47 phox and mediates respiratory burst activity in human neutrophils. J Immunol 170:5302–5308. https://doi.org/10.4049/jimmunol.170.10.5302

Chen J, He R, Minshall RD et al (2007) Characterization of a mutation in the Phox homology domain of the NADPH oxidase component p40 phox identifies a mechanism for negative regulation of superoxide production. J Biol Chem 282:30273–30284. https://doi.org/10.1074/jbc.M704416200

Clemens RA, Lowell CA (2019) CRAC channel regulation of innate immune cells in health and disease. Cell Calcium 78:56–65. https://doi.org/10.1016/j.ceca.2019.01.003

Cross AR (2000) p40(phox) participates in the activation of NADPH oxidase by increasing the affinity of p47(phox) for flavocytochrome b558. Biochem J 349:113–117. https://doi.org/10.1042/0264-6021:3490113

Cross AR, Segal AW (2004) The NADPH oxidase of professional phagocytes-prototype of the NOX electron transport chain systems. Biochim Biophys Acta Bioenerg 1657:1–22. https://doi.org/10.1016/j.bbabio.2004.03.008

Dahan I, Issaeva I, Gorzalczany Y et al (2002) Mapping of functional domains in the p22phox subunit of flavocytochrome b559 participating in the assembly of the NADPH oxidase complex by "peptide walking". J Biol Chem 277:8421–8432. https://doi.org/10.1074/jbc.M109778200

Dana R, Leto TL, Malech HL, Levy R (1998) Essential requirement of cytosolic phospholipase A 2 for activation of the phagocyte NADPH oxidase. J Biol Chem 273:441–445. https://doi.org/10.1074/jbc.273.1.441

DeCoursey TE, Ligeti E (2005) Regulation and termination of NADPH oxidase activity. Cell Mol Life Sci 62:2173–2193. https://doi.org/10.1007/s00018-005-5177-1

DeLeo FR, Burritt JB, Yu L et al (2000) Processing and maturation of flavocytochrome b 558 include incorporation of Heme as a prerequisite for heterodimer assembly. J Biol Chem 275:13986–13993. https://doi.org/10.1074/jbc.275.18.13986

Demaurex N, Saul S (2018) The role of STIM proteins in neutrophil functions. J Physiol 596:2699–2708. https://doi.org/10.1113/JP275639

Dewitt S, Laffafian I, Hallett MB (2003) Phagosomal oxidative activity during beta2 integrin (CR3)-mediated phagocytosis by neutrophils is triggered by a non-restricted Ca^{2+} signal: Ca^{2+} controls time not space. J Cell Sci 116:2857–2865. https://doi.org/10.1242/jcs.00499

Dewitt S, Francis RJ, Hallett MB (2013) Ca2+ and calpain control membrane expansion during the rapid cell spreading of neutrophils. J Cell Sci 126:4627–4635. https://doi.org/10.1242/jcs.124917

Dinauer MC, Pierce EA, Erickson RW et al (1991) Point mutation in the cytoplasmic domain of the neutrophil p22-phox cytochrome b subunit is associated with a nonfunctional NADPH oxidase and chronic granulomatous disease. Proc Natl Acad Sci 88:11231–11235. https://doi.org/10.1073/pnas.88.24.11231

Dupré-Crochet S, Erard M, Nüße O (2013) ROS production in phagocytes: why, when, and where? J Leukoc Biol 94:657–670. https://doi.org/10.1189/jlb.1012544

Dupré-Crochet S, Erard M, Nüße O (2019) Kinetic analysis of phagosomal ROS generation. Methods Mol Biol 1982:301–312

El-Benna J, Dang PM-C, Gougerot-Pocidalo M-A et al (2009) p47phox, the phagocyte NADPH oxidase/NOX2 organizer: structure, phosphorylation and implication in diseases. Exp Mol Med 41:217. https://doi.org/10.3858/emm.2009.41.4.058

Ellson CD, Gobert-Gosse S, Anderson KE et al (2001) PtdIns(3)P regulates the neutrophil oxidase complex by binding to the PX domain of p40phox. Nat Cell Biol 3:679–682. https://doi.org/10.1038/35083076

Ellson C, Davidson K, Anderson K et al (2006) PtdIns3P binding to the PX domain of p40phox is a physiological signal in NADPH oxidase activation. EMBO J 25:4468–4478. https://doi.org/10.1038/sj.emboj.7601346

Erard M, Dupré-Crochet S, Nüße O (2018) Biosensors for spatiotemporal detection of reactive oxygen species in cells and tissues. Am J Physiol Regul Integr Comp Physiol 314:R667–R683. https://doi.org/10.1152/ajpregu.00140.2017

Faure MC, Sulpice J-C, Delattre M et al (2013) The recruitment of p47 phox and Rac2G12V at the phagosome is transient and phosphatidylserine dependent. Biol Cell 105:501–518. https://doi.org/10.1111/boc.201300010

Flannagan RS, Jaumouillé V, Grinstein S (2012) The cell biology of phagocytosis. Annu Rev Pathol Mech Dis 7:61–98. https://doi.org/10.1146/annurev-pathol-011811-132445

Foyouzi-Youssefi R, Petersson F, Lew DP et al (1997) Chemoattractant-induced respiratory burst: increases in cytosolic Ca^{2+} concentrations are essential and synergize with a kinetically distinct second signal. Biochem J 322:709–718. https://doi.org/10.1042/bj3220709

Fradin T, Bechor E, Berdichevsky Y et al (2018) Binding of p67 phox to Nox2 is stabilized by disulfide bonds between cysteines in the 369 Cys-Gly-Cys 371 triad in Nox2 and in p67 phox. J Leukoc Biol 104:1023–1039. https://doi.org/10.1002/JLB.4A0418-173R

Freeman JL, Abo A, Lambeth JD (1996) Rac "insert region" is a novel effector region that is implicated in the activation of NADPH oxidase, but not PAK65. J Biol Chem 271:19794–19801. https://doi.org/10.1074/jbc.271.33.19794

Goulden BD, Pacheco J, Dull A et al (2019) A high-avidity biosensor reveals plasma membrane PI(3,4)P 2 is predominantly a class I PI3K signaling product. J Cell Biol 218:1066–1079. https://doi.org/10.1083/jcb.201809026

Greenlee-Wacker M, DeLeo FR, Nauseef WM (2015) How methicillin-resistant Staphylococcus aureus evade neutrophil killing. Curr Opin Hematol 22:30–35. https://doi.org/10.1097/MOH.0000000000000096

Greenwald EC, Mehta S, Zhang J (2018) Genetically encoded fluorescent biosensors illuminate the spatiotemporal regulation of signaling networks. Chem Rev 118:11707–11794. https://doi.org/10.1021/acs.chemrev.8b00333

Groemping Y, Rittinger K (2005) Activation and assembly of the NADPH oxidase: a structural perspective. Biochem J 386:401–416. https://doi.org/10.1042/BJ20041835

Hammond GRV, Balla T (2015) Polyphosphoinositide binding domains: key to inositol lipid biology. Biochim Biophys Acta Mol Cell Biol Lipids 1851:746–758. https://doi.org/10.1016/j.bbalip.2015.02.013

Han CH, Lee MH (2000) Activation domain in P67phox regulates the steady state reduction of FAD in gp91phox. J Vet Sci 1:27–31

Han C-H, Freeman JLR, Lee T et al (1998) Regulation of the neutrophil respiratory burst oxidase. J Biol Chem 273:16663–16668. https://doi.org/10.1074/jbc.273.27.16663

Henríquez-Olguín C, Renani LB, Arab-Ceschia L et al (2019) Adaptations to high-intensity interval training in skeletal muscle require NADPH oxidase 2. Redox Biol 24:101188. https://doi.org/10.1016/j.redox.2019.101188

Honbou K, Minakami R, Yuzawa S et al (2007) Full-length p40phox structure suggests a basis for regulation mechanism of its membrane binding. EMBO J 26:1176–1186. https://doi.org/10.1038/sj.emboj.7601561

Hsu F, Mao Y (2015) The structure of phosphoinositide phosphatases: insights into substrate specificity and catalysis. Biochim Biophys Acta Mol Cell Biol Lipids 1851:698–710. https://doi.org/10.1016/j.bbalip.2014.09.015

Imlay JA (2008) Cellular defenses against superoxide and hydrogen peroxide. Annu Rev Biochem 77:755–776. https://doi.org/10.1146/annurev.biochem.77.061606.161055

Immler R, Simon SI, Sperandio M (2018) Calcium signalling and related ion channels in neutrophil recruitment and function. Eur J Clin Investig 48:e12964. https://doi.org/10.1111/eci.12964

Ito T (2001) Novel modular domain PB1 recognizes PC motif to mediate functional protein-protein interactions. EMBO J 20:3938–3946. https://doi.org/10.1093/emboj/20.15.3938

Kalyanaraman B, Hardy M, Podsiadly R et al (2017) Recent developments in detection of superoxide radical anion and hydrogen peroxide: opportunities, challenges, and implications in redox signaling. Arch Biochem Biophys 617:38–47. https://doi.org/10.1016/j.abb.2016.08.021

Kamen LA, Levinsohn J, Cadwallader A et al (2008) SHIP-A increases early oxidative burst and regulates phagosome maturation in macrophages. J Immunol 180:7497–7505. https://doi.org/10.4049/jimmunol.180.11.7497

Kami K (2002) Diverse recognition of non-PxxP peptide ligands by the SH3 domains from p67phox, Grb2 and Pex13p. EMBO J 21:4268–4276. https://doi.org/10.1093/emboj/cdf428

Kanai F, Liu H, Field SJ et al (2001) The PX domains of p47phox and p40phox bind to lipid products of PI(3)K. Nat Cell Biol 3:675–678. https://doi.org/10.1038/35083070

Karathanassis D, Stahelin RV, Bravo J et al (2002) Binding of the PX domain of p47phox to phosphatidylinositol 3,4-bisphosphate and phosphatidic acid is masked by an intramolecular interaction. EMBO J 21:5057–5068. https://doi.org/10.1093/emboj/cdf519

Kim GHE, Dayam RM, Prashar A et al (2014) PIKfyve inhibition interferes with phagosome and endosome maturation in macrophages. Traffic 15:1143–1163. https://doi.org/10.1111/tra.12199

Klebanoff SJ, Kettle AJ, Rosen H et al (2013) Myeloperoxidase: a front-line defender against phagocytosed microorganisms. J Leukoc Biol 93:185–198. https://doi.org/10.1189/jlb.0712349

Knaus UG, Heyworth PG, Kinsella BT et al (1992) Purification and characterization of Rac 2. A cytosolic GTP-binding protein that regulates human neutrophil NADPH oxidase. J Biol Chem 267:23575–23582

Koga H, Terasawa H, Nunoi H et al (1999) Tetratricopeptide repeat (TPR) motifs of p67 phox participate in interaction with the small GTPase Rac and activation of the phagocyte NADPH oxidase. J Biol Chem 274:25051–25060. https://doi.org/10.1074/jbc.274.35.25051

Kreck ML, Freeman JL, Abo A, Lambeth JD (1996) Membrane association of Rac is required for high activity of the respiratory burst oxidase †. Biochemistry 35:15683–15692. https://doi.org/10.1021/bi960206l

Kuribayashi F, Nunoi H, Wakamatsu K et al (2002) The adaptor protein p40 phox as a positive regulator of the superoxide-producing phagocyte oxidase. EMBO J 21:6312–6320. https://doi.org/10.1093/emboj/cdf642

Lambert TJ (2019) FPbase: a community-editable fluorescent protein database. Nat Methods 16:277–278. https://doi.org/10.1038/s41592-019-0352-8

Lapouge K, Smith SJM, Groemping Y, Rittinger K (2002) Architecture of the p40-p47-p67phox complex in the resting state of the NADPH oxidase. A central role for p67phox. J Biol Chem 277:10121. https://doi.org/10.1074/jbc.M112065200

Levin R, Grinstein S, Schlam D (2015) Phosphoinositides in phagocytosis and macropinocytosis. Biochim Biophys Acta Mol Cell Biol Lipids 1851:805–823. https://doi.org/10.1016/j.bbalip.2014.09.005

Levin R, Hammond GRV, Balla T et al (2017) Multiphasic dynamics of phosphatidylinositol 4-phosphate during phagocytosis. Mol Biol Cell 28:128–140. https://doi.org/10.1091/mbc.e16-06-0451

Li XJ, Tian W, Stull ND et al (2009) A fluorescently tagged C-terminal fragment of p47 phox detects NADPH oxidase dynamics during phagocytosis. Mol Biol Cell 20:1520–1532. https://doi.org/10.1091/mbc.e08-06-0620

Li XJ, Marchal CC, Stull ND et al (2010) p47 phox Phox homology domain regulates plasma membrane but not phagosome neutrophil NADPH oxidase activation. J Biol Chem 285:35169–35179. https://doi.org/10.1074/jbc.M110.164475

Lopes LR, Dagher MC, Gutierrez A et al (2004) Phosphorylated p40PHOXAs a negative regulator of NADPH oxidase. Biochemistry 43:3723–3730. https://doi.org/10.1021/bi035636s

Magalhaes MAO, Glogauer M (2010) Pivotal advance: phospholipids determine net membrane surface charge resulting in differential localization of active Rac1 and Rac2. J Leukoc Biol 87:545–555. https://doi.org/10.1189/jlb.0609390

Magnani F, Nenci S, Fananas EM et al (2017) Crystal structures and atomic model of NADPH oxidase. Proc Natl Acad Sci U S A 114:6764–6769. https://doi.org/10.1073/pnas.1702293114

Makni-Maalej K, Boussetta T, Hurtado-Nedelec M et al (2012) The TLR7/8 agonist CL097 primes N -formyl-methionyl-leucyl-phenylalanine-stimulated NADPH oxidase activation in human neutrophils: critical role of p47phox phosphorylation and the proline isomerase Pin1. J Immunol 189:4657–4665. https://doi.org/10.4049/jimmunol.1201007

Marion S, Mazzolini J, Herit F et al (2012) The NF-κB signaling protein Bcl10 regulates actin dynamics by controlling AP1 and OCRL-bearing vesicles. Dev Cell 23:954–967. https://doi.org/10.1016/j.devcel.2012.09.021

Martynov VI, Pakhomov AA, Deyev IE, Petrenko AG (2018) Genetically encoded fluorescent indicators for live cell pH imaging. Biochim Biophys Acta Gen Subj 1862:2924–2939. https://doi.org/10.1016/j.bbagen.2018.09.013

Masoud R, Serfaty X, Erard M et al (2017) Conversion of NOX2 into a constitutive enzyme in vitro and in living cells, after its binding with a chimera of the regulatory subunits. Free Radic Biol Med 113:470–477. https://doi.org/10.1016/j.freeradbiomed.2017.10.376

Massenet C, Chenavas S, Cohen-Addad C et al (2005) Effects of p47 phox C terminus phosphorylations on binding interactions with p40 phox and p67 phox. J Biol Chem 280:13752–13761. https://doi.org/10.1074/jbc.M412897200

Matute JD, Arias AA, Wright NAM et al (2009) A new genetic subgroup of chronic granulomatous disease with autosomal recessive mutations in p40phox and selective defects in neutrophil NADPH oxidase activity. Blood 114:3309–3315. https://doi.org/10.1182/blood-2009-07-231498

Meijles DN, Howlin BJ, Li JM (2012) Consensus in silico computational modelling of the p22phox subunit of the NADPH oxidase. Comput Biol Chem 39:6–13. https://doi.org/10.1016/j.compbiolchem.2012.05.001

Miller JL, Velmurugan K, Cowan MJ, Briken V (2010) The type I NADH dehydrogenase of mycobacterium tuberculosis counters phagosomal NOX2 activity to inhibit TNF-α-mediated host cell apoptosis. PLoS Pathog 6:e1000864. https://doi.org/10.1371/journal.ppat.1000864

Minakami R, Maehara Y, Kamakura S et al (2010) Membrane phospholipid metabolism during phagocytosis in human neutrophils. Genes Cells 15:409–424. https://doi.org/10.1111/j.1365-2443.2010.01393.x

Murillo I, Henderson LM (2005) Expression of gp91phox/Nox2 in COS-7 cells: cellular localization of the protein and the detection of outward proton currents. Biochem J 385:649–657. https://doi.org/10.1042/BJ20040829

Nault L, Bouchab L, Dupré-Crochet S et al (2016) Environmental effects on reactive oxygen species detection—learning from the phagosome. Antioxid Redox Signal 25:564–576. https://doi.org/10.1089/ars.2016.6747

Nauseef WM (2004) Assembly of the phagocyte NADPH oxidase. Histochem Cell Biol 122:277–291. https://doi.org/10.1007/s00418-004-0679-8

Nauseef WM (2007) How human neutrophils kill and degrade microbes: an integrated view. Immunol Rev 219:88–102. https://doi.org/10.1111/j.1600-065X.2007.00550.x

Nauseef WM (2019) The phagocyte NOX2 NADPH oxidase in microbial killing and cell signaling. Curr Opin Immunol 60:130–140. https://doi.org/10.1016/j.coi.2019.05.006

Niedergang F, Grinstein S (2018) How to build a phagosome: new concepts for an old process. Curr Opin Cell Biol 50:57–63. https://doi.org/10.1016/j.ceb.2018.01.009

Nigorikawa K, Hazeki K, Sasaki J et al (2015) Inositol polyphosphate-4-phosphatase type I negatively regulates phagocytosis via dephosphorylation of phagosomal PtdIns(3,4)P2. PLoS One 10:e0142091. https://doi.org/10.1371/journal.pone.0142091

Nisimoto Y, Freeman JLR, Motalebi SA et al (1997) Rac binding to p67(phox). Structural basis for interactions of the Rac1 effector region and insert region with components of the respiratory burst oxidase. J Biol Chem 272:18834–18841. https://doi.org/10.1074/jbc.272.30.18834

Nisimoto Y, Motalebi S, Han C-H, Lambeth JD (1999) The p67 phox activation domain regulates electron flow from NADPH to flavin in flavocytochrome b 558. J Biol Chem 274:22999–23005. https://doi.org/10.1074/jbc.274.33.22999

Nunes P, Cornut D, Bochet V et al (2012) STIM1 juxtaposes ER to phagosomes, generating Ca2+ hotspots that boost phagocytosis. Curr Biol 22:1990–1997. https://doi.org/10.1016/j.cub.2012.08.049

Nunes P, Demaurex N, Dinauer MC (2013) Regulation of the NADPH oxidase and associated ion fluxes during phagocytosis. Traffic 14:1118–1131. https://doi.org/10.1111/tra.12115

O'Neill S, Mathis M, Kovačič L et al (2018) Quantitative interaction analysis permits molecular insights into functional NOX4 NADPH oxidase heterodimer assembly. J Biol Chem 293:8750–8760. https://doi.org/10.1074/jbc.RA117.001045

Ohayon D, De Chiara A, Dang PM-C et al (2019) Cytosolic PCNA interacts with p47phox and controls NADPH oxidase NOX2 activation in neutrophils. J Exp Med 216:2669–2687. https://doi.org/10.1084/jem.20180371

Ostuni MA, Gelinotte M, Bizouarn T et al (2010) Targeting NADPH-oxidase by reactive oxygen species reveals an initial sensitive step in the assembly process. Free Radic Biol Med 49:900–907. https://doi.org/10.1016/j.freeradbiomed.2010.06.021

Pal R, Basu Thakur P, Li S et al (2013) Real-time imaging of NADPH oxidase activity in living cells using a novel fluorescent protein reporter. PLoS One 8:e63989. https://doi.org/10.1371/journal.pone.0063989

Pozzan T, Lew DP, Wollheim CB, Tsien RY (1983) Is cytosolic ionized calcium regulating neutrophil activation? Science 221:1413–1415. https://doi.org/10.1126/science.6310757

Price MO, McPhail LC, Lambeth JD et al (2002) Creation of a genetic system for analysis of the phagocyte respiratory burst: high-level reconstitution of the NADPH oxidase in a nonhematopoietic system. Blood 99:2653–2661. https://doi.org/10.1182/blood.V99.8.2653

Regier DS, Waite KA, Wallin R, McPhail LC (1999) A phosphatidic acid-activated protein kinase and conventional protein kinase C isoforms phosphorylate p22 phox, an NADPH oxidase component. J Biol Chem 274:36601–36608. https://doi.org/10.1074/jbc.274.51.36601

Roma LP, Deponte M, Riemer J, Morgan B (2018) Mechanisms and applications of redox-sensitive green fluorescent protein-based hydrogen peroxide probes. Antioxid Redox Signal 29:552–568. https://doi.org/10.1089/ars.2017.7449

Roos D (2019) Chronic granulomatous disease. Methods Mol Biol 1982:531–542. https://doi.org/10.1007/978-1-4939-9424-3_32

Rosales C, Uribe-Querol E (2017) Phagocytosis: a fundamental process in immunity. Biomed Res Int 2017:1–18. https://doi.org/10.1155/2017/9042851

Sarfstein R, Gorzalczany Y, Mizrahi A et al (2004) Dual role of Rac in the assembly of NADPH oxidase, tethering to the membrane and activation of p67 phox. J Biol Chem 279:16007–16016. https://doi.org/10.1074/jbc.M312394200

Schlam D, Bohdanowicz M, Chatilialoglu A et al (2013) Diacylglycerol kinases terminate diacylglycerol signaling during the respiratory burst leading to heterogeneous phagosomal NADPH oxidase activation. J Biol Chem 288:23090–23104. https://doi.org/10.1074/jbc.M113.457606

Schlam D, Bagshaw RD, Freeman SA et al (2015) Phosphoinositide 3-kinase enables phagocytosis of large particles by terminating actin assembly through Rac/Cdc42 GTPase-activating proteins. Nat Commun 6:8623. https://doi.org/10.1038/ncomms9623

Schrenzel J, Serrander L, Bánfi B et al (1998) Electron currents generated by the human phagocyte NADPH oxidase. Nature 392:734–737. https://doi.org/10.1038/33725

Schürmann N, Forrer P, Casse O et al (2017) Myeloperoxidase targets oxidative host attacks to Salmonella and prevents collateral tissue damage. Nat Microbiol 2:16268. https://doi.org/10.1038/nmicrobiol.2016.268

Shao C, Novakovic VA, Head JF et al (2008) Crystal structure of lactadherin C2 domain at 1.7 Å resolution with mutational and computational analyses of its membrane-binding motif. J Biol Chem 283:7230–7241. https://doi.org/10.1074/jbc.M705195200

Someya A, Nagaoka I, Yamashita T (1993) Purification of the 260 kDa cytosolic complex involved in the superoxide production of guinea pig neutrophils. FEBS Lett 330:215–218. https://doi.org/10.1016/0014-5793(93)80276-Z

Song ZM, Bouchab L, Hudik E et al (2017) Phosphoinositol 3-phosphate acts as a timer for reactive oxygen species production in the phagosome. J Leukoc Biol 101:1155–1168. https://doi.org/10.1189/jlb.1A0716-305R

Staerck C, Gastebois A, Vandeputte P et al (2017) Microbial antioxidant defense enzymes. Microb Pathog 110:56–65. https://doi.org/10.1016/j.micpath.2017.06.015

Stasia MJ, Li XJ (2008) Genetics and immunopathology of chronic granulomatous disease. Semin Immunopathol 30:209–235

Steinckwich N, Frippiat J-P, Stasia M-J et al (2007) Potent inhibition of store-operated Ca^{2+} influx and superoxide production in HL60 cells and polymorphonuclear neutrophils by the pyrazole derivative BTP2. J Leukoc Biol 81:1054–1064. https://doi.org/10.1189/jlb.0406248

Steinckwich N, Schenten V, Melchior C et al (2011) An essential role of STIM1, Orai1, and S100A8–A9 proteins for Ca^{2+} signaling and FcγR-mediated phagosomal oxidative activity. J Immunol 186:2182–2191. https://doi.org/10.4049/jimmunol.1001338

Suh C-I, Stull ND, Li XJ et al (2006) The phosphoinositide-binding protein p40 phox activates the NADPH oxidase during FcγIIA receptor–induced phagocytosis. J Exp Med 203:1915–1925. https://doi.org/10.1084/jem.20052085

Sumimoto H (2008) Structure, regulation and evolution of Nox-family NADPH oxidases that produce reactive oxygen species. FEBS J 275:3249–3277. https://doi.org/10.1111/j.1742-4658.2008.06488.x

Sumimoto H, Minakami R, Miyano K (2019) Soluble regulatory proteins for activation of NOX family NADPH oxidases. Methods Mol Biol 1982:121–137

Sun J, Singh V, Lau A et al (2013) Mycobacterium tuberculosis nucleoside diphosphate kinase inactivates small GTPases leading to evasion of innate immunity. PLoS Pathog 9:e1003499. https://doi.org/10.1371/journal.ppat.1003499

Taylor WR, Jones DT, Segal AW (1993) A structural model for the nucleotide binding domains of the flavocytochrome b -245 β-chain. Protein Sci 2:1675–1685. https://doi.org/10.1002/pro.5560021013

Thomas DC (2018) How the phagocyte NADPH oxidase regulates innate immunity. Free Radic Biol Med 125:44–52. https://doi.org/10.1016/j.freeradbiomed.2018.06.011

Tian W, Li XJ, Stull ND et al (2008) Fc gamma R-stimulated activation of the NADPH oxidase: phosphoinositide-binding protein p40phox regulates NADPH oxidase activity after enzyme assembly on the phagosome. Blood 112:3867–3877

Tlili A, Dupré-Crochet S, Erard M, Nüße O (2011) Kinetic analysis of phagosomal production of reactive oxygen species. Free Radic Biol Med 50:438–447. https://doi.org/10.1016/j.freeradbiomed.2010.11.024

Tlili A, Erard M, Faure MC et al (2012) Stable accumulation of p67(phox) at the phagosomal membrane and ROS production within the phagosome. J Leukoc Biol 91:83–95. https://doi.org/10.1189/jlb.1210701

Ueyama T, Nakakita J, Nakamura T et al (2011) Cooperation of p40 phox with p47 phox for Nox2-based NADPH oxidase activation during Fcγ receptor (FcγR)-mediated phagocytosis. J Biol Chem 286:40693–40705. https://doi.org/10.1074/jbc.M111.237289

van Manen H-J, Verkuijlen P, Wittendorp P et al (2008) Refractive index sensing of green fluorescent proteins in living cells using fluorescence lifetime imaging microscopy. Biophys J 94:L67–L69. https://doi.org/10.1529/biophysj.107.127837

Vareechon C, Zmina SE, Karmakar M et al (2017) Pseudomonas aeruginosa effector ExoS inhibits ROS production in human neutrophils. Cell Host Microbe 21:611–618.e5. https://doi.org/10.1016/j.chom.2017.04.001

Westman J, Grinstein S, Maxson ME (2019) Revisiting the role of calcium in phagosome formation and maturation. J Leukoc Biol 106:837–851. https://doi.org/10.1002/JLB.MR1118-444R

Wientjes FB, Reeves EP, Soskic V et al (2001) The NADPH oxidase components p47phox and p40phox bind to moesin through their PX domain. Biochem Biophys Res Commun 289:382–388. https://doi.org/10.1006/bbrc.2001.5982

Wilson MI, Gill DJ, Perisic O et al (2003) PB1 domain-mediated heterodimerization in NADPH oxidase and signaling complexes of atypical protein kinase C with Par6 and p62. Mol Cell 12:39–50. https://doi.org/10.1016/S1097-2765(03)00246-6

Winterbourn CC, Hampton MB, Livesey JH, Kettle AJ (2006) Modeling the reactions of superoxide and myeloperoxidase in the neutrophil phagosome. J Biol Chem 281:39860–39869. https://doi.org/10.1074/jbc.M605898200

Wrona D, Siler U, Reichenbach J (2017) CRISPR/Cas9-generated p47 phox -deficient cell line for chronic granulomatous disease gene therapy vector development. Sci Rep 7:6–11. https://doi.org/10.1038/srep44187

Yeung T (2006) Receptor activation alters inner surface potential during phagocytosis. Science 313:347–351. https://doi.org/10.1126/science.1129551

Yeung T, Grinstein S (2007) Lipid signaling and the modulation of surface charge during phagocytosis. Immunol Rev 219:17–36. https://doi.org/10.1111/j.1600-065X.2007.00546.x

Yeung T, Gilbert GE, Shi J et al (2008) Membrane phosphatidylserine regulates surface charge and protein localization. Science 319:210–213. https://doi.org/10.1126/science.1152066

Yeung T, Heit B, Dubuisson J-F et al (2009) Contribution of phosphatidylserine to membrane surface charge and protein targeting during phagosome maturation. J Cell Biol 185:917–928. https://doi.org/10.1083/jcb.200903020

Yu L, Quinn MT, Cross AR, Dinauer MC (1998) Gp91phox is the heme binding subunit of the superoxide-generating NADPH oxidase. Proc Natl Acad Sci 95:7993–7998. https://doi.org/10.1073/pnas.95.14.7993

Zhao X, Carnevale KA, Cathcart MK (2003) Human monocytes use Rac1, Not Rac2, in the NADPH oxidase complex. J Biol Chem 278:40788–40792. https://doi.org/10.1074/jbc.M302208200

Zhao X, Xu B, Bhattacharjee A et al (2005) Protein kinase Cδ regulates p67phox phosphorylation in human monocytes. J Leukoc Biol 77:414–420. https://doi.org/10.1189/jlb.0504284

Zhen L, King AA, Xiao Y et al (1993) Gene targeting of X chromosome-linked chronic granulomatous disease locus in a human myeloid leukemia cell line and rescue by expression of recombinant gp91phox. Proc Natl Acad Sci 90:9832–9836. https://doi.org/10.1073/pnas.90.21.9832

Ziegler CS, Bouchab L, Tramier M et al (2019) Quantitative live-cell imaging and 3D modeling reveal critical functional features in the cytosolic complex of phagocyte NADPH oxidase. J Biol Chem 294:3824–3836. https://doi.org/10.1074/jbc.RA118.006864

Conclusions and the Futures of Phagocytosis

10

Maurice B. Hallett

Abstract

Although we know a wealth of detail about the molecular and cell biology of phagocytosis, there are many unsolved mysteries remaining. In this final chapter, some important may be tangential) questions are raised, that the bulk of researchers are not really addressing. In this chapter, some suggestions are given for this type of "blue skies" future work. These include new approaches to understanding phagocytosis and the possibility that this new knowledge may provide a solution to anti-microbial resistance. This future phagocytosis research would have an impact, not only on our understanding of phagocytosis, but potentially on the future of human health.

Keywords

Future Questions · Antibiotic Resistance · Universal Nature of Phagocytosis · Bacterial Anti-phagocytic Defences · Bacterial toxins

Endings and Beginnings

In this book, a wealth of detail about current thinking and experimental evidence has been given. There are many papers and books referenced in these chapters. However, these are just a tip of an information iceberg. In 2017 alone, there were 1983 papers on phagocytosis listed in Web of Science, with an accumulated number of papers from 1960 being over 37,000. It would be impossible for anyone to have read all these papers (properly) and to keep pace with new papers. It is therefore inevitable that many papers have been missed or inadvertently omitted. However, there would be no point in simply referring to every paper in this vast array and the purpose of reviews and books like this is to make sense of the jumble of data and find those needles in the information haystack which are of most importance. It could also be argued that, the purpose of reviews and books like this is to reduce the amount of reading for a newcomer to the field so that they can quickly carry the field forward and not have to re-discover what is already known.

Figure 10.1 shows that there is also an accelerating increase in the number of papers published on phagocytosis each year (Fig. 10.1). If all the answers to all the mysteries of phagocytosis were

M. B. Hallett (✉)
School of Medicine, Cardiff University, Cardiff, UK
e-mail: hallettmb@cf.ac.uk

© Springer Nature Switzerland AG 2020
M. B. Hallett (ed.), *Molecular and Cellular Biology of Phagocytosis*, Advances in Experimental Medicine and Biology 1246, https://doi.org/10.1007/978-3-030-40406-2_10

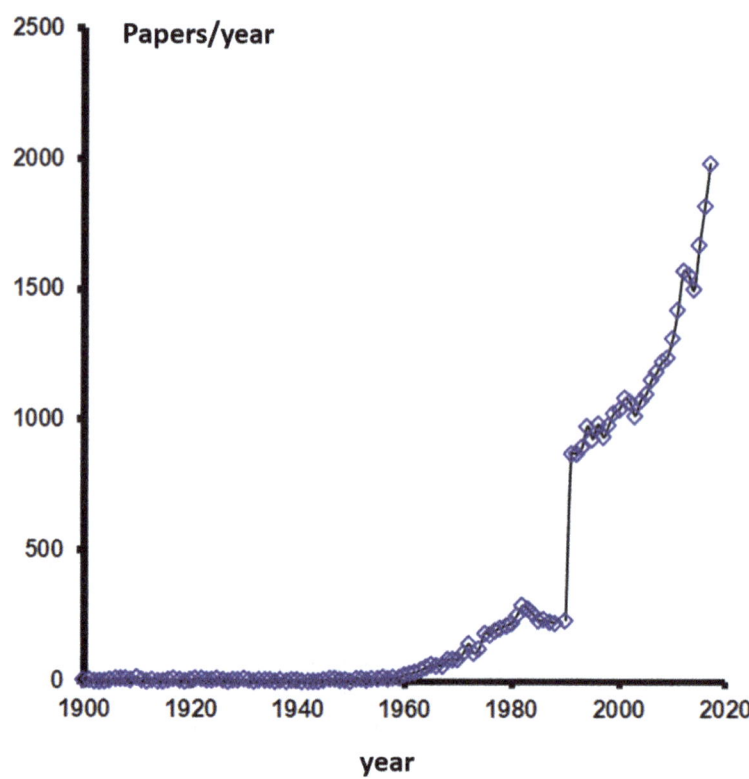

Fig. 10.1 Phagocytosis papers (1900–2019). The graph show the increase in phagocytosis papers over the period 1900–2019 reported in Web of Science Core Collection (https:// wok.mimas.ac.uk). The suddenly increase in the number of papers published from 236 papers in 1990 to 871 papers in 1991 may be due to feature of the database rather than a sudden increase in phagocytosis research activity

known, the number of papers published would decrease. For example, before the double helix structure of DNA was solved there were many research papers and speculative reviews and opinions. However, once its structure was known and accepted, the need for further research in this area obviously declined. There is no sign of a decline in the number of phagocytic papers appearing, so presumably there are still plenty of research questions to be tackled. Some are of the "i" dotting and "t" crossing type, of course, but many are focussed on more fundamental questions. In each field represented by chapters in this book, it may be apparent what these fundamental questions are. However, it may be less apparent how to answer them.

Questions for the Future

There is, of course, no agreement as to what the fundamental unanswered questions about phagocytosis are. However, here are a few, which many will disagree with, but could be considered.

Primitive and Universal Nature of Phagocytosis

How did phagocytosis originate and then evolve into the complex cell biological process that is studied today? This may be an unanswerable question but one that I feel is worth pursuing. Firstly, it is an important scientific question; not only because it relates the cell biological evolution and especially to the prokaryote/eukaryote question. The question is worth an answer because, it is the curiosity-led questions which usually lead to a fuller understanding. The question with limited scope usually only provide answers with limited scope.

What Is the Minimum Required for Phagocytosis?

The complexity of phagocytosis, especially of neutrophils and macrophages, overlays what is the essence of phagocytosis. Is it worth finding or rediscovering simpler systems which do not have this overlay. Defining the minimum requirement for phagocytosis would seem to be an important contribution, as once defined, the overlay of evolutionary specialisation (niche selection etc) can be added and, perhaps, sense made of it. This is a bottom up approach. The current situation is top down. We know a lot about the detail of complex phagocytosis at its most complex level ie the top and yet, the minimum requirement for phagocytosis (ie the bottom) remains obscure. Is it obscured by the wealth of detail, or will it be possible to see through the fog?

Bacterial Anti-phagocytic Defences as a Phagocytosis Signalling Map

Many fungi and bacteria which infect other species, eg humans, have evolved defence mechanism, often by inhibiting the mechanism underlying phagocytosis. For example, cytochalasins, which inhibit actin polymerisation, are a fungal product. Presumably this is an effective phagocytosis inhibitory tactic used by fungi, like *Candida albicans* (a commonly found commensal yeast) since actin polymerisation lies at the heart of the production of phagocytic pseudopodia. Many bacterial toxins, also inhibit steps in the phagocytic pathway. Since these have been produced by the evolutionary route of "exploring" random molecular possibilities, many have exquisite selectivity for proteins of key importance for phagocytosis and cell shape change. For example, the bacteria *Bordetella pertussis* produces pertussis toxin which, on entering the phagocyte cytosol, interferes with G-protein-coupled signalling, specifically by inhibition of the α subunit of heterotrimeric G_i proteins via ADP-ribosylation. Bacterial botulinum toxins also have anti-phagocytosis properties. Botulinum C2 toxin inhibits polymerization of G-actin by ADP-ribosylation of monomeric G-actin on (arginine 177) but not polymerised F-actin. This inhibits phagocytosis which depends on the ability to polymerise actin for the formation of phagocytic pseudopodia. Botulinum C3 toxin has ADP-ribosylation specificity for the small GTP-binding proteins RhoA, RhoB, and RhoC, all of which are implicated in cell shape change and phagocytosis. Without knowledge of the signalling pathways involved in phagocytosis, in may be inferred by reverse logic, therefore, that small and trimeric G-proteins, together with actin polymerisation were crucial parts of the mechanism for phagocytosis. Anti-phagocytosois toxins may thus provide a negative imprint of the signalling pathway. The toxins studied mostly have pathological "side effects" and cause disease. Is there a possibility that many non-pathological bacteria also produce other toxins, aimed at phagocytic inhibition. By increasing our knowledge of bacterial "toxins", we may therefore increase our knowledge of key steps in the phagocytic pathway. Would the complete bacterial toxin list of targets include the complete list the key proteins for phagocytosis?

Antibiotic Resistance and Human Phagocytosis

If we survive the impending climate crisis, and perhaps before, there is an impending antibiotic resistance crisis. Over the last 70 years, antibiotics have transformed some killer infections into easily treatable problems. Yet, there are only a few antibiotics (with many variations) and there are now emerging bacterial strains which are resistant to these. As bacterial resistance grows, the prospect of returning to pre-antibiotic medicine looms and simple infections may once again pose a lethal threat. There is an urgency about the problem, but no new antibiotic has yet emerged. Given that phagocytic immune cells are the first (and second) line of defence against infection, the question could be asked, whether phagocytic efficiency can be increased. Is it possible to pharmacologically switch on the "boosters" of the immune phagocytes and prevent or overcome infection without the use of antibiotics? If it were, bacterial resistance to antibiotics would no longer be a threat.

One way that nature has evolved to boost the efficiency of phagocytosis, is by opsonins, complement or antibody, which coat the infections. However, in infection overload, this does not seem to be enough. The possibility that the way in which opsonins accelerate phagocytosis could be replicated without the need for additional opsonins, is therefore attractive. As we understand more about the molecular mechanism which accelerate phagocytosis (see Chap. 9), it is a future hope that this knowledge could be exploited for combatting bacterial infection and be added to the armoury of anti-microbial medicine.

Beginnings and Endings

This is not a conclusion. That would imply an ending. The research on phagocytosis has progressed our understanding immeasurably since phagocytosis was first described in the eighteenth century (see Chap. 2). However, this is not the end of further research. It is not even the beginning of the end of phagocytosis research. There is an exciting prospect that new approaches, new minds and new technologies will push our understanding further. It may be that we have just reached the end of the beginning.

Index

A

Abstracted model, 58, 59
Accelerator of phagocytosis, 145
Acetoxymethyl ester (AM), 118
Actin network, 6, 45, 59, 60, 75, 94–96, 133, 134, 146, 150, 151
Actin nucleation-promoting factors, 45
Actinophrys, 14, 23
Actinosphaerium eichhorni, 33
Actin polymerisation (Brownian Ratchet), 100, 101, 151
Actin rings, 75, 111
Adenosine diphosphate ribose (ADPR), 109, 110, 113, 117, 118
Adenosine triphosphate (ATP), 108, 109, 113, 120–122, 147, 150
Adhesion, 96, 118, 135
ADP, *see* Adenosine diphosphate (ADP)
ADPR, *see* Adenosine diphosphate ribose (ADPR)
Alizarin, 36, 39
Alkanet *(tincture of alkanna),* 32
All-inclusive models, 58
ALLN, 137, 142, 143
AM, *see* Acetoxymethyl ester (AM)
2-Aminoethoxydiphenyl borate (2-APB), 115
Amoeba, v, 3, 6, 11–17, 19, 30, 31, 36–40, 94, 98
AMTB, *see* (N-(3-Aminopropyl)-2-[(3-methylphenyl)methoxy]-N-(2-thienylmethyl)benzamide hydrochloride) (AMTB)
ANO6, *see* Anoctamin 6 (ANO6) (also called TMEM16F)
Anoctamin 6 (ANO6) (also called TMEM16F), 121
2-APB, *see* 2-Aminoethoxydiphenyl borate (2-APB)
Apparent expansion, cell surface area, 85, 86
Arachidonate-regulated channel (ARC), 114
Araignee aquatique ("the fat water spider"), 20, 22, 23
ARC, *see* Arachidonate-regulated channel (ARC)
Aspartyl/asparaginyl beta-hydroxylase (ASPH), 110
ASPH, *see* Aspartyl/asparaginyl beta-hydroxylase (ASPH)
ATP, *see* Adenosine triphosphate (ATP)
ATP-driven SERCA pumps, 108
Auto-inhibitory region (AIR), 164

B

Bacteria (long), 74, 75, 82
Balfour Biological Laboratory for Women, 31
BAPTA-AM, *see* 1,2-Bis(2-aminophenoxy)ethane-N,N,N′,N′-tetraacetic acid tetrakis(acetoxymethyl ester) (BAPTA)
5-BDBD, *see* 5-(3-Bromophenyl)-1,3-dihydro-2H-Benzofuro[3,2-e]-1,4-diazepin-2-one (5-BDBD)
$\beta 2$ integrins, 110
Biophysics, 57
Bipinnaria, 25, 29
1,2-Bis(2-aminophenoxy)ethane-N,N,N',N'-tetraacetic acid tetrakis(acetoxymethyl ester) (BAPTA-AM), 111, 150
Bleb (membrane), 74, 76–80, 83, 96, 147
Bone marrow stromal cell antigen 1 (BST1), 113
Borrelia burgdorferi, 66
Bottleneck effect, 62
Bromocresol green and bromophenol blue, 33
Bromocresol purple, 33
5-(3-Bromophenyl)-1,3-dihydro-2H-Benzofuro[3,2-e]-1,4-diazepin-2-one (5-BDBD), 122
Bromothymol blue, 33
Brownian fluctuations (at membrane), 100, 133
Brownian Ratchet, 98–101, 133, 151
BST1, *see* Bone marrow stromal cell antigen 1 (BST1)
Burkholderia pseudomallei, 52

C

CAD, *see* Calcium release-activated channel activation domain (CAD)
Ca^{2+}-dependent phosphatase calcineurin, 112
cADPR, *see* Cyclic adenosine diphosphate ribose (cADPR)
Caged Ca^{2+}, 147
Caged IP_3, 121, 136, 147

© Springer Nature Switzerland AG 2020
M. B. Hallett (ed.), *Molecular and Cellular Biology of Phagocytosis*, Advances in Experimental
Medicine and Biology 1246, https://doi.org/10.1007/978-3-030-40406-2

Ca^{2+} hotspots, 111, 115, 119, 147, 149
Ca^{2+} ions, 34–40, 137, 148–150
Calcineurin, 112, 134, 136
Calcium (Ca^{2+}), 3, 34, 45, 89, 107, 135, 159
Calcium release-activated channel activation domain
    (CAD), 114
Calcium release-activated channel regulator (CRACR2B)
    (also called EFC4A), 115
Calcium release-activated channel regulator A
    (CRACR2A) (also called EFC4B), 115
Calcium signaling, 105–123
Calmodulin, 111, 121, 134, 135, 146
Calpain, 6, 96, 98, 102, 110, 120, 133–151
μ-Calpain, 136, 145–147
Calpain 1, 136, 141, 145–148
Calpain 2, 136, 137, 146, 147
Calpain 4, *see* Common subunit of calpain
Calpain (structure), 135, 145–146
Calpain activation inhibitors, 137, 138, 140
Calpain activity "burst", 144, 145
Calpain1 null cells, 136, 145
Calpastatin, 137, 146
Calpeptin, 136, 137, 142, 143
Ca^{2+} microdomains, 112
Capacitive Ca^{2+} entry, 109
Capn 4-/-, 136
Ca^{2+} pumps, *see* ATP-driven SERCA pumps
Carbonyl cyanide-4-(trifluoromethoxy)phenylhydrazone
    (FCCP), 117
Ca^{2+} release, 108–110, 113–116, 119, 121–123
Carmine, 16–18, 28, 29
Cav-1, *see* Caveolin-1 (Cav-1)
Caveolin-1 (Cav-1), 116
C3bi, 2, 113
CD38, 109, 113
Cdc42, 45
C2 domain, 146, 170
Cell drinking, 1
Cell shape change, 134–137, 150, 185
Cell spreading, 90, 91, 93, 101, 135, 136, 140, 141, 145,
    147, 148
Cell surface (wrinkled), 88–90, 94, 95, 98, 149
Cell surface area, 4, 13, 86–92
Cell surface area (direct measurement), 90
Cell surface topography, 88, 89, 92
CFTR, *see* Cystic fibrosis transmembrane conductance
    regulator (CFTR)
Chambers, R., 33, 36–38
Channels, 6, 108–110, 113–123, 147–150, 168
Charge, 3, 6, 36, 43–52, 117, 118, 120, 122, 169, 170,
    173
Chemokinesis, 136
Chinese hamster ovary cells, 136
Chlamydia trachomatis, 52
CHO cells, *see* Chinese hamster ovary (CHO) cells
Cholesterol, 44, 47
Chronic Granulomateous Disease (CGD), 158, 163, 173
Claparède, E., 12, 13, 24
Claus, Carl Friedrich Wilhelm, 11

C2-like domain, 146
Closure, 2, 12, 13, 46, 75, 82, 86, 89, 91–93, 112, 135,
    138, 139, 141, 145, 160, 167, 168, 170, 174
Common subunit of calpain, 146
Complement component C3bi, 2
Complement receptor (CR), 168
Computational model, 57–59, 68
Confocal microscopy, 99
Congo red and phenol red, 32
Connexins (Cx), 109, 110, 121
Cortical actin network, 6, 94–96, 98, 100, 134, 146, 150,
    151
COS7 cell line, 160
cPLA2, *see* Cytosolic phospholipase A2 (cPLA2)
CR, *see* Complement receptor (CR)
CRACR2A, *see* Calcium release-activated channel
    regulator A (CRACR2A) (also called EFC4B)
CRACR2B, *see* Calcium release-activated channel
    regulator (CRACR2B) (also called EFC4A)
Cross-linking protein, 51, 95
Cx, *see* Connexins (Cx)
Cyclic adenosine diphosphate ribose (cADPR), 108–110,
    113
Cystic fibrosis transmembrane conductance regulator
    (CFTR), 117
Cytokine production, 107
Cytoskeletal proteins, 95, 147
Cytoskeletal remodelling, 44, 45, 107
Cytoskeleton, 2, 60, 62, 70, 93, 94, 96, 97, 147, 164,
    165
Cytosolic Ca^{2+}, 3, 6, 10, 36, 38–40, 89, 92, 97, 98,
    108–110, 117, 119, 133, 135, 136, 138–142, 144,
    147, 150
Cytosolic phospholipase A2 (cPLA2), 111, 172
Cytosolic subunits (oxidase), 158
Cytovilln, 96

D
DAG, *see* Diacylglycerol (DAG)
Damage-associated molecular pattern (DAMP), 121
DAMP, *see* Damage-associated molecular pattern
    (DAMP)
Darwin, C., 11, 28, 29
DC, *see* Dendritic cell (DC)
Decision-making, 6
Dendritic cell, 43, 115, 150
Deoxyribonucleic acid (DNA), 3, 49, 113, 184
Diacylglycerol (DAG), 45, 46, 48, 109, 113, 116, 117,
    168, 170, 172
Dichlorodihydrofluorescein (DCFH2), 174
Dictyostelium, 73–76, 82
Dictyostelium discoideum, 73
Diffusion, 33, 38, 60–64, 69, 91, 92, 100, 112, 143, 145,
    147, 148, 160, 174, 175
Diffusion constant, 61, 69, 145, 148
Directed phagocytosis, 89
Dissociation constant (Kd), 118, 146, 148
3D model of the cytosolic complex, 161, 168
DNA, *see* Deoxyribonucleic acid (DNA)

E

EF hand, 114, 137, 146, 158
Ehrlich, P., 29
Electrical charge, 44
Electro-microinjection, 134
Electrostatic interactions, 47, 48, 50, 51
Endocytic pathway, 44
Endocytosis, 58, 60–63, 66, 68, 70, 86
Endoplasmic reticulum (ER), 47, 52, 108–110, 112–117, 119, 121–123, 168
Englemann, T., 30
Eosinophils, 134, 135
ER, *see* Endoplasmic reticulum (ER)
ER Ca^{2+} release, 108, 109, 113, 115, 116, 121, 123
Esyt, *see* Extended synaptotagmin (also called FAM62) (Esyt)
Exocytosis, 13, 86–91, 101, 114, 119, 147
Exocytotic addition to the plasma membrane, 86–88
Expansion, cell surface area, 13
Extended synaptotagmin (Esyt), 123
Ezrin, 5, 6, 85–102, 133, 134, 137, 146, 149–151
Ezrin phosphomimetic, 96
Ezrin-Radixin-Moesin (ERM), 95, 96, 137, 146

F

F-actin, 95, 96, 98, 99, 185
FAM62, *see* Extended synaptotagmin (Esyt)
Fc, *see* Fragment crystallizable (FC)
FcγR, *see* Fc gamma receptor (FcγR)
FcγRIIA, *see* Fc gamma receptor IIA (also called FCGR2A, or CD32) (FcγRIIA)
FCCP, *see* carbonyl cyanide-4-(trifluoromethoxy)phenylhydrazone (FCCP)
Fc gamma receptor (FcyR), 113
FcR-mediated phagocytosis, 111
Fibroblast, 86, 88, 91, 115, 135–137
Flavocytochrome b$_{558}$, 111, 161, 166
Fluorescence Loss Induced by Photobleaching (FLIP), 160
Fluorescence Recovery After Photobleaching (FRAP), 92, 160
Fluorescent Ca2+ probes fura-2 (FFP-18), 148
Fluorescent proteins, 159–161, 170, 172
FM 1-43, 89, 90
fMLP, *see* N-formylmethionyl-leucyl-phenylalanine (also called fMLF) (fMLP)
Formylated peptide, 113, 136, 142
Formylmethionyl-leucyl-phenylalanine, 113
Free Ca2+, 35, 38–40, 89, 135, 138, 139, 143, 146–150
Frustrated phagocytosis, 86, 89, 90, 136, 140
2FYVE decoration, 75, 81

G

Gelsolin, 110
Gene transcription, 107
Geometry, 4, 6, 86, 94, 145, 149
GFP-2FYVE, 75, 77, 78, 81, 82

GFP-myosin-II heavy chain, 79
Gleichen, 17, 18
Goeze, J.A.E., 19–24
GPCR, *see* G-protein coupled receptor (GPCR)
Gp91^{phox}, 158
G-protein coupled receptors (GPCRs), 117, 120, 121
Greenwood, M., 31–33
Greifenstein castle, 17, 18

H

Haarwanzen (hairy bug), 20
Hairy bug (Haarwanzen), 20, 22, 23
Hamburger, H.J., 34, 35, 135
HEK, *see* Human embryonic kidney cells (HEK)
Heliozoans, 12–14, 16, 17, 19, 23, 24, 31, 33
Heme oxygenase 1 (HMOX-1), 118
HL60 neutrophils, 91, 101
HMOX-1, *see* Heme oxygenase 1 (also called HO-1)
HO-1, *see* heme oxygenase 1 (HO-1)
H_2O_2, 118, 158, 174, 175
HOCl, 158, 175
Howland, R.B., 31, 33, 36, 38
Human embryonic kidney cells (HEK), 121, 122
Hydrophobic interactions, 50, 52

I

IgG, *see* Immunoglobulin G (IgG)
Immunoglobulin G (IgG), 64, 160, 168, 176
Immunoreceptor tyrosine-based activation motifs, 113
Infusoria, 22, 23
Inositol trisphosphate (IP$_3$), 45, 46, 108, 109, 113, 116, 121, 136, 140, 147, 168, 170, 172
Inositol trisphosphate receptor (also called ITPR) (IP$_3$R), 108–110, 112–114, 116, 120, 122
Inside the phagosome, 157, 174
Integrins, 110, 113, 135, 168
Intraphagosomal Ca^{2+}, 118, 123
Intra-wrinkle Ca^{2+}, 148, 149
Ion channels, 6, 110, 116–123
Iontophoretic release, 141
IP$_3$, *see* Inositol trisphosphate (IP$_3$)
IP$_3$R, *see* Inositol trisphosphate receptor (also called ITPR) (IP$_3$R)
IP$_3$ Receptor (IP$_3$R), 108–110, 112–114, 116, 120, 122, 147
ITAM, *see* Immunoreceptor tyrosine-based activation motifs (ITAM)

J

Joblot, L., 20–23
Junctate, 109, 110, 114–116, 122

K

Kd, *see* Dissociation constant (Kd)
Kölliker, R.A., 13–17, 23, 24
KRas, 48, 50, 51
Krukenberg, 32

L

Lamellipodia, 46, 48, 91, 136, 137
Laplace law, 92
Leeuwenhoek, 10, 11
Legionella pneumophila, 52, 66
Leidy, M., 11
Leukotriene-regulated channel (LRC), 114
LFG3, *see* Transmembrane Bax inhibitor motif (TMBIM)
Lipid microdomains, 44
Lipids, 3, 4, 6, 43–45, 47, 49, 51, 52, 86, 89, 90, 95, 107, 108, 110, 123, 134, 150, 158, 162, 165, 166, 169–173, 176
Lipopolysaccharide (LPS), 113, 118, 120, 121
Litmus, 30–32
Localised electroporation, 141
Long bacteria, 74, 75, 82
LPS, *see* Lipopolysaccharide (LPS)
LRC, *see* Leukotriene-regulated channel (LRC)
Lymphocytes, 92, 93, 96, 136
Lysosomes, 2, 33, 44, 47, 107–112, 114, 118, 119, 121, 122

M

Macrophage, 2, 32, 33, 43, 51, 82, 83, 86, 88–90, 97–100, 111, 113, 115–112, 145, 150, 160, 165, 169, 170, 172, 175, 184
Madin-Darby canine kidney cells (MDCK cells), 97
MARCKS, *see* Myristoylated alanine-rich C kinase substrate (MARCKS)
Mathematical modelling, 58, 148, 149
m-calpain, 146, 147, 152
MCOLN, *see* Transient receptor potential mucolipin (TRPML)
MCS, *see* Membrane contact site (MCS)
MCU, *see* Mitochondrial calcium uniporter (MCU)
MDCK cells, *see* Madin-Darby canine kidney cells (MDCK cells)
Mechanical bottleneck, 62
MEF, *see* Mouse embryonic fibroblast (MEF)
Meissner, M., 31, 32
Membrane, 1, 13, 43, 57, 74, 85, 107, 133, 158
Membrane-associated proteins, 146
Membrane capacitance, 90, 91
Membrane contact sites (MCS), 110, 112, 114, 115, 123
Membrane remodelling, 50
Membrane reservoir, 88, 94, 95, 101, 140
Membrane ruffling, 1
Membrane shape, 60, 62, 123
Membrane subunits (Oxidase), 158, 159, 161, 167
Membrane supply and demand, 101
Membrane surface charge, 43–52
Membrane tension, 6, 74, 83, 85–102, 137
Membrane tension hypothesis, 86
Mercaptoacrylate calpain inhibitors, 137, 138
Mesodermic cells, 28, 30
Messina, 28, 29
Metchnikoff, 11, 25, 26, 29, 30, 32
Metchnikov, O., 11, 26, 29, 30
Microinjection, 9, 21, 31, 33, 36–38, 97, 134, 141, 143

Micropinocytosis, 1, 5
Micropipette, v, 31, 33, 36, 64, 89, 92, 100, 138, 139, 142
Micropipette aspiration, 92
Micropipette delivery phagocytic stimulus, 138
Microridge, 88, 90, 92, 99, 101, 148, 149
Microvilli, 95–97
Micrurgery, 36
Migration, 107, 118, 136, 166
Mitochondrial calcium uniporter (MCU), 109
Mobile Ca^{2+} buffers, 148, 149
Modelling, 6, 57–59, 65–70, 148, 149
Moesin, 95, 96, 98, 137, 146, 164
Mouse embryonic fibroblast (MEF), 111
Mouse fibroblasts, 91
Müller, O.F., 19, 22–24
Myeloperoxidase, 175
Myosin-II, 75, 76, 79
Myristoylated alanine-rich C kinase substrate (MARCKS), 51

N

NAADP, *see* Nicotinic acid adenine dinucleotide phosphate
NAD, *see* Nicotinamide adenine dinucleotide
NADPH, *see* Nicotinamide adenine dinucleotide phosphate
NADPH oxidase, 6, 111, 112, 157–175
Na^+/H^+ exchange regulatory cofactor 1 (also called SLC9A3R2) (NHERF2), 116
N-(3-Aminopropyl)-2-[(3-methylphenyl)methoxy]-N-(2-thienylmethyl)benzamide hydrochloride) (AMTB), 119
NCX, *see* Sodium calcium exchanger
Near membrane Ca2+, 148, 149
Neuropeptide tachykinin receptor (also called substance P receptor, or TACR1) (NK1), 120, 125
Neutral red, 32, 33
Newnham College Cambridge, 31
Newt eosinophils, 134, 135
NFAT, *see* Nuclear factor of activated T cells
N-formylmethionyl-leucyl-phenylalanine (also called fMLF) (fMLP), 113, 117, 142
NHERF1, *see* Na+/H+ exchange regulatory cofactor 1 (also called SLC9A3R1)
NHERF2, *see* Na+/H+ exchange regulatory cofactor 1 (also called SLC9A3R2)
Nicotinamide adenine dinucleotide, 111
Nicotinamide adenine dinucleotide phosphate, 111
Nicotinamide adenine dinucleotide phosphate oxidase, 111
Nicotinamide adenine dinucleotide phosphate oxidase 2 (NOX2), 111, 158–159, 161–168, 174–176
Nicotinic acid adenine dinucleotide phosphate (NAADP), 9, 10, 13
NIH 3T3 fibroblasts, 136
Ni^{2+} ions, 138, 140, 142, 150
Nipple mouth (papilla oris), 23
Nir, *see* Pyk2 N-terminal domain-interacting receptor 1 (also called PITPNM)

Nitrous oxide (NO), 120
NK1, *see* Neuropeptide tachykinin receptor (also called substance P receptor, or TACR1)
NO, *see* Nitrous oxide
Nobel Prize, 11, 25, 29
NOX2, *see* Nicotinamide adenine dinucleotide phosphate oxidase 2
NOX2, phagocyte NADPH oxidase, 158
NSC668394, 95, 97
Nuclear factor of activated T cells (NFAT), 121

O

OAG, *see* 1-oleoyl-2-acetyl-sn-glycerol
1-oleoyl-2-acetyl-sn-glycerol (OAG), 117
Opsonin, 2, 135, 150, 185, 186
ORAI family, 109
ORP, *see* Oxysterol-binding protein-related protein (also called OSBPL)
ORP5, *see* Oxysterol-binding protein-related proteins
ORP8, *see* Oxysterol-binding protein-related proteins
ORP1L, *see* Oxysterol-binding protein-related protein 1L
OSBP, *see* Oxysterol-binding protein-related protein
Oxysterol-binding protein-related protein 1L (ORP1L), 47
Oxysterol-binding protein-related proteins (ORP), 47, 123

P

PA, *see* Phosphatidic acid
Pannexins (Panx), 109, 110, 121
Panx, *see* Pannexins
Paramecium, 13, 14, 19, 20, 32, 33
PARP, *see* Poly (ADP-ribose) polymerase
Particle binding, 2, 45, 48, 110, 113, 120
Particle recognition, 2, 5, 111
Particle signalling, 3
Particle stiffness, 69
Partner of stromal interaction molecule 1 (POST), 115
Pasteur, 25, 29, 32
Pasteur Institute, 25, 29
PB1 domain, 164, 165
PC, *see* Phosphatidyl choline
PD150606, 138, 140
Pendeloques, 18–20, 22
Periphagosomal Ca^{2+} signals, 108, 111
Phagocyte ruffling, 51
PhagocyteS, 2–5, 9, 11, 29, 30, 32, 34, 38, 43, 45, 46, 51, 58, 60, 62, 64, 65, 69, 73–75, 81, 82, 93, 99, 101, 109, 110, 141–143, 157, 158, 161, 163–165, 168, 173–176, 185
Phagocyte theory, 11
Phagocytic cup, 2, 3, 12, 13, 16, 17, 23, 24, 45, 46, 48, 51, 60–62, 64, 69, 70, 72, 77, 79, 80, 82, 88, 89, 92, 93., 97, 98, 100, 110, 114, 119, 135, 138–140, 144, 145, 149
Phagocytic cup formation, 2, 3, 16, 93, 119, 135, 138, 139, 145
Phagocytic pseudopodia, 3, 89, 93, 98–101, 141, 149

Phagocytic vacuole, *see* Phagosome
Phagocytosing archaeon, 3
Phagolysome, 2
Phago-lysosome fusion, 114, 118, 121
Phagosomal closure, 112
Phagosomal ionic homeostasis, 120
Phagosomal membrane, 2, 44, 47, 75, 78, 80, 99, 117, 118, 120, 123, 160, 163, 165, 167, 170, 174, 176
Phagosome, 2, 10, 44, 50, 75, 78, 86, 107, 135, 157, 165
Phagosome (inside), 79, 157, 169, 174, 176
Phagosome formation, 3, 44, 46, 138
Phagosome maturation, 46, 47, 111, 115, 119, 121, 174
Phenol red, 32, 33
pH indicators, 32
pH indicators, neutral red, congo red and phenol red, phenol red, bromothymol blue, bromocresol purple, bromocresol green and bromophenol blue
pH of the phagosome, 30
3-phosphatase, 45, 47
5-phosphatases, 45, 46
Phosphatidic acid (PA), 45, 113, 169
Phosphatidylcholine, 44, 120
Phosphatidyl choline (PC), 113
Phosphatidylethanolamine (PtdEth), 44, 46
Phosphatidylinositides, 44
Phosphatidylinositol 3,5-bisphosphate (PI(3,5)P2), 109, 119
Phosphatidylinositol 4,5-bisphosphate (PI(4,5)P2), 95, 97, 109, 113, 116–119, 170, 173
Phosphatidylinositol 3-kinases (PtdIns3K), 45, 46,–47
Phosphatidylinositol 3-phosphate (PI(3)P), 46, 47, 75, 160, 165
Phosphatidylinositol-3-phosphate 5-kinase, 47
Phosphatidylinositol-phosphate-5-kinase, 45, 46
Phosphatidylinositol 3,4,5-trisphosphate (PI(3,4,5)P2), 117
Phosphatidylserine, 44, 146, 169, 170
Phosphoinositide interacting regulator of TRP (PIRT), 118
Phospholipase C gamma (PLCγ), 113, 170, 172, 173
Phospholipase D (PLD), 109, 113, 159
Phospholipid bilayer, 4, 87, 98, 100, 134
Phosphomimetic (ezrin), 96
Phox proteins, 159–161
PI3K, *see* Posphatidylinositol 3 kinase
PIKFYVE, *see* Phosphatidylinositol-3-phosphate 5-kinase
PIP3, 80, 82, 98, 170, 171, 173
PI(3,4,5)P2, *see* Phosphatidylinositol 3,4,5-trisphosphate
PI(3,5)P2, *see* Phosphatidylinositol 3,5-bisphosphate
PIRT, *see* Posphoinositide interacting regulator of TRP
pKa, 33, 44, 148
PKC, *see* Protein kinase C
Plasma membrane, 1, 2, 4–6, 43–52, 74, 76, 85–88, 91, 92, 94–101, 109, 110, 113, 114, 116–118, 120, 122, 123, 132, 134, 146–151, 166–168, 170, 172, 173, 176
Plasma membrane calcium ATPase (PMCA), 108, 109
Platelets, 136

PLB-985 cells, 60, 173
PLCγ *see* Phospholipase C gamma
PLD, *see* Pospholipase D
PMCA, *see* Pasma membrane calcium ATPase
Pollack, H., 33, 36, 37
Poly (ADP-ribose) polymerase (PARP), 109, 113
POST, *see* Partner of stromal interaction molecule 1 (also
    called TMEM20 or SLC35G1)
p22^{phox}, 158–164, 166–168, 172, 173, 176
p40^{phox}, 111, 158, 160, 162, 164–176
p47^{phox}, 158, 160–169, 171–173
p67^{phox}, 158–162, 164–168, 172–174
Prediction, 59, 63
Prenylation, 49, 166
Primitive phagocytosis, 3–5
Pritchard, A., 14–17, 19, 24
Progression of phagocytosis, 59, 60, 92
Protein kinase C (PKC), 111, 134, 159, 164, 168, 169,
    172, 173
Proteolytic site (of calpain), 137
PRR region (region rich in prolines), 163–165
Pseudopodia /pseudopods, v, 2, 3, 12, 14, 17, 24, 36,
    38–40, 44, 45, 51, 89, 91–94, 97–101, 133–136,
    138–141, 145, 146, 149–151, 185
PtdIns(3,4,5)P$_3$, 49, 50, 4446
PtdIns4P, 45, 50–52
PtdSer, *see* Phosphatidylserine
PTEN, 45, 174
Purinergic P2X receptor, 121
P2X Channels, 121
PX domain, 164–167, 171–173, 177
Pyk2 N-terminal domain-interacting receptor 1 (also
    called PITPNM) (Nir), 123

R
Rab7, 47
Rac, 45, 158, 159, 162, 165–169, 172–174
Radixin, 95, 137, 146
RAW 264.7 macrophages, 89, 170
(CBZ-Ala Ala)$_2$ R110 calpain indicator, 143, 144
Reactive oxygen species (ROS), 107, 109, 113, 115, 117,
    118, 121, 122, 157–161, 163–169, 173–176
Receptor, 2, 43, 45, 49, 57–70, 107, 109–111, 113, 114,
    116, 117, 120–122, 136, 139, 150, 159–161, 168,
    170, 172–174
Receptor diffusion, 61, 63, 69
Receptor model, 57–70
RECS1, *see* Transmembrane Bax inhibitor motif
Redox sensitive fluorescent proteins, 161
Release of a particle, 74, 75, 79
Releasing after phagocytosis, 4
Rho-GTPases, 45
RING finger protein 24 (RNF24), 116
RNF24, *see* RING finger protein 24
ROS, *see* Reactive oxygen species
Rosco, *see* (R)-roscovitine (also called Seliciclib)
ROS detection, 174
R-roscovitine (Rosco), 117
Ruffling, 5, 51

Rustizky, 30
Ryanodine receptors (RyR), 108, 109, 113
RyR, *see* Ryanodine receptors

S
S100A8/A9, 112, 158, 169
Salmonella enterica, 52
SARAF, *see* store-operated calcium entry associated
    factor (SARAF) (also called TMEM66)
Sarco/endoplasmic reticulum calcium ATPase (SERCA),
    108, 109
Sarcoma kinase (Src), 113, 120
S100 Ca2+-binding protein, 111
Scanning electron microscopy (SEM), 88, 92, 98
sdFRAP, *see* subdomain Fluorescence recovery after
    photobleaching (sdFRAP)
SEC14 domain and spectrin repeat-containing protein 1
    (SESD1), 107
Second messenger, 107, 110, 113, 168
Secretion, 31, 32, 107, 175
Seliciclib, *see* (R)-roscovitine
SEM, *see* scanning electron microscopy (SEM)
SERCA, *see* Sarco/endoplasmic reticulum calcium
    ATPase *(SERCA)*
SERCA pumps, 109
SESD1, *see* SEC14 domain and spectrin repeat-containing
    protein 1 (SESD1)
SH3 domain, 164, 165
Shigella flexneri, 52
SHIP1, 45
SHIP2, 45
Simple lipid assisted microinjection (SLAM), 134
Simultaneous uptake and release of particle, 79–81, 83
SK, *see* Sphingosine kinase (SK)
Slack membrane, 89
SLC35G1, *see* Partner of stromal interaction molecule 1
    (SLC35G1)
Small regulatory subunit of calpain, *see* common subunit
    of calpain
SNARE, *see* soluble N-ethylmaleimide-sensitive-factor
    attachment protein receptors (SNARE)
SOAR, *see* STIM1 Orai1-activating region (SOAR)
SOCE, *see* Store-operated calcium entry (SOCE)
Sodium calcium exchanger (NCX), 106, 108, 109
Soluble N-ethylmaleimide-sensitive-factor attachment
    protein receptors (SNARE), 107, 110
S1P, *see* sphingosine-1-phosphate($S1P$)
Spheroid (oblate), 65–68
Spheroid (prolate), 65–68
Sphingolipids, 44, 46, 123
Sphingosine kinase (SK), 107, 113
Sphingosine-1-phosphate (S1P), 107–109, 113, 116, 123
Spinning disk microscope, 160
Spleen tyrosine kinase (Syk), 107, 113, 120, 159, 168
Spreading, 45, 86, 90, 91, 93, 99, 101, 102, 135–137,
    140, 141, 145, 147, 148
Src, *see* sarcoma kinase (Src)
Staphylococcus aureus, 158
Starfish, 3, 25, 28–30

Starfish larva, 28, 29
Stefan problem, 61, 62
Stiffness, 69, 70, 97, 121
STIM, *see* Stromal interaction molecule (STIM)
STIM-depleted cells, 116
STIM1 Orai1-activating region (SOAR), 107, 114, 115
Store-operated calcium entry (SOCE), 109, 110, 113–117, 119, 120, 150, 168, 174
Store-operated calcium entry associated factor (SARAF), 107, 115
Stromal interaction molecule (STIM), 107, 109, 110, 114–117, 122, 135, 138, 150
Structure of calpain, 135, 145, 146
Subdomain Fluorescence recovery after photobleaching, 89
Surface charge, 43–52, 169, 170, 173
Sweet Briar College, 31, 33
Syk, *see* Spleen tyrosine kinase (Syk)
Synaptotagmins, 110
Synaptotagmin VII (SytVII), 107, 119
SytVII, *see* Synaptotagmin VII *(SytVII)*

T
Target radius, 63, 64
Target shape, 6, 57–70
Tension, 6, 74, 76, 83, 85–102
Thorn (in starfish), 10, 25, 29, 30
TMBIM1, *see* transmembrane Bax inhibitor motif (also called LFG3, or RECS1)
TMEM20, *see* partner of stromal interaction molecule 1
TMEM66, *see* store-operated calcium entry associated factor
Total internal reflection fluorescence (TIRF) microscopy, 148
TPC, *see* two-pore channel (TPC)
Transient receptor potential (TRP), 105, 107, 110, 116, 117, 119, 120
Transient receptor potential canonical (TRPC), 107, 109, 110, 114–117, 122, 123
Transient receptor potential melastatin (TRPM), 107, 109, 110
Transient receptor potential mucolipin (TRPML), 107, 109, 110, 114, 119, 120, 122
Transient receptor potential polycystic 1 (TRPP1)(also called PKD2), 107–109, 122
Transient receptor potential vanilloid (TRPV), 107, 109, 110, 114, 120, 121
Transmembrane Bax inhibitor motif (TMBIM1), 107, 109
TRPC, *see* transient receptor potential canonical (TRPC)

TRPC channels, 109, 114, 119
TRPM, *see* transient receptor potential melastatin (TRPM)
TRPM channels, 109
TRPM2 channels, 113
TRPML, *see* Transient receptor potential mucolipin (TRPML) (also called MCOLN)
TRPML channels, 109, 119
TRPP1, *see* transient receptor potential polycystic 1 (TRPP1) (also called PKD2)
TRPV Channels, 120, 121
Tubular phagosome, 76–80
Two-pore channel (TPC), 107, 109, 113

U
Uncaging Ca2+, 140
Uncaging IP3, 140
University of Groningen, 34
Uptake of long particle, 75

V
vacuolar-type H+-ATPase (v-ATPase), 107, 117, 118
VAMP7, *see* vesicle-associated membrane protein 7 (VAMP7) (also called synaptobrevin-like protein 1 SYBL1)
v-ATPase, *see* vacuolar-type H+-ATPase (v-ATPase)
Vesicle-associated membrane protein 7 (VAMP7)(also called synaptobrevin-like protein 1 SYBL1), 107, 119
VGCC, *see* voltage-gated calcium channel (VGCC)
Voltage-gated calcium channel (VGCC), 107, 109
von Gleichen-Russworm, Baron Friederich Wilhelm, 18

W
WASp, *see* Wiskott-Aldrich Syndrome protein (WASp)
Wiskott-Aldrich Syndrome protein (WASp), 94, 99, 100, 134, 151
Wrinkled cell surface, 88, 94, 95, 98, 149
Wrinkles, 4, 5, 87–99, 101, 102, 140, 148–151

Y
Yersinia pseudotuberculosis, 52

Z
Zebrafish, 110
Zymosan, 99, 100, 112, 139, 140, 144, 145, 172

CPSIA information can be obtained
at www.ICGtesting.com
Printed in the USA
LVHW061105170520
655852LV00005B/261